The Economics of Managing Biotechnologies

Economy & Environment

VOLUME 22

The titles published in this series are listed at the end of this volume.

The Economics of Managing Biotechnologies

Edited by

Timothy Swanson
Department of Economics and School of Law,
University College London, United Kingdom

SPRINGER-SCIENCE+BUSINESS MEDIA, B.V.

A C.I.P. Catalogue record for this book is available from the Library of Congress.

ISBN 978-94-017-3950-4 ISBN 978-0-306-47659-4 (eBook)
DOI 10.1007/978-0-306-47659-4

Printed on acid-free paper

Table of Contents

Part C: Managing Biology: The Economics of Managing Biological Resistance

Part D: Managing Industry: The Economics of Managing Industrial Impacts

Part E: Policy Conclusions

Preface

Why do biotechnologies raise new problems of economic interest? Are there legitimate policy concerns to be voiced about problems concerning the rate or direction of technological change? Why do biological contexts raise distinct forms of management problems? How should policy makers respond to these problems?

This volume is the result of a workshop on Environmental Policy, Agriculture and Biotechnology held in Rome, Italy in May 2000 that focused on these general questions in the context of recent agricultural biotechnology changes. This workshop was attended by a large group of economists interested in the issues that exist at the interface between the biological, industrial and technological worlds. It also attracted the interest and involvement of a number of scholars and policy-makers expert in these individual fields. There were presentations that covered the basics of the biotechnologies currently appearing in agriculture, and presentations on the essence of the policy discussions about these biotechnologies. Then there were sessions where the economists discussed their various approaches to thinking about the problems that lie at the interface between the biological and industrial worlds.

The first part (Part A) of this volume includes a number of survey papers devised to 'set the scene' concerning the problem of agricultural biotechnologies and their management implications. The first paper by Charles Spillane and Yvonne Pinto covers a lot of the agricultural grounding for this issue. It lists the range and forms of biotechnological changes occurring in that field, and the various types of benefits that might be anticipated to result. After indicating the range of benefits, it then catalogues the various types of risks and limitations inherent within the technologies. Finally the paper then moves into the policy realm, indicating the wide variety of regulatory forums and regimes that are currently dealing with these problems. Thus the paper provides a very broad and general introduction to the benefits, risks and policies applicable to these technologies.

The second paper in the volume is by a pioneer in the field of biological management, Uri Regev. It was Professor Regev and his co-authors who first identified and discussed the economic nature of the problem of resistance nearly thirty years ago. This problem concerns the management of the capability of biological organisms to adapt to prevailing technologies by reason of natural selection. Thus any new pest management technology that is extremely (but not absolutely) effective sows the seeds of its own demise by reason of the enhanced reproductive success of those pests naturally resistant to the technology. His early work on the intertemporal externalities inherent within pest management technologies is here extended to the case of biotechnological innovations. That work had demonstrated that the biological context introduced an intertemporal externality amongst joint pest managers (e.g. farmers) of the nature of a common pool problem. Each pest manager would share in the future benefits from reduced selection pressure, and hence the trait of resistance is like that of a common pool resource. It will be subject to overexploitation if there are a large number of managers sharing the resource.

The third paper of the volume is a neat economic survey of the new biotechnologies, analogous to the first chapter, but applied to the case of forestry. Roger Sedjo places the benefits, risks and policy issues applicable to biotechnologies in terms that economists understand by assessing the costs and benefits of their introduction into forestry plantations.

Together the papers of Part A are intended to introduce the problems, principles and policies inherent within the new agricultural biotechnologies. This first part makes clear that the problems of biotechnologies within agriculture are well-known and much-discussed. There are many different venues that have considered these problems for many years, on account of the on-going international discussions over agriculture and the long-standing problems with pest management.

However, as indicated above, there are several relatively new lines of economic enquiry raised by the questions of biotechnological advance. While Part A raises surveys the issues within a very specific context, it is also possible to abstract from this particular context and to consider the general problems involving technological changes introduced within biological systems. This is what we do in the ensuing parts of the volume.

The first general issue raised by biotechnology concerns the question of directing technological change down specific avenues through appropriate incentive systems. Part B contains three papers considering the direction of technological change within the biological context. Chapter 4 is a survey of the literature on the problem of pest resistance and a consideration of the capacity for a decentralised industry to address it. As the evolutionary model in Goeschl and Swanson make clear, the nature of the resistance problem is the directed evolution of pests in response to the application of the technology – evolution ensures that the pests reproducing best in the presence of the technology are precisely those which predominate within future populations. This introduces the common pool problem earlier noted by Regev, but it also introduces a problem at the level of the technological innovator. Since innovations must be rewarded through their use across time horizons during which they are marketed, the in-built depreciation through use reduces the incentives for innovation. A decentralised industry, rewarded through use of its innovations over time, has little incentive to solve such problems.

Chapter 5 is a specific application regarding the problem of directing technology to a desirable outcome. Here the problem being considered by O'Shea and Ulph concerns the externality inherent in the use of an innovative technology by reason of its extreme novelty within the biological environment. Biotechnologies represent the capacity to bridge across vast distances within the genetic pool, and this implies new biological innovations of an extreme nature. Such innovations, when introduced within a biological world, may not fit in well with already-existing organisms. This implies the possibility of pest management, but also has the unintended effect of removing other layers of the trophic pyramid. Hence, severe impacts on wide varieties of insects might translate into losses of bird life as well. O'Shea and Ulph examine this problem and analyse the policies required to internalise the externality inherent in such forms of technological change.

Chapter 6 is a paper on the economics of risk management in the context of uncertainty and irreversibility. Wesseler examines the problem of technological management when there is extreme uncertainty about the possible effects from introducing the innovation. Biotechnologies involve such uncertainties precisely because they involve the generation of organisms across greater distances than nature has attempted, and because the existing system of organisms will respond in unpredictable ways to such innovations. Under standard theories of option pricing, he assesses the 'hurdle rate' for biotechnologies, i.e. the additional benefits required of these technologies to compensate for their inherent uncertainties.

Thus, Part B of the volume assesses the general problem of managing technological change for its environmental/biological consequences. The three forms of analysis consider various aspects of the problem of managing biotechnologies: resistance; unintended effects; and, irreversible uncertainties. This part of the volume demonstrates that an unregulated market may not generate the direction of technological change that society desires.

Part C of the volume considers the management of biotechnological change from a different perspective: how should drastic technological change be introduced within the context of biological resources? The specific policy issue of concern is the optimal size and extent of refugia retained to dampen the effects of dramatic technical change. The paper by Laxminarayan and Simpson in Chapter 7 analyses this issue from the economic perspective. It assesses the costs of retaining reserves and balances them against the benefits of reduced resistance. Reserves are a buffer against enhanced resistance effects because they retard the rate at which successful pests are able to diffuse across space (the epidemiological benefit) and because they provide a source of diverse pests and pathogens (the selection benefit). Of course these benefits come at the cost of both reduced production and increased pestilence, and the analysis by Laxminarayan and Simpson assesses the economically optimal level of reserves given these trade-offs.

Chapter 8 by Hurley et al. considers the same issues but looks at them solely from the agricultural perspective, i.e. what is the optimal level of reserves for containing the development of resistance? This is an important policy question because laws already exist that specify minimum reserve requirements for the introduction of biotechnological products. This paper analyses whether those policies are correct, when assessed solely against their own objective, i.e. the avoidance of enhanced resistance effects.

Together the papers demonstrate that the management of dramatic technological change must involve the managed introduction of the products of those technologies. When the adverse effects are known and expected, then the phased introduction of the technology may enable the dampening of those impacts. In a biological setting, phased introduction accords with managed introduction across both time and across space.

Part D of the volume turns to the industrial, as contrasted with the environmental, impacts of new technologies. New technologies or rapid rates of technological change often have distributive as well as efficiency effects. This is because such technologies often imply highly concentrated industries. The reasons for this vary. Sometimes it is attributable to the complex of patents and other restrictions that develop around the new technologies. Sometimes it is because the capital structures or development capacities of certain parts of the world render them incapable of assimilating or applying the new technologies. In either event, fields of rapid technological change frequently display the characteristics of heavily concentrated industries. Chapter 9 by Munro assesses the industrial economics of the biotechnology industries, determining the conditions under which these new technologies might result in greater industrial concentration. One of the more straightforward routes by which this is occurring in the field of biotechnology is through the advance of new use restriction technologies. These are technologies that inhibit the reproduction of products sold previously, e.g. the development of sterile seed technologies. Such technologies enable the enhanced appropriation of the rents from innovation, by disallowing any unlicensed transfers or uses.

Chapter 10 by Goeschl and Swanson assesses the distributive impacts of such use restriction technologies. The efficiency effects should be straightforward: greater appropriation of the benefits of innovation should result in greater investment in innovations, wherever they might be applied. However, the distribution of the benefits from innovation will depend in part on: (a) the investment by the innovator in technologies applicable in all parts of the world; and (b) the diffusion of the information embedded in technologies across all parts of the world by the innovator. In theory the innovator should have incentives to pursue both activities to distribute benefits, but in practice Chapter 10 demonstrates that this has not always occurred. For this reason the benefits of rapidly advancing technologies are often concentrated near that technological frontier.

Part E of the volume comprises two papers with some policy reflections and conclusions. Chapter 11 by Dale states the obvious benefits from biotechnologies and the clear-cut risks associated with them. The chapter lists the elements of the necessary risk assessment that must be conducted before new biotechnologies are introduced. This is one of the clear messages from the analysis contained within this volume: new technologies imply new management obligations.

Chapter 12 by Crompton and Tzotzos gives a less obvious but equally important set of conclusions. New technologies also imply new needs for international cooperation. This is because the implications of the technologies vary significantly for different parts of the world, and these varying needs must be taken into account when managing safety implications and when managing technological diffusion and distribution.

In sum, the volume sets forth a range of issues concerning technological change, its impacts (environmental and industrial) and its management implications. The papers have been selected to provide a survey of the issues concerned rather than in-depth analysis of a single topic. The volume is best read in its entirety in order to take the message from the Rome Workshop.

Acknowledgements

This volume has its origins within a project sponsored by the European Science Foundation (ESF) entitled 'Environmental Policy, Agriculture and Biotechnology' granted in 1998. We are grateful to that foundation for their support, and to the Scientific Officer, Aad Correlje for his participation in these activities. The ESF provided funding for two workshops on the topic of biotechnology management, the first occurring at the University College London School of Public Policy in April 1999. We are grateful to the Administrator of the School, Sally Welham, for her assistance in that undertaking. Thereafter a general call for papers went out for a Workshop on Biotechnology Management to be held in Rome, Italy, May 2000. This Workshop was hosted by the International Plant Genetic Resources Institute with considerable grace and aplomb. We are grateful to Pablo Eyzaguirre for his assistance there. Finally the Workshop was expanded to include not only the environmental but also the industrial impacts of biotechnologies, and for this we are grateful to the UK Department for International Development (DfID) for their support. We are especially grateful to Robert Carlisle of that office for his encouragement of the Workshop.

I also would like to thank all of the participants at the Rome Workshop for their excellent contributions and enthusiastic involvement. The quality of this volume is primarily attributable to their efforts.

Finally I would like to thank the editorial staff at Kluwer for their advice and assistance in completing the volume.

Timothy Swanson
University College London

List of Contributors

Babcock, B.A., Department of Agricultural Economics, Iowa State University, Ames, IA, USA

Crompton, T., Ty Mawr Erwallo, Dolywern, Llangollen LL20 7AG, Wales

Dale, P.J., John Innes Centre, Colney, Norwich NR4 7UH, UK

Goeschl, T., Department of Land Economy, University of Cambridge, Cambridge CB3 9EP, UK

Hellmich, R., United States Department of Agriculture, Agriculture Research Service, Iowa State University, Ames, IA, USA

Hurley, T.M., Department of Applied Economics, University of Minnesota, 231 Classroom Office Building, 1994 Buford Avenue, St. Paul, MN 55108-6040, USA

Laxminarayan, R., Resources for the Future, 1616 P. Street NW, Washington, DC 20036, USA

Munro, A., School of Economic and Social Studies, University of East Anglia, Norwich NR1 2BQ, UK

O'Shea, L., Department of Economics, The University of Southampton, Highfield, Southampton SO17 1BJ, UK

Pinto, Y., Cereals Research Department, John Innes Centre, Norwich, Norfolk NR4 7UH, UK

Regev, U., Department of Economics, Ben-Gurion University, 84105 Beer-Sheva, Israel

Secchi, S., Department of Agricultural Economics, Iowa State University, Ames, IA, USA

Sedjo, R., Resources for the Future, 1616 P. Street NW, Washington, DC 20036, USA

Simpson, R.D., Resources for the Future, 1616 P. Street NW, Washington, DC 20036, USA

Spillane, C., Room P2-39, Institut für Pflanzenbiologie, Universität Zürich, Zollikerstrasse 107, CH-8008 Zürich, Switzerland

Swanson, T., Faculty of Laws & Department of Economics, University College London, Gower Street, London WC1E 6BT, UK

Tzotzos, G.T., Biodiversity Unit, UNIDO, SES/PEM, Vienna International Centre, Wagramerstr. 5, A-1400 Vienna, Austria

Ulph, A., Department of Economics, The University of Southampton, Highfield, Southampton SO17 1BJ, UK

Wesseler, J., Department of Environmental Economics and Natural Resources, Wageningen University, Hollandseweg 1, 6706 KN Wageningen, The Netherlands

PART A

SURVEYING THE ISSUES: TECHNOLOGIES, BIOLOGY AND ECONOMICS

1. Biosafety in Agricultural Biotechnology: Balancing Social and Environmental Impacts

CHARLES SPILLANE and YVONNE PINTO

1. Introduction

Agriculture, grazing, forestry and other human managed ecosystems are highly important economic sectors that are geographically extensive covering at least two thirds of the terrestrial surface of the planet. Although the percentage of people engaged in agriculture can vary widely between countries (e.g. 92% in Nepal, 2% in USA), on average almost 50% of the world's economically active people are engaged in agriculture. From a social perspective, over 1100 million farmers, in many different farming systems and environments, are economically active in agricultural production globally (about 50 million farmers in the developed countries and 1050 million in the developing countries).

The vast majority of the world's farmers have little access to productive resources but practice approximately 60% of global agriculture, producing 15–20% of the world's food. The low productivity of resource poor farmers tends to perpetuate rural poverty to the extent that of the more that 2,500 billion people in developing countries who live in rural areas, approximately 1000 million live below the poverty line: 633 million in Asia; 204 million in Africa; 27 million in the Near East and North Africa; and 76 million in Latin America. Even in 2010 over half of the people in the developing countries will look to farming for their livelihoods.

The challenge facing world agriculture in terms of improved food production in the next decades is enormous. As the world's population increases to 8.5 billion by 2025, world food production will need to increase by about 60% in the next 25 years in order to ensure adequate food supplies – especially from local production. By the year 2025, 83% of the expected global population of 8.5 billion will be living in developing countries. It has been estimated that in the next 25 years humanity will have to produce as much food as it has over the past 10,000 years. Even if yields are to be maintained at current levels there will have to be a sustained public-sector funding base for ongoing research to develop techniques and technologies which can limit the negative effects of pests and diseases, while improving the livelihood security of the world's farmers and consumers. This will have to be done through sustainable intensification of existing arable lands.

Timothy Swanson (ed.), The Economics of Managing Biotechnologies, 3–49.

Agricultural biotechnology comprises a collection of scientific techniques, including genetic engineering, that are used to modify and improve plants, animals and microorganisms for human benefit. Agricultural biotechnologies are not a substitute for conventional plant and animal breeding, but can be a powerful complement to improving the efficiency and sustainability of agricultural production. Agricultural biotechnologies are widely expected to change the nature of agriculture and food production over the coming decades. The ongoing debate between groups who are either advocates or opponents of agricultural biotechnologies is whether such change will be beneficial to society. Both groups have differing perceptions of the stakeholders in society that should be the key beneficiaries, regarding the application of modern biotechnologies to agriculture.

Advocates of agricultural biotechnology point to issues such as the limitation of conventional breeding approaches, rising demand for food (mainly urban based), the 'environmentally friendly' substitution of chemical inputs with renewable gene inputs, and the generation of beneficial options that were hitherto unavailable to humanity. The opponents of agricultural biotechnology raise issues such as perceived long-term health risks of transgenic foods, environmental risks of the release of transgenic organisms, increasing industrial concentration in the food supply sector, increasing property claims over 'life' products and processes, and the 'unnatural' decontextualisation of the code of life.

The practical application of agricultural biotechnology has become increasingly contentious largely because of conflicts between multiple stakeholder groups regarding the stringency of biosafety regulations. In such a polarised political climate, emerging biosafety assessment models require that risks, benefits and needs be given balanced assessment in relation to transgenic organisms. Indeed, many opponents of plant biotechnology cite biosafety as the key risk based issue for the more stringent regulation of transgenic organisms. At one end of the extreme, many powerful environmental groups call for a moratorium on the planting of 'genetically modified' crops. The other end of the extreme would be no biosafety regulations regarding transgenic or conventional organisms. The academies of sciences in Brazil, China, India, Mexico and the United States, the Third World Academy of Sciences in Trieste and the UK's Royal Society have collectively stated that concerns about safety and the possible environmental impacts of GM crops should be countered by research. The environmental effects of transgenic crops, if any different, should be measured against the effects of conventional agriculture and care taken to maintain a diversity of crops, conventional and transgenic (Spillane, 2000).

This review provides an introduction to current developments in agricultural biotechnology research, followed by a review of some of the biosafety issues that have been raised, mainly regarding transgenic crops, and concludes with a review of current policy developments in the area of biosafety of agricultural biotechnologies.

2. Scientific Developments in Agricultural Biotechnology

Agricultural biotechnology research is now generating a wealth of information of utility to crop improvement strategies. The basic research being undertaken at the level of genomics is making it possible to generate a large number of interesting applications of biotechnology for agriculture (see Section 3). In this section we survey some of these important fundamental scientific developments that underlie these new applications.

2.1. THE EMERGING IMPORTANCE OF AGRICULTURAL GENOMICS

Genomics is a recently coined term used to describe the development and application of large-scale, high throughput and parallel processing approaches to the functional analysis of entire genomes (or genetic systems) (Bouchez and Hofte, 1998). Genomics technologies can allow the identity and function of tens of thousands of different genes to be analysed simultaneously or in parallel. The science of genomics has arisen because the speed and scale at which the genomes of economically important organisms can be functionally analysed is increasing at a rapid pace (Broder and Venter, 2000). Through private and public sector initiatives, a first working draft of the complete sequence of the human genome was had been completed was announced in June 2000 (Dunham, 2000). Technology spillovers from such programmes will result in high throughput DNA sequencing and gene expression analysis being increasingly applied to the genomes of commercially important organisms such as human pathogens, crops or domestic animals. Efforts are underway to sequence the complete genome of a range of priority organisms. Whole genome sequencing is now being complemented by proteomics where a range of high throughput techniques for analysing the protein profile (or proteome) of an organism are under development (Pandey and Mann, 2000; Banks et al., 2000). Agricultural genomics research is currently underway for a range of important agricultural species (Timberlake, 1998). Large-scale DNA sequencing of some crop and plant genomes is now at an advanced stage, especially for *Arabidopsis* and rice (Rounsley et al., 1998).

Expressed sequence tags (ESTs) are small fragments of genes that have been sequenced and which can act as unique 'bar-codes' for each particular gene in an organism. Large-scale EST projects can rapidly generate unique ESTs for the majority of the genes in an organism in a very rapid and cost effective manner. For instance, ESTs which each uniquely identify more than 95% of the estimated 120,000 human genes are available on public databases on the Internet. Publically funded research has now identified ESTs representing the majority of genes in the *Arabidopsis* (International Arabidopsis Genome Initiative) and genomes (Rice Genome Project in Japan). The majority of such ESTs have been placed in the public domain via their publication in publically accessible EST databases on the

Internet. It is also known that there are several large-scale commercially funded EST projects for crops such as maize and soybean. However, to date much of this EST data is publically unavailable (Cohen, 1997).

Agricultural genomics initiatives are both generating genetic markers and identifying genes that can be used either for marker assisted breeding or the development of transgenics with improved agronomic properties (Phillips and Freeling, 1998). The application of molecular markers to genetic linkage maps of a wide range of crops is allowing the identification of the chromosomal (physical) locations of genes for improving yield and other complex traits important to agriculture (McCouch et al., 1997). The underlying genetics of a wide range of quantitative agronomic traits are being unravelled through the identification of quantitative trait loci (QTLs) using a combination of molecular markers and advanced statistical breeding procedures. A reflection of the wide range of agronomic traits and corresponding QTLs currently under investigation or improvement can be obtained from browsing through the abstracts of the annual Plant and Animal Genome Conferences (http://probe.nalusda.gov:8000/otherdocs/pg/pg6/pag6.html). As research progresses it is possible that a range of quantitative trait nucleotides (QTNs), which are responsible for traits of agricultural importance will be identified (Phillips, 1999).

Powerful 'functional genomics' systems to explore the function of genes through 'knockout' and 'monitoring' strategies have also been developed using existing transposon mutagenesis systems, which have traditionally been widely used for the genetic analysis of maize (Martienssen, 1998). The development of DNA 'chip' technology is also set to revolutionize the scale at which genetic experiments can be done and can potentially allow simultaneous and rapid analysis of all (e.g. 10,000–100,000) of the genes in any organism at any particular point in time or environment (Collins, 1999). If this technology becomes cost effective in the same manner which has led semiconductor technology to become widespread, it will have a very significant impact on how genetic improvement of crops and animals is conducted. There are some large-scale publically funded efforts to develop genomics technologies and tools which would be made available to the academic plant biology community with 'no strings attached' regarding intellectual property. In the US the *Arabidopsis* Functional Genetics Consortium is adopting this approach (Green et al., 1998).

The potential value of agricultural genomics has been recognised by the private sector judging by its current level of investment. High levels of public sector funding have also been approved for agricultural genomics initiatives in a few countries. For instance, the US government has approved $40 million to fund genomics initiatives for crops of national importance (Bennetzen et al., 1998), and the Japanese government has earmarked substantial financial support for a national Rice Genome Project (Briggs and Helentjaris, 1997). Some analysts consider that the USA's research capacity in agricultural genomics now outstrips that of all other countries and that the technology gap is widening in this area (Joly and Lemarie, 1998). The promotion of the French GenoPlante agricultural genomics consortium is one national initiative which is trying to bridge such gaps between regions (Joly and Lemarie, 1998). The Indian government is assessing proposals to strengthen India's capacity to conduct agricultural genomics research (Bagla, 1999).

Scientists participating in some of the plant genomics initiatives have stated that publicly available research tools and data are a priority if broader scientific, social and economic spillovers are to be promoted. It is felt by some in the scientific community that international co-operation is critical to such large-scale genome projects, both because genome projects are typically too large for any one country and the information forthcoming should be of benefit to the world and not just the teams involved (see http://www.staff.or.jp). International co-operation is essential to ensure that overly protectionist proprietary positions are not taken which would inhibit the availability of potentially useful public domain information.

2.2. Comparative Genomics: Unifying Crop Genepools

An important discovery that is accelerating genetic mapping research has been the discovery that the genomes of distantly related crop species may be quite similar to each other in terms of gene order and structure (Gale and Devos, 1998). For instance, a gene on the chromosome of one grass species can be anticipated to be present in a predicted location on a specific chromosome of other grass family species. For this reason, many cereal geneticists now view the grass family as a single genetic system whereby genomic information from one cereal species can be used to understand much of the genetics of other cereal species (McCouch, 1998).

Because it is relatively easy to determine the DNA sequence of species with smaller genomes, a number of species (e.g. rice, *Arabidopsis*) are emerging as the 'anchor genomes' which are expected to act as the 'Rosetta Stones' for understanding the larger genomes of other related species (Messing and Llaca, 1998). A number of major multi-partner efforts are underway to determine the sequence and structure of major cereal crop genomes which will provide valuable information for understanding the genomes of other cereals. Examples include the International Grass Genome Initiative, International Triticeae Mapping Initiative, the International Rice Genome Sequencing Project (http://rgp.dna.affrc.go.jp/), the Japanese Rice Genome Project (Sasaki, 1998) and the US National Corn Genome Initiative (Coe, 1998). The entire *Arabidopsis* genome has now been sequenced by a multi-partner international effort and made publically available (Anon, 2000). Such genome data is already providing valuable information regarding the genetics of economically important crops both within the Brassicaceae and many other tribes (Lagercrantz and Lydiate, 1996). The *Arabidopsis* genome has already been sequenced by at least one company, and a rough draft of the rice genome has been sequenced by the private sector. Until recently, because of the cost of generating genomic sequences, the sequence data for plant genomes sequenced by the private sector has typically not been made unconditionally available to the academic community, or to competitor companies. However there are signs that such corporate policies are beginning to change. In July 2000 Monsanto established a web site (http://www.rice-research.org) where researchers may register and gain access to the Monsanto Rice Genome Sequence data at no charge, as part of Monsanto's new strategy to support global agricultural research. Since important scientific advances are aided by the dissemination and exchange of information Monsanto is encouraging the users of this rice genome sequence database to publish their research results widely.

2.3. Biotechnologies to Increase the Accessibility and Use of Genepools

Crop wild relatives stored in the world's genebanks are valued as a unique source of genetic variation, but they have rarely been used for the genetic improvement of quantitative traits. Indeed it is widely acknowledged that exotic germplasm is infrequently used by breeders (Duvick, 1996). A recent development of significance has been a powerful molecular marker based methodology whereby QTLs conferring complex traits such as yield and organ size can now be effectively transferred from wild relatives into crop varieties (Tanksley and McCouch, 1997). This powerful technique has now been demonstrated for rice (Xiao et al., 1998) and tomato (De Vicente and Tanksley, 1993) and is being tested in other crop species for which suitable molecular maps are available. This discovery suggests that the innovative use of molecular maps and markers will increase the accessibility and realisable value of wild and exotic germplasm.

3. Biotechnology Applications in Agriculture

Plant biotechnology research is best realised when it is integrated with conventional breeding or crop improvement approaches to pursue important agronomic objectives. In many countries, plant biotechnology research is being integrated in this way wherever the technical and financial resources for such integration are available. Among the scientific community it is often highlighted that plant biotechnologies could generate agricultural and environmental benefits, especially where renewable genetic inputs can be effectively used to substitute agro-chemical inputs.

A vast range of applications through biotechnology are being pursued. These include:

- Micropropagation and plant tissue culture technology (e.g. the vegetative propagation of disease-free plantlets of staple crops such as cassava, potato, sweet potato, taro, etc.).
- Improved fermentation technologies.
- Improved technologies for generating biomass derived energy.
- Improved vegetable oils for cooking that maintain texture at raised temperatures, reduce processing energy costs and create healthier products.
- Generation of higher micronutrient levels (e.g. vitamin A, iron, essential amino acids) in micronutrient deficient staple crops such as rice and other cereals.
- Reduced toxin levels (e.g. nutraceuticals, reduced mycotoxin contamination) and reduced allergenic potential (e.g. hypo-allergenic peanuts).
- Crops with better storage and transport characteristics through delayed ripening and fungus/pest protection.
- Multiple marker assisted selection strategies for improving agronomic traits in animal and plant varieties/breeds, including yield potential (e.g. rice, tomato).
- Development of abiotic stress tolerance genes (e.g. Aluminium and manganese tolerant crops which can grow in acidic soils, salt tolerance, drought tolerance).
- Vaccines against human and animal diseases (e.g. cholera-toxin B, hepatitis B, herpes, rabies, bovine rhinotracheitis, Salmonella, East Coast Fever, liverflukes).
- Insect resistance (e.g. against corn borer).
- Bacterial and fungal disease resistance (e.g. against powdery mildew).
- Virus resistance (e.g. against cassava mosaic virus).
- Better crop digestibility for animals and humans (e.g. reduced or modified lignin content).
- Delayed over-ripening of fruits and vegatables (e.g. to reduce post-harvest losses).
- Herbicide tolerant seeds (e.g. to combat broomrape and witchweed infestations).

3.1. Generating New Options for Pest and Disease Resistance

Breeding for resistance to pests and diseases is a major ongoing activity for the vast majority of crops. There are many pests and pathogens against which conventional breeding has failed to provide crop varieties with durable resistance. There are also serious pests and pathogens which integrated pest management (IPM) approaches have not yet managed to address. A significant proportion of plant biotechnology is targeted at developing new strategies for pest and disease control for which there are currently few options, although most such work is concentrated on pests and diseases of major commercial or export crops.

The genetic basis of gene-for-gene interactions in plant pathogen interactions is being rapidly elucidated. Plant biotechnology research is identifying and isolating a vast range of genes and QTLs conditioning resistance against a wide range of pests and pathogens (Hammond-Kosack and Jones, 1997; Bent, 1996). Powerful techniques have been developed for 'scanning' crop genomes for related resistance genes and subsequent cloning of such genes (Leister et al., 1996; Chen et al., 1998). The effective transfer of resistance genes from one crop species to another has been demonstrated as a new option for resistance breeding. Some transgenic approaches have demonstrated higher levels of broad-spectrum resistance against pathogenic viruses (Tacke et al., 1996), bacteria (Cao et al., 1998) and fungi (Shah, 1997).

Plant biotechnology could have significant substitution effects in the global insecticide market. The latter is estimated to be worth approximately US$ 8100 million per annum (James, 1997). The development of genes (conditioning resistance against insect pests) which may replace insecticides could have a significant impact both on the environment and sales of insecticide. The source of many such genes has been the bacterium *Bacillus thuriengiensis* (Bt) although many other organisms are now being screened for useful genes. Bt has been used for decades as an insecticide spray but has had limited use outside of organic agriculture and Canadian forestry, in total accounting for less than 1% of the insecticide market. It is now estimated that US$ 2700 million of chemical insecticide applications could be replaced with Bt biotechnology applications, either as improved sprays or through expression in transgenic crops (Krattiger, 1997). Bt transgenic crops were planted on an estimated 20 millions acres in the USA in 1997. It has been proposed that if Bt corn, crop rotations, monitoring of pest populations and other IPM related strategies were used in corn production, in some instances insecticide use in corn production could be reduced by more than 50% (Pimentel et al., 1993). It is considered that there are many environmental and other benefits that can be derived from the widespread use of Bt transgenic crops (Betz et al., 2000).

It has required the advent of transgenic single gene resistance approaches to warrant legal requirements on farmers to engage in large-scale resistance gene 'deployment' or 'recycling' (McGaughey et al., 1998). The planting of large contiguous crop areas to varieties which are relying on monogenes to condition resistance to important pathogens has long been recognised to be unwise because some monogenes can exert a selection pressure for resistance breaking strains of the pathogen to evolve (Kiyosawa, 1982; Mundt, 1994). Research is underway to determine in the case of Bt crops whether resistant pest biotypes are likely to be selected for (Gould et al., 1997; Tabashnik et al., 2000) and to develop effective crop management strategies to maintain the efficacy of resistance genes (e.g. Shelton et al., 2000). There have been few public policy instruments developed to encourage that low selection pressures on pathogen populations are maintained in agricultural production systems.

The first generation of transgenic plants expressing the insecticidal *Bacillus thuriengiensis* (Bt) protein now require a resistance gene 'deployment' strategy to limit selection for an insect population that is insensitive to the particular Bt protein (Gould, 2000; Shelton et al., 2000). Because Bt transgenic seed is proprietary it can be sold with a legal requirement that a certain percentage of the cropping area (i.e. refugia or mixed plantings) be planted to varieties which lack Bt transgenes to limit selection for Bt insensitive pests. Also Bt gene 'recycling' strategies are being established whereby a range of transgenic varieties each containing a different type of Bt protein are released over time. These will effectively combat the potential ability of the insect pest to overcome any particular Bt resistance gene, and hence minimize the conventional 'boom-bust' cycles in variety-pathogen co-evolution.

Unless consideration is given to how monogenic resistance genes (whether transgenic or not) are spatially and temporally deployed against important plant pests and pathogens (especially airborne fungi) it is likely that the familiar 'boom-bust' cycle will be perpetuated and valuable genetic resources will have been wasted (Gould, 1998). Multiline and mixture strategies for resistance (trans)gene deployment may become more feasible because of the increasing number of resistance genes being isolated and the ease with which transgenic crops can be established.

It is suggested that past failures with multiline and mixture strategies may have been partly due to a lack of control by breeders over the accurate deployment of resistance genes by farmers. In addition that stronger property rights (IPRs or user contracts) over varieties or resistance genes would allow a requirement to be placed on the variety user to ensure that optimal gene deployment or recycling practices are followed (Pink and Puddephat, 1999). Less coercive resistance gene deployment strategies are reportedly under development or in use in some IPM programmes, including the FAO's Farmer Field School Approach. Modern biotechnologies could be supportive of integrated pest management (IPM) approaches. However, the lack of fruitful interaction between farmer participatory IPM approaches and transgenic or biotechnological approaches to pest/disease management has been noted (Carozzi and Koziel, 1997; Brown, 1998). It is suggested that there is a need for more collaborative interaction between experimenting farmers and scientists, especially regarding pest and disease problems that cannot be solved by local level research alone (Loevinshon and Meijerink, 1998; Thro and Spillane, 2000).

Barriers to inter-disciplinary approaches to agricultural research may stem from a lack of constructive communication channels between 'upstream' public sector agricultural biotechnology researchers and 'downstream' on farm researchers and farmers groups. There are currently no mechanisms for effective translation of farmers' expressed needs into research action through appropriate 'participatory problem transfer' (Thro and Spillane, 2000). Most public sector bodies which either fund or conduct agricultural biotechnology research have no incentive mechanisms which would ensure that agricultural biotechnology research might be targetted to the needs of poorer farmers or social groups. In essence, many public bodies involved in agricultural biotechnology research with a poverty alleviation mandate may experience 'mission creep' whereby public sector research agendas are more targetted to commercially lucrative markets or clients, rather than social markets or poorer clients. This is a public policy problem that can only be addressed by governments, their institutions and publically funded researchers.

3.2. Labour Saving Approaches to Weed Management

While dangerous pesticides have been over-used in many developing countries, herbicide use has been very low even though most herbicides are far less toxic than pesticides. Yet weed control is a major time, labour and resource-consuming task especially for resource poor farmers with limited access to inputs such as affordable herbicides, etc. It is estimated that these farmers spend more than 60% of their time weeding. Much of the weeding is done by unpaid women and children (Halos, 1992). There is a convincing case to be made that herbicide-resistant crops could in the near term offer significant, economically accessible, advantages to many farmers in developing countries, in particular poorer farmers with limited labour availability (Gressel et al., 1996a).

In particular, there are many weed problems faced by farmers in developing countries with no effective control measures in existence, with or without the application of herbicides. These include the parasitic broomrapes and witchweeds (*Striga* spp). The areas infested with such weeds are vast and expanding. For example, a survey of 18,000 square kilometres in Nigeria found that 70% of fields were infested with witchweed seeds (Hartman and Tanimonure, 1991). In the seven agro-ecological zones of sub-Saharan Africa witchweeds are generally listed as the worst pests affecting agriculture (Akobundu, 1991). Witchweeds infest the grain crops of more than 100 million people in sub-Saharan Africa and Asia, reducing yields by 50%, and by more in drought years. Labour intensive weeding is largely ineffective against such weeds. Crop yields could potentially be doubled if such weeds could be controlled. In addition labour spent in weeding could be released for other more productive activities, such as increasing literacy and schooling for children.

Control of such weeds has proven to be a formidable challenge for IPM based approaches. ICIPE (Kenya) has recently reported success in the control of stemborers, termites and the parasitic weed striga using a push-pull intercropping strategy whereby one inter-crop repels the stemborers (the push) and another attracts (pulls) the pests. In addition to its stemborer repelling effect, the leguminous silverleaf (*Desmodium uncinatum*) inter-crop has the added advantage of suppressing *Striga* infestations when it is inter-cropped with maize (http://www.icipe.org/agriculture/maizepest.html).

In recent years it was also discovered that it is possible to control *Striga* spp using imadizoline-resistant maize (Abayo et al., 1996). One such strategy of potential utility to developing country farmers is being developed whereby only the transgenic seed is treated with high levels of a systemic imadizoline with a resultant excellent control of *Striga*. Using $5 of herbicide gave $100 of increased maize yield per hectare in *Striga* infested areas in Kenya (Gressel et al., 1996a). Such strategies would require resistance management measures to ensure that *Striga* resistance to the herbicide would not evolve (Gressel et al., 1996b). Herbicide-resistant crops could form part of integrated weed management systems, where resistance management strategies are used to ensure that herbicide tolerance does not develop in the weed flora. However it would seem that current biosafety regulations will limit African farmers access to herbicide tolerant crops in the near term, even for crops such as maize which have no weedy wild relatives in Africa.

The FAO held a workshop on regulating herbicide tolerant crops in 1998 and is reported to be in the process of publishing Guidelines for the Regulation of Herbicide Tolerant Crops. The Nuffield Council on Bioethics has expressed some concern about the labour displacement potential of herbicide tolerant crops, especially if used to promote no-till farming (Naylor, 1994; Lipton, 1999). Questions have also been raised regarding the differential gender impact of herbicide tolerant crops and/or the utility of herbicide tolerant crops for polycropping and intercropping systems, where the secondary crops are also of value.

3.3. Other Notable Improvements in Desirable Crop Traits

Hidden hunger, such as protein and micronutrient deficiencies, is a widespread and endemic problem for the worlds poorest people, especially women and children. A range of transgenic approaches have now been developed to nutritionally improve the amino acid profile of crop protein. This is achieved either by transferring genes encoding more nutritious proteins from other species (Molvig et al., 1997), or by manipulation of crop biosynthetic pathways to increase the nutritional profile of endogenous proteins (Karchi et al., 1993). Where transformation protocols have been developed, many important legumes (whether landraces or modern varieties) such as peanut, beans, clover, etc., could feasibly be nutritionally improved through transferring methionine-rich protein genes from a species such as sunflower (Molvig et al., 1997; Khan et al., 1996). Sunflower seeds, unlike Brazil nuts, are not known to cause any allergic reactions.

Insufficient intake of dietary vitamin A is implicated in the death of approximately 1–2 million children annually (Humphrey, 1992). In South-East Asia, every year an estimated 5 million children develop the eye disease xerothalmia (Sommer, 1988). Unfortunately, many staple crops such as rice are deficient in dietary vitamin

A. In addition, the vitamin A containing tissues of rice (embryo, aleurone layer) are removed during milling. Genetic engineering approaches have now developed milled rice which accumulates vitamin A to provide one additional means of facilitating increased dietary intake of vitamin A (Burkhardt et al., 1997). If technology transfer of such vitamin A rich crops can reach its intended clients, it is likely that the transgenes used to increase vitamin A production could be applied to other crop species or varieties in locations where vitamin A deficiency is a medical problem. Similar approaches to combating micronutrient deficiencies by increasing both the content and availability of iron in transgenic rice are showing much promise (Goto et al., 1999).

The biosynthetic pathways which produce commercially interesting compounds in plants and other species can now be manipulated by 'metabolic engineering' of the pathway so that higher levels of desirable compounds may be obtained. For example, seed oil content has been increased using this strategy (Zou et al., 1997). Such strategies are currently the focus of much commercial biotechnology interest (Murphy, 1999). It is now becoming possible to tailor the specifications for the modification of vegetable oils in transgenic plants that more specifically meet end user needs (Knauf, 1995). One example is the production of higher levels of laurate in rapeseed (*Brassica napus*) (Voelker et al., 1996). Metabolic engineering approaches are also being explored for the transfer of the key biochemical components of C4 photosynthesis to those crops which rely on the less energy efficient C3 photosynthesis (Ku et al., 1999).

Similar antisense or gene-suppression technologies approaches can be used to 'switch off' or control the timing of the production of undesirable (or desirable) compounds (Senior, 1998). For instance, the Flavr Savr tomato developed by Calgene uses anti-sense technology to suppress one of the genes responsible for ripening, so that tomatoes remain on the vine longer and become sweeter without going soft. Major improvements in the efficiency of such procedures have recently been developed (Waterhouse et al., 1998). The introduction of genes which delay ripening or spoilage, could help to reduce the post-harvest losses. Some perishable vegetables and fruit markets may gain, especially in situations where poor farm-to-farm-to-market roads, inadequate transportation and storage facilities exist to exacerbate post-harvest losses. Similar types of transgenic approaches are being extended from tobacco (Gan and Amasino, 1995) to other crops such cassava.

A number of transgenic methods have been developed for changing the molecular structure of plant structural or storage compounds so that the crops are more digestible for either humans or domestic animals (Halpin et al., 1994). For instance, lignin is a plant compound that adversely affects pulp and paper production processes and which also lowers the nutritional value of animal feeds. Mutant lines of maize, sorghum and pearl millet have been described that have a reduced lignin content and improved digestibility (Cherney et al., 1990). A number of transgenic strategies for the manipulation of lignin quantity and quality have also now emerged (Chapple and Carpita, 1998).

Human disease is a major constraining factor to labour availability in many agricultural projects and to socio-economic development in general. Lack of effective cold storage facilities limits the efficacy of linear supply chains for many vaccines. Vaccine production can also be expensive. Production of effective oral vaccines against major tropical diseases in transgenic plants may be an extremely appropriate and low technology means of decentralising both vaccine production and distribution in developing countries (Walmsley and Arntzen, 2000). The potential feasibility of producing oral vaccines in transgenic plants has now been demonstrated for diseases such as cholera-toxin B (Arakawa et al., 1998) and hepatitis B (Mason et al., 1992; Richter et al., 2000). If they are made widely accessible, such transgenic plants may be of major utility to hospitals and medical centres in providing a reliable and cost effective supply of heat-stable vaccines and other protein based pharmaceuticals (Yu and Langridge, 2000).

3.4. Efficiency in Domestic Animal Production

Population growth combined with increasing incomes and urbanisation primarily drive the demand for animal products and feed grains. This is especially so for poultry, pig and milk production. There have been significant increases in the production efficiency of most livestock products in the industrial world, primarily through development and use of improved technologies in genetics, health, nutrition and management. With limited resources, it is critically important that growth in efficiency rather than in numbers should be the dominant factor in the doubling of global output of livestock products that is expected to take place in the next 25 years. Improvements in efficiency arise from the development, spread and adoption of improved technologies for breeding, feeding, management, and healthcare of animals. The new molecular technologies and *in vitro* reproductive technologies have been responsible for significant progress in research on livestock genetics, physiology and health, in the context of improved production efficiency. Agricultural biotechnologies are expected to play an increasing role in developing environmentally friendly animal production systems (Cunningham, 1999; FAO, 2000b).

In vitro reproduction technologies have had major impacts on animal production, especially in the intensive meat and dairy sectors. For instance, in the OECD countries Artificial Insemination (AI) has now been a practical technology in dairy cattle for over 50 years. Long-term semen and embryo storage, without loss of viability, is a valuable technique for assisting conservation in endangered breeds of most of the more important farm animal species. Embryo storage and transfer (ET) techniques have been developed to increase reproductive rates of selected cows so that genetically outstanding cows can contribute more to the breeding program. Because the cost of ET is high, use is almost confined to the trade in high value pedigree animals. The application of AI and ET in a manner which does not displace local breeds and varieties is an important policy issue regarding the conservation of animal genetic resources.

Genetic marker technologies, such as marker-assisted selection, parentage identification, and gene introgression can equally be applied to livestock selection programs (Davis and DeNise, 1998; Axford et al., 1999). Highly saturated genetic maps are now available for cattle, swine, and sheep to provide the genetic framework for developing marker assisted selection (MAS) programs. Clonal propagation of genotypes of crops such as potato and cassava is the norm for such crops, and somatic embryogenesis in plants was first demonstrated in the 1950s. In the animal arena, the advent of the cloning of farm mammals such as the sheep named 'Dolly' has only now become at least a technical possibility within domestic animal breeding programmes (Ashworth et al., 1998).

Partly because the vast majority of agronomic traits in livestock improvement are quantitative, transgenic technologies are currently not widely used in animal improvement programmes for agricultural purposes (Cunningham, 1999). The impact of transgenic animals on animal breeding and production is presently very limited as there is a dearth of single gene traits in livestock and the propagation process of a transgene in an animal population is relatively slow (Cunningham 1999). Transgenic animals have mainly been developed for 'niche' markets such as production of high value pharmaceutical proteins where transgenic plants cannot fulfil the same function (Rudolph, 1999).

Vaccines against brucellosis, encephalitis, liverfluke, hepatitis, etc., have been developed for domestic farm animals and poultry. It is likely that some of these vaccines can also be produced in transgenic plants using the same processes as developed for human vaccines. A range of new vaccines have been or are under development for animals diseases of importance to commercial animal production. There have also been notable successes with DNA vaccines, which if continued may represent a more cost-effective means of both developing and distributing vaccines in many resource poor situations (Beard and Mason, 1998; Krishnan, 2000).

3.5. MICROBIAL BIOTECHNOLOGIES

Microorganisms are essential components of agricultural ecosystems. For instance, beneficial soil microorganisms such as rhizobia and ectomycorrihizae contribute greatly to agricultural productivity. Conversely, most crop and animal diseases are caused by pathogenic microorganisms. Microbial biotechnology is proceeding at a more rapid pace than other biotechnology sectors simply because the genomes of many scientifically or commercially important microorganisms are typically smaller in size and hence can be easier to analyze. Because of their importance and the relatively higher levels of funding for medical biotechnology research, most of the microbial genomes currently being sequenced are human pathogens (Saunders and Moxon, 1998). In 1995, *Haemophilus influenzae* became the first free-living organism to have its entire genome sequence published (Jenks, 1998).

By 1999, the sequence of the genomes of 20 different microbial organisms had been completed and it is known that the sequencing of an additional 69 different microbial genomes is at an advanced stage in both public and private sector research institutions (http://www.tigr.org/tdb/mdb/mdb.html). In 1996 the cyanobacterium *Synechocystis* became the first completely sequenced photosynthetic organism, of critical importance to plant biology for functional and evolutionary comparisons (Kaneto et al., 1996). To date few microbial genomes of direct importance to agricultural productivity are being completely sequenced. A notable exception has been the sequencing of the genome of the plant pathogen *Xylella fastidiosa* by a decentralised consortium of Brazilian laboratories (Lambais et al., 2000). Important advances are being made regarding the underlying biology of symbioses between nitrogen fixing bacteria and leguminous crop species (Bladergroen and Spaink, 1998).

3.6. The Utility of Plant Tissue Culture and Micropropagation

The rapid propagation of many desirable plant varietal genotypes using plant micropropagation technology is a relatively low technology 'appropriate' biotechnology which is now delivering tangible benefits to many farmers in both developed and developing countries (Sasson, 1998; Van Uyen and Van der Zaag, 1993; Bryan, 1988). In addition to its rapid propagation advantages, such tissue culture can also be used to generate disease-free planting materials. There are numerous examples of micropropagation initiatives which are delivering disease-free planting materials to poorer farmers. These include local level micropropagation work on taro in Samoa, *Musa* spp and multipurpose trees in Kenya, potato in Vietnam (Van Uyen and Van der Zaag, 1993) and cassava in Colombia.

Micropropagation techniques have now been developed for a wide range of crops. For instance, China now has developed micropropagation technology for more than 100 crop species (Sasson, 1998). In the Guangdong province, 3–4 million micropropagated banana plantlets are produced annually and 1 million exported. Micropropagated bananas are reported to have been successfully adopted by poorer farmers in Guangxi and Guangdong in China. In 1994 it was estimated that the farmers in Guangxi received an extra income of about $723,000 as a result of adoption of approximately 600,000 disease-free plantlets. Similarly, 10% of the area of China planted to potatoes was derived from micropropagated virus-free material in the early 1990s and yields are reported to have increased by up to 100–200%.

Micropropagation of both food and export crops is also now routine in many Latin American and Caribbean countries (Sasson, 1998). Large-scale micropropagation is conducted for crops such as coffee, banana, plantain, taro, cocoa, cocoyam, sweet potato, apple, blueberry, raspberry, pineapple, citrus, grapes, papaya, mango, guava, potato, kiwi, cherry, pear, ornamentals, and yams. The existing and potential benefits of adoption of such micropropagated plants by farmers have been substantial as testified by the following selected examples (Sasson, 1998):

- In Mexico about 2.6 million people are dependent on coffee cultivation and production, However coffee rust affects 90% of coffee plantations. The Mexican government is facilitating the distribution of disease-free plantlets both to protect labour in this area and export earnings.
- Costa Rica is the world's second largest banana producer. The national banana corporation, Corbana, a semi-public company is undertaking large-scale micropropagation of nematode-free plantlets. The combination of a fallow period with nematode-free plantlets has eliminated the need for the use of nematicides in Corbana's plantations since 1987.
- At the Federal University of Rio de Janeiro, Brazil meristem culture of sweet potato cultivars to eliminate viruses and pathogens, followed by field trials around Rio de Janeiro resulted in the raising of yields from 9 to 19 tonnes.
- In Argentina, a company called Tecnoplant SA provides a service whereby it will 'clean' varieties delivered by customers from viruses and other pathogens, through tissue culture methods. It does this under contract for large-scale producers and planters. With appropriate subsidies such initiatives could be extended to producing disease-free planting material of the locally adapted varieties of low income farmers.
- At the Biotechnology Institute, Santa Clara, Cuba clonal propagation of banana following *in vitro* micropropagation produces about 5 million plantlets annually. This is done using relatively low tech facilities at 25–30°C with the result that yield from tissue culture derived plantlets was 30% higher than from conventional planting material.

Micropropagation capacity is less well developed in most African countries, yet it represents a technology which if better integrated with ongoing efforts in seed/planting material production and supply, could yield significant agronomic benefits to farmers. There are few links between genetic resources conservation initiatives and micropropagation initiatives for the more rapid supply of a wide range of healthy planting materials to farmers. Plant tissue culture and micropropagation capacity if effectively coupled to local seed/planting material delivery channels (private sector, state sector and informal sector) could generate benefits for many resource poor farmers. The application of such biotechnologies to local varieties and landraces of root and tuber crops could generate disease-free plantlets while helping to both boost yields and also limit 'genetic erosion'.

3.7. The Development of Apomixis Technology

Apomixis is a naturally occurring phenomenon whereby some plant species can produce seeds without fertilisation. While apomixis has been described in over 400 different plant species it is only found in a few crop species. The harnessing of apomixis genetics for heterosis breeding and general crop improvement could have significant implications for agricultural research. One potential benefit is that it may be possible to develop true breeding hybrids which retain their yield advantages over generations. Indeed, the long lists of potential agronomic benefits that could be derived from apomictic systems which are easy to use in crop improvement suggests that apomixis may be one of the most important targets for concerted international research efforts (Jefferson, 1994; Jefferson and Bicknell, 1996). In theory apomixis could allow both plant breeders (and possibly farmers) to genetically adapt plants to specific micro-environments, rather than the current practice of adapting the overall cultivation environment to the crop plants requirements. Hence, through the use of apomixis plant breeding could become extremely rapid and responsive to specific micro-environments, cropping conditions and markets. This in turn could stimulate diverse strategies for more sustainable agro-ecosystem management and could have profound implications for biological resource management within agricultural systems. The development of apomictic technology is one biotechnology that could provide major food and livelihood security benefits to farmers in developing countries. Some advances have been made in the application of biotechnology for the development of apomictic crops and a number of promising research approaches are now underway (Grossniklaus et al., 1998). If any of the existing research approaches are successful, it is thought that apomictic crops may be developed within the next decades. However broad social benefits are only likely to occur if apomixis technology can be made accessible to developing countries and in particular resource-poor scientists and farmers.

4. Transgenic Biotechnologies and Biosafety Concerns

Since the first developments of transgenic organisms (GMOs, LMOs) through recombinant DNA technology there has been a vast body of scientific research undertaken on risk assessment regarding the use of different types of transgenic organisms. Scientific risk assessment procedures regarding transgenic organisms are now an active and specialised area of scientific research. UNIDO and UNEP maintain a roster of scientists who have recognised expertise in biosafety related risk assessment regarding GMOs and a list of national contacts (http://irptc.unep.ch/biodiv/).

Living organisms (whether transgenic or not) retain the ability to disperse to new environments, colonize and multiply there. All transgenic organisms are assessed for risks associated with human health including food toxicity/safety, the presence of gene products conferring allergies, long-term effect of antibiotic resistance genes . In addition there are also perceived environmental risks such as the increased fitness or invasiveness of organisms containing transgene products, the effects of transgene products on non-target organisms, the effect of transgenic organisms on current agronomic practice and instability in the expression of some early-generation transgenes. There are also impacts which may require assessment at the level of the agricultural system including the potential for the development of 'superweeds' (Gressel, 1999), alteration in nutritional value of crops and the loss of beneficial biodiversity. General concerns regarding transgenic organisms (especially for consumption) include a loss of familiarity, and ethical issues such as labelling.

To avoid technology prejudice, all risk assessments on transgenic organisms should be compared with substantially equivalent non-transgenic organisms from whence the transgenic was derived. For instance, risk assessment of transgenic maize plants expressing the *Bacillus thuriengiensis* insecticidal protein should be comparative to conventionally bred insect-resistant and insect-susceptible maize varieties, and also compared with the risks from the use of Bt-protein containing insecticidal sprays. These very necessary base-line studies are being established for conventional agriculture against which risk analysis of transgenic crops can be conducted.

The basic features of risk assessments include risks to animal and human health, risks to the environment, risks to agriculture and general risks. Whether a particular type of risk has to be assessed in a particular location or context will largely depend on the risk itself. Hence, there may be possibilities for international harmonisation of food safety including testing of most transgenic foods, whereby if they have not proved uniformly detrimental to human health in one country, they are unlikely to do so in another country. Such risks can then simply be assessed based on what is already known about human dietary biology regarding conventional foods (e.g. allergies and food intolerances that may be specific to certain individuals or cultural groups). However, other types of risks such as transfer of genes which confer a selective advantage in a particular ecological niche, or adverse secondary effects on non-target organisms may have to be assessed within the particular context or agro-ecosystem which they are to be applied. It will be possible to develop hierarchical procedures for international harmonisation of biosafety regulations.

On a crop by crop basis, many studies have now been done of pollen dispersal from transgenic crops and gene transfer from transgenic crops to wild relatives. Such 'transgene independent' plant based studies have shown that the overall likelihood of gene-flow from the cultigen can be estimated for any particular plant species at any particular location. Hence, while gene flow from transgenic potatoes to its wild relatives is virtually impossible in most of Europe it is more probable in the centres of diversity of the potato in the Andean region (FAO, 1995).

The scientific consensus emerging from the vast range of biosafety studies of transgenic plants is that each case should ideally be evaluated on a case by case basis. Hence biosafety decisions might differ according to the particular type of transgene, crop, environment and end-use involved. Useful analysis tools for such evaluations can be the general concepts of 'substantial equivalence' (OECD, 1993b) and 'familiarity' (OECD, 1993b). The concept of substantial equivalence was first introduced in 1993 by the OECD and was subsequently endorsed by the WHO, the FAO and many governments including Japan, Canada, USA, UK, and the European Community (Banati, 2000). In devising process-based biosafety legislation there is now a conflict regarding what emphasis to place on two quasi-legal conceptual principles: the Precautionary Principle and the Principle of Familiarity, both of which are proposed by opposing groups as primary guidance for risk assessment (Matthee and Vermersch, 2000). Many scientists are questioning the objectivity of a 'no risk' Precautionary Principle which applies only to genetically engineered traits/varieties and does not compare or apply the same Principles to the possible health and environmental effects of conventionally bred traits/varieties. A range of points and counter-points which illustrate the wide range of interpretation of precautionary principles can be found at: http://www.cid.harvard.edu/cidbiotech/.

However, there is little doubt that biosafety regulations should exist for introduction of any new living organisms (whether transgenic or not) to an ecological niche and should be based on scientific analysis of the risks and benefits from such introductions. The application of the science of ecology to risk assessment is to be welcomed because of past knowledge gained from exotic species introductions.

4.1. The Origins of a Need for Transgenic Approaches to Crop Improvement

Many discussions on the merits or drawbacks of modern plant biotechnologies rarely separate the vast range of technologies commonly referred to as 'modern plant biotechnologies'. Generalisations regarding the utility (or not) of biotechnologies as a generic category to different groups of farmers are usually not very meaningful or informative.

From a public funding and regulatory (biosafety) perspective it is worth considering that not all biotechnologies generate transgenic or so-called 'genetically modified' organisms (GMOs). Modern biotechnologies such as vaccine development, antibody production, immunodiagnostics, molecular genetic mapping, marker assisted breeding and plant tissue culture are also highly useful technologies which can be applied within any particular crop genepool to generate improved varieties/breeds which are not transgenic and hence outside the increasingly onerous restrictions of current biosafety legislation.

Conventional plant breeding has been extremely successful and increased financial support for plant breeding (and basic research to further improve our knowledge of plant genetics) will be necessary if plant breeding is to both maintain and improve agricultural yields (Cooper et al., 2000). However, there are some limitations inherent in conventional plant breeding such as the lack of practical access to useful germplasm due to sexual incompatibility barriers or undesirable linkage blocks and concomitant time lags in incorporating useful genes into existing varieties (Spillane and Gepts, 2000).

Most crop genepools are typically depauparate in a range of agronomically useful traits (e.g. protein quality, abiotic stress tolerance, virus or viroid resistance, etc), that may be available in the genepools of other crops or species. For instance, resistance to softrot (*Erwinia carotovara*) is lacking in the potato genepool and causes crop losses estimated at $100 million per year worldwide. Another example is resistance to rice yellow mottle virus (RYMV) which is lacking in many popular rice varieties, and sources of resistant germplasm are only found in some African rice landraces, from which it is difficult to transfer by conventional breeding crosses (Pinto, 2000). As there were no conventional solutions to this disease problem, a transgenic approach for control of RYMV disease has been developed (Pinto et al., 1999). Indeed, many transgenic approaches to crop improvement arise from a lack of suitable conventional approaches available to deal with a particular agronomic problem (e.g. rice sheath blight, cassava mosaic virus, potato leaf roll virus, black sigatoka in plantains). For many serious pests or pathogens, transgenic approaches may provide new control options where current options are lacking in their efficacy or existence. Transgenic approaches can therefore be of use for a range of crops and areas where there are limited options available through conventional breeding e.g. nuclear male sterility, improved heterosis breeding, reducing toxic compounds, herbicide tolerance, generating novel resistance genes.

Transgenic approaches have considerably broadened the range of genepools which are now accessible for crop improvement purposes (Flavell, 1999). For instance, the application of useful gene transfer from microorganisms through genetic engineering techniques range from the introduction of vaccine antigen genes (Mason et al., 1996; Arakawa et al., 1998) to aluminium tolerance genes (De la Fuente et al., 1997) to food plants. Isolated plant genes (such as those conferring resistance against pests and pathogens) can now be usefully transferred between sexually incompatible crop plant species (Whitham et al., 1996; Wilkinson et al., 1997; Molvig et al., 1997). Transgenic technologies are currently not widely used in animal improvement programmes for agricultural purposes. Transgenic animals have mainly been developed for 'niche' markets such as production of high-value pharmaceutical proteins where transgenic plants cannot fulfil the same function.

In the context of ongoing debates regarding transgenic crops, public funding agencies should not forget that many modern agricultural biotechnologies such as plant tissue culture, molecular genetic mapping and marker assisted selection could still have a major impact on any conventional crop improvement approaches which decide to limit themselves to the genetic variation accessible within the primary to secondary genepools.

4.2. Biodiversity and Plant Biotechnology – Gene Flow Considerations

There is currently no concrete evidence either way to suggest that transgenic crops or agricultural biotechnology innovations per se would either decrease or increase biodiversity in agricultural or 'natural' ecosystems. Indeed, any tendency towards crop monocultures was well established before any transgenic varieties existed, and was also well in evidence before the era of the 'Green Revolution' varieties (Smale, 1997; Wood and Lenne, 1997). A diversity of crop varieties and species in agro-ecosystems may have an intuitive and popular appeal (e.g agriculture in nature's image), to those seeking greater stability in agricultural systems. However, there is a paucity of ecological evidence supporting the diversity equals stability hypothesis in either agricultural or wild ecosystems (Wood, 1996, 1998; Wood and Lenne, 1999). For instance, there are many natural ecosystems that are highly productive and stable despite being species poor or genetically uniform pure stands (e.g. mangrove, papyrus, bamboo, water hyacinth, *Dipterocarp* forests, karite trees, bracken, *Spartina*, *Phragmites*) (Wood, 1996, 1998).

Within agricultural systems, plant biotechnology research could be applied to either increasing or decreasing genetic diversity of cultivated plants depending on the research objectives and the agro-ecosystem in question. For instance, the wild relatives of crops, although a major genetic resource, are actually rarely used in the breeding of plant varieties, because of practical difficulties in using such exotic germplasm in breeding programmes. With some modern biotechnological methods the use of such resources may increase (Tanksley and McCouch, 1997). Recent advances in agricultural genomics, marker assisted breeding and transgenesis suggest that useful genetic diversity is actually becoming more accessible to crop researchers. This has the potential that aggregate increases in genetic diversity within crop genepools could now practically be achieved through increased use of genes from wild relatives and other species (Tanksley and McCouch, 1997). Novel transgenic traits can add to the diversity of traits that are available in any crop genepool (Spillane and Gepts, 2000).

Plant micropropagation can generate many clones of a particular variety in an analogous manner to vegetative propagation of root and tuber crops. While plant tissue culture and micropropagation might (contingent on its objectives) possibly increase the propensity for monocultures, such techniques may also be used to generate and multiply healthy plantlets of diseased locally adapted varieties which without such intervention are likely to be abandoned by farmers. This tissue- culture approach has been used for the maintenance of traditional landraces of Andean potatoes (Iriarte et al., 2000). Reductions in broad spectrum pesticide applications through the substitution effects of resistance genes conferring specific resistances against agronomic pests may contribute to an increase in beneficial insect biodiversity in agricultural systems. If escalating demand for food due to population pressure is to be met without an agricultural expansion into natural areas containing high levels of biodiversity (e.g. tropical forests) then yields in high potential areas must be significantly increased. FAO has estimated that two thirds of the required agricultural productivity will have to occur through intensified use of lands already under cultivation. Plant biotechnology which substitutes renewable genetic inputs for chemical inputs is likely to be one source of the potential yield increases required for high-potential agricultural areas (FAO, 1999).

Invasive exotic species (such as zebra mussels, kudzu, water hyacinth, etc.) are major and well known environmental and agricultural problems world wide (Williamson, 1996a). In the context of biodiversity, the severe environmental and economic damage that can be caused by such 'genetically unmodified' exotic species introductions are likely to pose a much greater threat to biodiversity and ecosystems than transgenic crops *per se* (Barton et al., 1997). Yet, transboundary movements of exotic species which, on rare occasions result in the emergence of an invasive pest are unlikely under current phytosanitary legislation to be subjected to the same level of intense scrutiny or biosafety regulation as transgenic organisms (Williamson, 1996b).

Gene flow to related wild plants is a natural long-term phenomenon and not a novelty of transgenic plants – such geneflow is essential for species evolution. Crop wild relatives only account for a small proportion of the world's genebank accessions and it is generally agreed that the *in situ* conservation of such wild resources is preferable to the many difficulties in maintaining them under long-term *ex situ* conditions. Any slight risk potential of mono-transgene gene flow contributing to the genetic erosion of sympatric wild relatives should be assessed relative to other factors which are known to contribute to genetic erosion of wild relatives. Base line studies are being established which will allow any risks from transgenes to wild relatives to be assessed. Recent studies have shown that despite a century of geneflow from sugar beet production fields to adjacent wild sea beet populations in the Italian Po-valley no negative effects on the intra-species genetic diversity was measured.

Concern has been voiced about the perceived risks of transgene 'escape' to wild relatives of crops with the potential for creation of weeds with additional selective advantages exclusively conferred by a transgene (Rissler and Mellon, 1996). For any crop, the risk of any such geneflow will differ considerably according to a range of factors including the nature of the transgene and whether the crop is cultivated in a region where there are sympatric weedy wild relatives (Ellstrand and Hoffmann, 1990). Unlike transgenic food safety analysis, any such risk assessment has to be region or country specific and will differ widely according to the reproductive and agricultural harvesting biology of the crop in question (Dale and Irwin, 1995).

Risk assessment will also be contingent on whether the transgene in question can confer any selective advantage on the wild relative, either within or outside of agricultural ecosystems (Dale, 1994). In considering what would happen if gene flow of a transgene to weedy wild relatives there are many additional issues to be considered before any level of risk can be assessed. For instance, any long-term effects will be contingent on whether the transgene can confer a selective advantage and on the likelihood of persistence and spread of the transgene in weedy of natural populations (Linder and Schmitt, 1994; Mikkelsen et al., 1996). A detailed risk analysis methodology for assessing the risks of geneflow from herbicide tolerant crops to their weedy wild relatives has concluded that no generalisations can be made as to whether genetic engineering would either exacerbate or alleviate herbicide resistance (Gressel, 1997b).

Nonetheless, to maintain the efficacy of herbicides it is important to limit any potential for herbicide-resistant transgene flow into weedy wild relatives. Any such risk assessment should be done based on what is known about weed biology and be assessed relative to the risks and problems encountered with conventional herbicide tolerant weeds. The evolution of herbicide tolerance in weeds which are related to crop plants is very well documented (Gressel et al., 1996a, 1996b). The more than 10 million hectares of herbicide-resistant weeds that have appeared in the past 30 years all result from selection of naturally occurring herbicide-resistant mutants among weed populations. This has occurred for legume weeds in soybean, Abutilon in cotton and bromes in wheat, and is still occurring for major crops such as rice and wheat (Gressel, 1997). In most instances, the use of genetic isolating mechanisms such as male sterility and/or maternally inherited expression systems (Daniell and Varma, 1998) will substantially reduce any risks of transgene flow into sympatric weedy wild relatives (Gressel, 1999). However, some scientists question whether such risk minimisation strategies will be sufficient to appease those who are *per se* opposed to transgenic crops (Chamerlain and Neal Stewart, 1999).

In general, risks transgenic crops pose to biodiversity should ideally be assessed relative to other non-transgene related factors. These may include urbanisation, agriculture and land use changes, exotic plant introductions, conventional weeds, etc., which are likely to more drastically reduce the geographic ranges of useful crop wild relatives or biodiversity in general. Many risk assessment studies regarding transgenic crops fail to do comparative studies to assess each particular risk comparative to the levels of risk from other factors. Ideally risk assessment studies using transgenic plants should be conducted in parallel (using the same risk analysis indicators and criteria) with risk assessments of conventional farming, and organic farming.

4.3. NON-TARGET SPECIES

Another area of concern regarding transgenic biotechnologies and biodiversity is a potential impact of these technological changes on non-target species. One theoretical possibility is the potentially direct impact of introduced characteristics on non-target species (Pimentel and Raven, 2000). For example, a laboratory based study from Cornell University (USA) was published in 1999 which suggests that pollen from transgenic maize which produces an insect-killing Bt toxin from *Bacillus thuriengiensis* may have toxic effects on the larvae of the monarch butterfly (Losey et al., 1999). The monarch butterfly is not an endangered species in the USA. The larvae of the monarch butterfly feed on the milkweed plant which may grow in proximity to maize fields in some parts of the USA. Pollen from Bt maize was artificially dusted onto milkweed leaves and the pollen-dusted leaves used to feed monarch larvae, most of which died. The initial results of the study have since been exaggerated and extrapolated in the popular media and are viewed by anti-biotechnology groups as vindicating their fears about Bt transgenic maize.

However, the scientists who performed the initial studies of the effects of Bt toxin on Monarch butterflies (a non-target organism) have stressed that their lab-based studies cannot be extrapolated to predict the situation that might (or not) occur in farmer's fields. The study raised no new concerns regarding the susceptibility of butterfly species to Bt proteins because the target pests of most Bt microbial insecticides (e.g. such as Dipel(r) which is used in organic farming) are members of the same insect order (*Lepidoptera*) that includes moths and butterflies. Reductions in use of broad-spectrum insecticides through the use of Bt maize has been reported for control of the European corn borer in the USA. Because not all monarch butterflies are killed due to the proximity of their host plant to maize fields, such reductions may overall be of benefit to the Monarch butterfly and other insect species. Hence more comprehensive studies on biodiversity impacts must be completed, before the risks of direct, non-target impacts can be assessed (Hodgson, 1999; Wraight et al., 2000).

Although Bt corn pollen under certain circumstances has the potential of adversely affecting the population levels of monarch butterflies and other non-target *Lepidoptera*, these impacts are considered to be minimal when compared with habitat loss and the widespread use of pesticides throughout the ecosystem (Pimental and Raven, 2000). Field trials conducted since 1999 have not supported the suggestions of laboratory studies that concluded that Monarch or swallowtail butterflies would be threatened under natural field conditions by Bt-containing corn pollen. The US EPA recently concluded that the currently available evidence does not support the conclusion that registered (biotech plants) may cause unreasonable adverse effects on the environment (http://www.epa.gov/scipoly/).

There is always a possibility of unforeseen and unintended indirect impacts from any technological changes. For pest-resistant crops, these should effectively be raised during resistance management strategies employed to plan effective field tests (Shelton et al., 2000). Indirect impacts due to changes in agricultural crops (or cropping practices) might occur through large-scale changes in the pest population within agricultural environments, e.g. by reason of the loss of certain classes of insect life. Such losses might indirectly result in the loss of bird and other wildlife, themselves dependent on the continuing existence of particular forms of insects at given times of the year. Recent studies on herbicide-resistant crops and farmland biodiversity are gathering evidence on the dynamics of pest populations and their natural enemies (Woiwod et al., 2000; Cui and Xia, 2000). Changes in the field ecology of agriculture may result indirectly in changes in the biodiversity of crop pests and their natural enemies. Of course, the existence of such indirect effects is possible as much for other technological changes (chemical, cultivational) as it is for biotechnological changes. Once again it is important that the net effect of the technological change be considered, not just any discernible impacts. For example, the efficacy of using refuges in Bt-resistant crops as a management strategy to prevent the build up of Bt-resistant pests can be compared against other strategies to maintain the long-term efficacy of Bt insecticidal sprays (Gould, 2000).

4.4. Human Health and Transgenic Foods

Despite much searching, there is currently no scientifically accepted evidence to suggest that transgenic crops per se are any more or less toxic or allergenic than their conventionally bred counterparts (Ruibal-Mendieta and Lints, 1998). Indeed, genetic engineering approaches and other research approaches are underway to develop 'functional foods' or 'nutraceuticals' which would contain lower levels of allergens and toxins, or higher levels of beneficial compounds, than conventional foods (Kottke, 1998; Weiss, 1997; Astwood and Fuchs, 1996; Knauf and Facciotti, 1995). There is now a substantial body of research addressing the issue of whether transgenic foods have any negative effects on mammalian digestion. The emerging evidence is that the transgenic food available on the market for human or animal consumption are substantially equivalent to their conventionally bred counterparts and are safe for consumption (e.g. Burks and Fuchs, 1995; Berberich et al., 1996; Nida et al., 1996; Padgette et al., 1996; Hammond et al., 1996; Harrison et al., 1996; Brake and Vlachos, 1998; List et al., 1999; Pusztai et al., 1999; Taylor et al., 1999; Sidhu et al., 2000). There has been much misinformation circulated in the popular media regarding perceived dangers to human health from use of transgenic foods or 'Frankenfoods'. As a result of such misinformation, the majority of the public in many countries are not aware of the fact that any scientific evidence that transgenic foods are inherently unsafe is largely non-existent, despite intensive research to assess its safety. Changes in consumer attitudes over recent years have provided worldwide evidence of the effect of this campaign of misinformation (Lappe et al., 2000).

Many crops and foods have potential health risks associated with them, especially if they are not correctly prepared (e.g. cooked) for eating. There are dozens of crops that are toxic unless treated, or if taken in quantity (D'Mello et al., 1991). Many naturally occurring plant proteins and compounds can be anti-nutrients, toxic or allergenic. In addition, in humid areas a range of carcinogenic compounds (e.g. fumonisin and aflatoxins) can be produced by fungal growth on grain crops. A significant number of crop species are toxic if not cooked or prepared properly to reduce or inactivate such compounds. A low percentage of the human population experiences some form of food allergy (Bruijnzeel-Koomen et al., 1995; Hefle et al., 1996). In most instances, standard procedures for assessing toxicity (LD50) and allergenicity (*in vitro* test, skin prick tests) can equally be applied to conventional and transgenic varieties to identify those transgenics which are substantially equivalent to conventional varieties (Lehrer and Reese, 1997). Such standard testing procedures were sufficient to identify that a methionine-rich 2S albumin protein from the Brazil nut (*Betholletia excelsa*) was allergenic (Nordlee et al., 1996) to some people and hence was not as good a candidate. An alternative was a non-allergenic methionine-rich sunflower seed albumin gene which may be used to improve the nutritional content of legumes (Nestle, 1996).

Selectable marker genes (e.g. antibiotic resistance or herbicide tolerance genes) are used in constructing transgenic plants containing an associated transgene of interest, but are usually not required once the transgenic plants are produced. Existing biosafety regulations have been stringent enough to disallow corn-borer tolerant maize in Europe because of the extremely low risk of antibiotic tolerance spreading in bacteria in the rumen of cattle (Williamson, 1996b, Estruch et al., 1997). Considerable research on the widely used nptII selectable marker gene has shown that it is safe for both the environment and the consumer (Fuchs et al., 1993). Even though most selection markers in constructing transgenics are likely to pose little danger either to humans of the environment because of perceived consumer concern it is likely that future generations of final product transgenic plants will not contain selectable marker genes (Nap et al., 1992; Goldsbrough, 1992). This is because a number of quite efficient systems have now been developed for the development of 'marker-free' transgenics (Komari et al., 1996). More innocuous marker systems which are not based on antibiotic or herbicide resistance genes are also being developed (e.g. mannose-6-phosphate isomerase) (Joersbo et al., 1998). Hence it is possible that biosafety considerations regarding such genes may gradually become less of an issue as improved 'marker-free' transformation systems become available.

Any risk assessment regarding pest-resistant plants should be comparative with existing approaches to dealing with the pest in question while maintaining productivity levels (e.g. insecticides, cropping strategies, IPM, conventionally bred resistant varieties, transgenic varieties, no control measure). The use of transgenic crops expressing the Bt insecticidal protein can be assessed relative to the widespread use of insecticidal sprays composed of the entire bacterium (the active ingredient of which is the insecticidal toxin). For instance, it has been proposed that there may be certain health risks associated with the use of Bt insecticidal sprays that are not associated with the use of Bt transgenic plants (MacKenzie, 1999).

4.5. THE NEED TO UNPACK THE REASONINGS BEHIND THE LABELLING OF 'GM FOODS'

There is no scientific evidence that transgenic derived foods are any more injurious to human health than food derived from conventionally bred crops. While the labelling of transgenic foods may be seen as a right of consumers, in the absence of any known risks to the consumer it is a technology-prejudiced labelling system. In most cultures, food taboos exist which often are historically associated with perceived undesirable effects of a particular food in a particular cultural context (O'Laughlin, 1974; Odebiyi, 1989; Simoons, 1995; Harris, 1998). Strategic anti-biotechnology lobbyist activity against 'GM foods' has had an intended effect of inculcating novel food taboos among consumers regarding the consumption of such foods. The manufacturing of dissent regarding transgenic foods (in the absence of any unique health or environmental risks) is evidently considered by politicians to have created an increased demand for labelling and segregation of transgenic foods.

However the extent of such demand for non-transgenic produce can simply be assessed via voluntary rather than mandatory labelling requirements, i.e. if there is a major demand for segregated non-transgenic foods then voluntary labelling schemes will be adopted by food companies in response to such consumer demands. Most current certification and labelling standards for 'organically produced' food require that transgenic technology was not used on the production of the organic food, and currently provides the best example of a 'non-GM food' label. Indeed, for many people there are valid belief-based reasons for deliberately choosing certain types of food (or food preparations), for which they may pay a premium. The most well known and widely adopted examples are probably veganism or vegetarianism. Other religion- based examples are Kosher food (foods which are acceptable according to Jewish dietary laws) and Halal/Haram food (food categories which are lawful [Halal] or unlawful [Haram] according to the Holy Quran). Certification systems have been developed for all these classes of foods. For instance, there are Kosher certification authorities such as the London Beth Din or the Federation of Synagogues which give kosher certification to food commodities.

In the absence of known risks to human health from 'GM foods' the costs of segregation of transgenic foods will ultimately be borne by the consumer and will push transgenic food production towards larger-scale farmers where economies of scale regarding the transaction costs of segregation can be achieved (Barefoot et al., 1994). In 1999, the consultancy firm KPMG was commissioned to analyze the cost of compliance of mandatory 'GM food' labelling for Australia, and found that the cost of mandatory labelling would be about $3 billion or 7% of the Australian industry's $42 billion in sales (J. Bishop Grewell, pers. comm.). The KPMG consultancy report prepared for health ministers also warned that a stringent labelling regime could reduce key Australian food exports by up to 20%. Requests for traceability of segregated food lines derived from either transgenic *versus* non-transgenic sources (e.g. as proposed by the European Union's revised Directive 90/220) will add further costs, which will ultimately be borne by the consumers. It is estimated that food production costs could rise by between 6 and 17% under the new labelling regime for genetically modified food within the EU. Public policy makers should consider whether mandatory labelling of foods derived from transgenic crop varieties makes sense if such 'GM foods' have no health risks associated with them that are any different to foods derived from conventionally bred crop varieties.

5. International Policy Making Regarding Biosafety Concerns

Policy decisions taken in regard to biosafety regulations may have long-term implications for the sustainability of agriculture and food security. In particular, policy makers should realise that long-term negative implications for agriculture and food security can equally arise from having biosafety regulations which are either too lax or too stringent. If any countries expect over the long term to benefit from modern biotechnologies in their agriculture and food sectors, they will have to give serious consideration to the drafting of biosafety regulations which are tailored to meet their socio-economic needs.

Biosafety risk assessment procedures can vary widely between countries. While many OECD countries have functional biosafety systems, many developing countries do not. Many countries are now establishing national biosafety committees and biosafety regulations regarding the use of GMOs. There are also initiatives to harmonise biosafety regulations at the regional level. For instance, the South Asian Association for Regional Cooperation recently developed an agreement between seven countries on germplasm exchange and the future development of biosafety regulations (Jayaraman, 1999). The Biosafety Information Network and Advisory Service (BINAS) is a service of UNIDO monitoring global developments in regulatory issues in biotechnology (see http://binas.unido.org/binas/). The BINAS maintains an on-line database of the state of development of national biosafety legislation worldwide.

5.1. THE BIOSAFETY PROTOCOL TO THE BIODIVERSITY CONVENTION

At the international policy level there are 171 governments which are Parties to the Convention on Biological Diversity (CBD), see http://www.biodiv.org/biosafety/. Article 19.3 of the CBD requested Parties to consider a legally binding international Protocol for Biosafety, recognising the potential risks posed to biodiversity by living modified organisms (LMOs) resulting from biotechnology. The proposed Biosafety Protocol was intended to specify obligations for international transfer of LMOs and set out means of risk assessment, risk management, advance informed agreement, technology transfer and capacity building regarding biosafety.

In 1995 the Conference of Parties (COP) to the Convention on Biological Diversity (CBD) established a negotiation process to develop, in the field of the safe transfer, handling and use of living modified organisms, a protocol on biosafety, specifically focusing on transboundary movement of any living modified organism (LMO). The primary focus of these negotiations was to be the adverse effects of LMOs on the conservation and sustainable use of biological diversity. The inter-governmental negotiations of the draft Biosafety protocol reached deadlock in Cartagena, Colombia, in February 1999 (Masood, 1999). At that point issues still under negotiation were the scope of the protocol (only LMOs, or LMOs and products thereof), trade implications, the introduction of socio-economic considerations, liability, and the treatment of non-parties. Informal consultations on reviving the negotiations took place at meetings in Montreal (June 1999) and Vienna (September 1999).

The Cartagena Protocol on Biosafety was agreed by 130 governments in Montreal in January 2000 and represents a legally binding international agreement. LMOs are now legally defined as 'any living organism that possesses a novel combination of genetic material obtained through the use of modern biotechnology'. The objective of the Protocol is 'to contribute to ensuring an adequate level of protection in the field of the safe transfer, handling and use of living modified organisms resulting from modern biotechnology that may have adverse effects on the conservation and sustainable use of biological diversity, taking also into account risks to human health, and specifically focusing on transboundary movements'. The Protocol recognises that modern biotechnology has great potential for human well-being if developed and used with adequate safety measures for the environment and human health. The Protocol refers to the use of the precautionary approach contained in Principle 15 of the Rio Declaration on Environment and Development. The following paragraphs outline some of the key elements of the Protocol, based on the text of the Protocol.

5.1.1. *Scope*
LMOs covered by the Protocol are to include, for instance, various food crops that have been genetically modified for greater productivity or nutritional value, or for resistance to pests or diseases. The Protocol will not apply to the transboundary movement of living modified organisms which are pharmaceuticals for humans.

5.1.2. *AIA Procedures*
Under the Protocol, governments are supposed to signal whether or not they are willing to accept imports of agricultural commodities that include LMOs by communicating their decision to the world community via an Internet-based Biosafety Clearing House. The specific mode of operation of the Biosafety Clearing-House is to be decided by the Parties to the Protocol. An advance informed agreement procedure (AIA) procedure is outlined for transboundary movement of LMOs. However, the AIA procedures will not extend to transit or contained use of LMOs (e.g. within appropriate research facilities), and less stringent AIA procedures are to apply to LMOs intended for direct use as food or feed, or for processing. Stricter AIA procedures are to apply to LMOs (e.g. seeds, live fish) that are to be intentionally introduced into the environment.

Under the AIA procedure, exporters will be legally required to provide extremely detailed and accurate information to the competent authority in each importing country in advance of the first shipment. The Party of import then has to acknowledge receipt of the notification, in writing, to the exporter within 90 days of its receipt. The Party of import has then to provide in writing, within 270 days, to the exporter and to the Biosafety Clearing-House their decision as the whether the intentional transboundary movement may proceed. Failure of the Party of import to acknowledge receipt of a notification or provide a decision will not imply its consent to an intentional transboundary movement. Lack of scientific certainty due to insufficient relevant scientific information and knowledge is not supposed to prevent an importing Party from taking a decision, one way or the other. The Parties to the Protocol are now supposed to decide upon appropriate procedures and mechanisms to facilitate decision-making by Parties of import. A Party that makes a final decision regarding domestic use, including placing on the market, of an LMO that may be subject to transboundary movement for direct use as food or feed, or for processing is required within 15 days of making that decision to inform and provide detailed and legally accurate information to the other Parties through the Biosafety Clearing-House. However, this provision shall not apply to decisions regarding field trials.

If the 130 governments who are currently Parties to the Protocol formally agree that certain LMOs are not likely to have adverse effects biological diversity or human health, they can exempt such LMOs from the AIA procedure. Similarly, an importing Party may specify in advance to the Biosafety Clearing-House that imports of LMOs to it to be exempted from the AIA procedure. Parties may only enter into bilateral, regional and multilateral agreements regarding trans-boundary movement of LMOs if such agreements do not result in a lower level of protection than that provided for by the Protocol.

5.1.3. *Risk Assessment Procedures*

The Protocol also makes some general provisions regarding risk assessment procedures. It proposes that risk assessment should be carried out in a scientifically sound and transparent manner, taking into account recognised risk assessment procedures and/or expert advice of relevant international organisations. According to the Protocol, lack of scientific knowledge or scientific consensus should not necessarily be interpreted as indicating a particular level of risk, an absence of risk, or an acceptable risk. Risks associated with LMOs or products thereof, are supposed to be considered in the context of the risks posed by the non-modified recipients or parental organisms in the likely potential receiving environment. Risk assessment is to be carried out on a case-by-case basis. The Protocol outlines a six-step risk assessment methodology. The importing Party is required to ensure that risk assessments are carried out and may require the exporter to carry out and/or bear the cost of the risk assessments. The Protocol also makes provisions for risk management.

5.1.4. *Labelling*

The Protocol aims to ensure that shipments of commodities that may contain LMOs are to be clearly labelled. Each Party is required to take measures to require that relevant LMOs that are subject to intentional transboundary movement are handled, packaged and transported under conditions of safety, taking into consideration relevant international rules and standards. There are requirements for minimum but detailed documentation, which include:

- for LMOs intended for direct use as food or feed, or for processing to use the words ‘may contain’ LMOs. The detailed requirements for this are to be negotiated by governments over the next two years;
- for relevant LMOs destined for contained use to be clearly identified as LMOs;
- for LMOs that are intended for intentional introduction into the environment of the importing to be clearly identified as LMOs, in addition to more detailed documentation requirements.

5.1.5. *Other Provisions*

The Protocol also raises questions of liability and redress. According to the Protocol, the Parties are to adopt a process for the elaboration of international rules and procedures in the field of liability and redress for damage resulting from transboundary movements of LMOs over the next four years.

Under the agreement, the Protocol and the WTO are to be mutually supportive with a view to achieving sustainable development. At the same time, the Protocol is not to affect the rights and obligations of governments under any existing international agreements.

The Protocol will legally enter into force for its member governments shortly after 50 countries have ratified it. From that point on all Parties to the Protocol will be legally bound to take appropriate legal, administrative and other measures to implement their obligations under the Protocol. However, the protocol is supposed to be evaluated for its effectiveness by the Parties to the Protocol on a five yearly basis.

5.2. THE WORLD TRADE ORGANISATION

The World Trade Organisation is an important forum for the legal settlement of disputes between trading blocks. The issue of whether some biosafety-related regulations may constitute barriers to trade is likely at some stage to be considered by the dispute settlement mechanism of the WTO. The resolution of such disputes will likely depend as much on questions of political economy as on the state of knowledge within the biological sciences.

In particular, the WTO's Agreement on Sanitary and Phytosanitary (SPS) Measures is likely to be of increasing importance regarding explicit requirements for transparent, science-based risk assessment of material for import. In the case of beef hormones the USA alleged an infringement of the SPS by the EU[1] and the subsequent WTO Dispute Panel finding stated that risk assessment should not involve social value judgements made by political bodies (WTO, 1997). Upon appeal by the EU, the Appelate Body supported the Panel's decision and stated further that the precautionary principle did not override the requirements of the SPS Agreement to take into account relevant scientific evidence (WTO, 1998).

The scope of the SPS Agreement covers measures in trade that are intended to protect human, animal, and plant health or life. Although other WTO Agreements refer to international standards, only the SPS Agreement identifies specific organisations to be used for international harmonisation of standards under the Agreement. For instance, the FAO/WHO Codex Alimentarius Commission is named for human health and food safety while the FAO's International Plant Protection Convention (IPPC) is responsible for plant health. There may also implications for trade in transgenics in standard-setting under the Agreement on Technical Barriers to Trade (TBT Agreement). This Agreement covers a large number of technical measures concerning human, animal and plant life and health not covered by the SPS Agreement, and the environment.

The WTO's SPS Agreement concerns sanitary and phytosanitary measures (a) to protect animal or plant life or health from risks arising from the entry, establishment or spread of pests, diseases, disease-carrying organisms or disease-causing organisms; (b) to protect human or animal life or health from risks arising from additives, contaminants, toxins or disease-causing organisms in foods, beverages or feedstuffs; (c) to protect human life or health risks arising from diseases carried by animals, plants or products thereof, or from the entry, establishment or spread of pests; or (d) to prevent or limit other damage from the entry, establishment or spread of pests.

The SPS Agreement is based upon international standards, guidelines and recommendations:

(a) for *food safety*, the standards, guidelines and recommendations established by the FAO/WHO Codex Alimentarius Commission relating to food additives, veterinary drug and pesticide residues, contaminants, methods of analysis and sampling, and codes and guidelines of hygienic practice;
(b) for *animal health and zoonoses*, the standards, guidelines and recommendations developed under the auspices of the Office of International Epizootics (OIE); In particular, OIE guidance includes internationally agreed principles and methods for risk analysis with specific applications in the evaluation of risk and measures for animal diseases.
(c) for *plant health*, the international standards, guidelines and recommendations developed under the auspices of the Secretariat of the International Plant Protection Convention (IPPC) in cooperation with regional organisations operating within the framework of the IPPC; and
(d) for *matters not covered by the above organisations*, appropriate standards, guidelines and recommendations promulgated by other relevant international organisations open for membership to all Members, as identified by the Committee.

The SPS Agreement is based on several important principles. These include, for example, the principles of sovereignty, necessity, harmonisation, transparency, and equivalence. The same principles are reflected in the work programmes for global harmonisation under Codex and the IPPC. These principles provide the foundation for the elaboration of standards, guidelines and recommendations to be used in implementing the SPS Agreement. Measures based on the SPS principles and conforming to corresponding standards from the designated standard setting organisations are deemed acceptable without further justification. In instances where measures deviate from established standards, or where measures are established in the absence of standards, the SPS Agreement requires justification based on scientific principles and evidence. Risk analysis methods elaborated by Codex and the IPPC are to provide systematic frameworks for this purpose.

5.3. The WHO/FAO Codex Alimentarius Commission

In the specific context of labelling of transgenic foods the WHO/FAO Codex Alimentarius Commission (CAC) is of increasing international importance. Its current membership is 163 countries. Since 1962, the Codex Alimentarius Commission has been responsible for developing standards, guidelines and other recommendations on the quality and safety of food to protect the health of consumers and to ensure fair practices in food trade (Randall and Whitehead, 1997). Codex standards, guidelines and recommendations are based on current scientific knowledge including assessments of risk to human health. The risk assessments are carried out by FAO/WHO expert panels of independent scientists selected on a worldwide basis.

Codex standards, guidelines and other recommendations are not binding on Member States, but are a point of reference in international law (General Assembly Resolution 39/248; Agreement on the Application of Sanitary and Phytosanitary Measures; Agreement on Technical Barriers to Trade). The CAC is presently developing Recommendations for the Labelling of Foods obtained through Biotechnology (CAC, 1998). The CAC is also considering the development of a general standard which would apply basic food safety and food control disciplines to foods which are derived from biotechnology. The advice of prior FAO/WHO expert consultations on biotechnology and food safety will be used as guidance for the conditions required for foods prepared from biotechnology. The FAO states that foremost among these are consideration of potential allergenicity, possible gene transfer from LMOs, pathogenicity deriving from the organism used, nutritional considerations and labelling (FAO, 1999).

The 23rd Session of the Codex Alimentarius Commission (June/July 1999) established an *Ad Hoc* Intergovernmental Task Force on Foods derived from Biotechnology. The Task Force will submit a preliminary report to the Codex Alimentarius Commission in 2001, and a full report in 2003. The Codex Committee on Food Labelling is also working on the development of recommendations for the labelling of foods obtained through biotechnology. It is envisaged by FAO that Codex standards will apply to all types of foods, and, for this reason, Codex will need to deal with foods of plant, animal and fish origin. The impact of feeding GMO plants to animals, and the nature of the resulting foods from these animals will also be addressed.

5.4. The FAO International Plant Protection Convention (IPPC)

The IPPC was adopted by the FAO Conference in 1951 and came into force in 1952. It is recognised as the primary instrument for international cooperation in the protection of plant resources from harmful pests. There are currently 107 governments that are contracting parties to the IPPC. The IPPC's purpose is common and effective action to prevent the introduction and spread of pests of plants and plant products, and the promotion of appropriate control measures. It covers both cultivated and wild plants; the direct and indirect effects of pests; and the prevention of the introduction and spread of weeds, and their control. The IPPC also covers the movement of biological control agents, and other organisms of phytosanitary concern claimed to be beneficial. The IPPC provides the global standard setting mechanism for phytosanitary measures. It may be concerned with evaluating the potential 'pest' characteristics (including weediness) of GMOs, that is, whether a GMO may be detrimental to plant life or health. The IPPC allows parties to take phytosanitary measures, i.e. any legislation, regulation or official procedure having the purpose to prevent the introduction and/or spread of pests. These cover the pest concerned and may also cover any plant, plant product, storage place, packaging, conveyance, container, soil and any other organism, object or material capable of harbouring or spreading pests that are deemed to require phytosanitary measures.

The IPPC calls for phytosanitary measures to be based on a pest risk analysis, which covers both economic and environmental factors including possible detrimental effects on natural vegetation. The Convention also allows for the prohibition or restriction of the movement of biological control agents and other organisms of phytosanitary concern claimed to be beneficial into the territories parties. Any transgenic organism that can be considered a pest of plants falls within the scope of the IPPC and is likely to be subject to the provisions of the Convention.

The interpretation of areas such as risk assessment and the precise legal definition of terms such as injurious, economic and phytosanitary concern will have a bearing on the level of application of the IPPC to the regulation of some transgenic organisms. Where certain classes of transgenic organisms fall under the auspices of the IPPC, this will allow for their regulation under national phytosanitary legislation and providing quarantine services with the authority to take measures.

At the second meeting of the Interim Commission on Phytosanitary Measures (ICPM) in October 1999, a number of governments gave high priority to standard setting in relation to GMOs in particular to risk assessment and testing and release of GMOs. They indicated that this could be addressed within the framework of the IPPC. Others advocated a more cautious approach while some indicated the need to give sufficient priority to development of standards for plant quarantine. The ICPM decided that an exploratory working group would address the issues of biosafety in relation to GMOs and of invasive species and report back to the 3rd meeting of the ICPM in April 2001.

6. Conclusions

Scientific methodologies for risk assessment regarding the biosafety of transgenic organisms (e.g. crops, animals, biocontrol agents) in agriculture have been under development since it first became apparent that transgenic organisms could have a useful role to play in agricultural and other anthropocentric production systems. As scientific risk assessment has evolved it has become clear that most biosafety assessments regarding transgenics can only be meaningful when compared to the best or worst of conventional practices involving their non-transgenic counterparts (e.g. conventionally bred plant varieties or animal breeds). Many of the initial studies focussing only on the effects of transgenics failed to include adequate controls to allow comparative assessment with the effects of their non-transgenic equivalents. The emerging picture seems that there is no clear evidence for environmental risks regarding transgenic crops which are generically over and above those observed for conventional crops.

While many scientists (and their representative organisations) do not perceive any additional risk inherent in transgenic *versus* conventional crop varieties per se, the anti-biotechnology lobbying via the popular media has resulted in the public having higher perceptions of risk associated with agricultural biotechnology. Unlike the pharmaceutical sector, the agricultural biotechnology sector has suffered from a failure to communicate to the public the low relative risks associated with agricultural biotechnology products (Spillane and Thro, 2000). In the domain of public policy, public perceptions can set reality as politicians try to address the fears of their constituents. Hence, in parallel to the evolution of scientific methodologies for risk assessment, a substantial body of national and international policy and legislation has emerged over the past decade concerning the regulation of transgenic organisms in production systems, especially in response to popular concerns regarding the biosafety of transgenic organisms. Such biosafety legislation has tended to focus only on the transgenics, even though non-transgenics which are substantially equivalent may have similar effects.

The exaggeration of the risks of transgenic crops and concomitant over-regulation has in some regions led to reductions in public sector funding support for any plant genetics research. For instance, support for plant science research within the European Union (EU) has been dropping sharply (Heselmans, 1999). In recent years, there has been a two-thirds reduction in the number of successful plant research proposals funded by the EU. Dutch authorities are in the midst of implementing a 30% cut of plant research funding over five years and have ended funding for all collaborative projects. All Danish programs in plant biotechnology will expire by 2002, and no new initiatives are planned (Hodgson, 1999).

Because of emerging biosafety legislative regulations, there are likely to be high initial costs for the introduction of transgenic crops or products (Tripp, 2000a). Traditionally the public sector has played a facilitating role for the introduction of novel useful innovations to farmers and consumers. In some instances, this has included reducing the cost of entry for private firms (Pray and Umali-Deiniger, 1998). However, in the case of agricultural biotechnology it is by no means clear who will bear the burden of the initial high costs of agricultural biotechnology entry. It is probable that such costs are likely to be borne by the earliest adopters of any benefits that transgenic crops or products confer. If current concerns driving the drafting of stringent biosafety models transpire to be exaggerated, then such over-regulation regarding biosafety will have amounted to no more than a technology-specific tax which pushes agricultural biotechnology research towards meeting the needs of richer clients.

Risk management regarding transgenics in agriculture will require any relevant risks (which may not be novel) to be weighed against the expected benefits of developing and deploying such transgenics. Much of the risk assessment procedures that have been developed to date mainly focus on biological effects of the deployment of transgenic organisms. However, most transgenics have been developed in response to a particular social need and hence any risks should ideally be weighed against the expected social benefits that could accrue from the use of the transgenic innovation. In this regard, there has been a lack of economic analyses which conduct comparative cost-risk-benefit studies of transgenic *versus* non-transgenic approaches to addressing particular human needs. In some instances there will undoubtedly be conventional non-transgenic approaches which are perfectly adequate to addressing particular agronomic needs or problems (Spillane and Thro, 2000).

There are also important questions of representation and accountability regarding the issue of the main actors are that are developing/influencing biosafety regulations and whose interests they are ultimately representing. Perceptions of risk will differ between those whose livelihoods depend on technological innovations reaching them (e.g. farmers, poorer consumers, patients) and those whose livelihoods will be largely unaffected whether certain technologies are developed or not. There is a need for a new era of good governance and democratisation regarding decision making over biosafety regulations. A move towards more needs-driven decision making regarding biosafety and biotechnology policy should involve membership based organisations (e.g. trade unions, farmers organisations, consumers organisations, scientific bodies) (Tripp, 2000b; Spillane, 2000). When we enter such an era it will become possible to weigh the risks and benefits of transgenics or other biotechnological products against their conventional counterparts, in the context of the needs of different societies or social groups.

Selected Biosafety-related Websites

- Biosafety Clearing House of the Convention on Biological Diversity
  http://www.biodiv.org/biosafe/
- Biosafety Information Network and Advisory Service UNIDO
  http://binas.unido.org/binas/index.php3
- National Focus Points for Biosafety
  http://www.unep.org/unep/program/natres/biodiv/irb/focpts.htm
- Biosafety: National Contacts
  http://irptc.unep.ch/biodiv/cont03.html

- Convention on Biological Diversity – Working Group on Biosafety
  http://www.iisd.ca/linkages/biodiv/bios
- Third World Network viewpoints on biosafety
  http://www.twnside.org.sg/bio.htm
- Biotechnology Advisory Commission (Stockholm Environmental Institute) – an outline of the function of this organisation
  http://www.sei.se/envtech.html#BiotechnologyAdvisoryCommission
- Organisation for Economic Cooperation and Development Biotrack Online – a database of transgenic trials
  http://www.oecd.org/ehs/service.htm
- European Commission: GMOs in food and environment: database of trials in member states
  http://food.jrc.it/gmo/gmo.asp
- IRRO Databases on Environmental Releases (Includes OECD)
  http://www.bdt.org.br/bdt/irro
- Biosafety – Bioline's electronic only journal
  http://www.bdt.org.br/bioline/by

Note

[1] WTO disputes WT/DS26 (USA *versus* EU) and WT/DS48 (Canada *versus* EU).

References

Abayo, G.O., Ransom, J.K., Gressel, J. and Odhiambo, G.D. (1996) Striga hermonthica control with acetolactate synthase inhibiting herbicides seed dressed to maize with target site resistance, in *Advances in Parasitic Weed Research*, Junta de Andalucia, pp. 762–768. *Adv. Drug Deliv. Rev.* **43**, 3–11.

Akobundu, I.O. (1991) Weeds in human affairs in sub-Saharan Africa, *Weed Technol.* **5**, 860.

Anon (2000) Analysis of the genome sequence of the flowering plant *Arabidopsis thaliana*. The Arabidopsis Genome Initiative, *Nature* **408**, 796–815.

Arakawa, T., Chong, D.K.X. and Langridge, W.H.R. (1998) Efficacy of a food plant-based oral cholera toxin B subunit vaccine, *Nature Biotechnol.* **16**, 292–297.

Ashworth, D., Bishop, M., Campbell, K., Colman, A., Kind, A., Schnieke, A., Blott, S., Griffin, H., Haley, C., McWhir, J. and Wilmut, I. (1998) DNA microsatellite analysis of Dolly, *Nature* **394**, 329.

Astwood, J. and Fuchs, R. (1996) Preventing food allergy: Emerging technologies, *Trends Food Sci. Technol.* **7**, 219–226.

Axford, R.F.E., Bishop, S.C., Nicholas, F.W. and Owen, J.B. (1999) *Breeding for Disease Resistance in Farm Animals*, 2nd edn., CABI Publishing, Wallingford, UK, 411 pp.

Bagla, P. (1999) India prepares research, policy initiatives, *Science* **283**, 309.

Banati, D. (2000) The development of European food law – From the 'principle of the free movement of goods' to the 'precautionary principle', Part 1, *Elelmezesi Ipar* **54**(2), 46–48.

Banks, R.E., Dunn, M.J., Hochstrasser, D.F., Sanchez, J.C., Blackstock, W., Pappin, D.J. and Selby, P.J. (2000) Proteomics: New perspectives, new biomedical opportunities, *Lancet* **356**, 1749–1756.

Barefoot, S.F., Beachy, R.N. and Lilburn, M.S. (1994) Labelling of food-plant biotechnology products, Council for Agricultural Science and Technology, CAST Issue Paper No. 4, July.

Barton, J., Crandon, J., Kennedy, D. and Miller, H. (1997) A model protocol to assess the risks of agricultural introductions, *Nature Biotechnol.* **15**, 845–848.

Beard, C.W. and Mason, P.W. (1998) Out on the farm with DNA vaccines, *Nature Biotechnol.* **16**, 1325–1327.

Bennetzen, J.L., Kellogg, E.A., Lee, M. and Messing, J. (1998) A plant genome initiative, *Plant Cell* **10**, 488–490.

Bent, A.F. (1996) Plant disease resistance genes: Function meets structure, *Plant Cell* **10**, 1757–1771.

Berberich, S.A., Ream, J.E., Jackson, T.L., Wood, R., Stipanovic, R., Harvey, P., Patzer, S. and Fuchs, R.L. (1996) Safety assessment of insect-protected cotton: The composition of the cottonseed is equivalent to conventional cottonseed,. *J. Agric. Food Chem.* **41**, 365–371.

Betz, F.S., Hammond, B.G. and Fuchs, R.L. (2000) Safety and advantages of *Bacillus thuringiensis*-protected plants to control insect pests, *Regul. Toxicol. Pharmacol.* **32**, 156–173.

Bladergroen, M.R. and Spaink, H.P. (1998) Genes and signal molecules involved in the *rhizobia-Leguminosae* symbiosis, *Curr. Opin. Plant Biol.* **1**, 353–359.

Bouchez, D. and Hofte, H. (1998) Functional genomics in plants, *Plant Physiol.* **118**, 725–732.

Brake, J. and Vlachos, D. (1998) Evaluation of event 176 "Bt" corn in broiler chickens, *J. Poultry Sci.* **77**, 648–653.

Briggs, S.P. and Helentjaris, T. (1997) Plant genomics moves into the limelight. *Genome Res.* **7**, 865–867.

Broder, S. and Venter, J.C. (2000) Whole genomes: The foundation of new biology and medicine, *Curr. Opin. Biotechnol.* **11**, 581–585.

Brown, J.K.M. (1998) How to feed the world, in two contradictory lessons, *Trends Plant Sci.* **3**, 409–410.

Bruijnzeel-Koomen, C., Ortolani, C., Aas, K., Bindslev-Jensen, C., Bjorksten, B., Moneret-Vautrin, D. and Withrich, B. (1995) Adverse reactions to food, *Allergy* **50**, 623–635.

Bryan, J.E. (1988) Implementation of rapid multiplication and tissue culture methods in Third World countries, *Amer. Potato J.* **65**, 199–207.

Burkhardt, P.K., Beyer, P., Wunn, J., Kloti, A., Armstrong, G.A., Schledz, M., Lintig, J.V. and Potrykus, I. (1997) Transgenic rice (*Oryza sativa*) expressing daffodil (*Narcissus pseudonarcissus*) phytoene synthase accumulates phytoene, a key intermediate of provitamin A biosynthesis, *Plant J.* **11**, 1071–1078.

Burks, A.W. and Fuchs, R.L. (1995) Assessment of the endogenous allergens in glyphosate tolerant and commercial soybean varieties, *J. Allergy Clin. Immunol.* **96**, 1008–1010.

Cao, H., Li, X. and Dong, X. (1998) Generation of broad-spectrum disease resistance by overexpression of an essential regulatory gene in systemic acquired resistance, *Proc. Natl. Acad. Sci. USA* **95**, 6531–6536.

Carozzi, N. and Koziel, M. (eds.) (1997) *Advances in Insect Control: The Role of Transgenic Plants*, Taylor & Francis, London.

Chamerlain, D. and Neal Stewart, C. (1999) Transgene escape and transplastomics, *Nature Biotechnol.* **17**, 330–331.

Chapple, C. and Carpita, N. (1998) Plant cell walls as targets for biotechnology, *Curr. Opin. Plant Biol.* **1**, 179–185.

Chen, X.M., Line, R.F. and Leung, H. (1998) Genome scanning for resistance gene analogs in rice, barley, and wheat by high-resolution electrophoresis, *Theor. Appl. Genet.* **97**, 345–355.

Cherney, D.J., Patterson, J.A. and Johnson, K.D. (1990) Digestibility and feeding value of pearl millet as influenced by the brown-midrib, low-lignin trait. *J. Anim. Sci.* **68**, 4345–51.

Codex Alimentarius Commission (CAC) (1998) Proposed draft recommendations for the labelling of foods obtrained through biotechnology, Codex Committee on Food Labelling, 26th session agenda papers, CS/FL 98/8.

Coe, E.H. (1998) Potentials of the National Corn Genome Initiative, *Proc. Natl. Acad. Sci. USA* **95**, 2029–2032; http://www.inverizon.com/ncgi.

Cohen, J. (1997) Please pass the data, *Science* **276**, 1961.

Collins, F.S. (1999) Microarrays and macroconsequences, *Nature Genet.* **21**, 2–3.

Cooper, D.C., Spillane, C. and Hodgkin, T. (eds.) (2000) *Broadening the Genetic Bases of Crop Production*, IPGRI/FAO, CABI Publishing, Wallingford, UK, in press.

Cui, J. and Xia, J.Y. (2000) Effects of Bt (*Bacillus thuringiensis*) transgenic cotton on the dynamics of pest population and their enemies, *Acta Phytophyl. Sinica* **27**, 141–145.

Cunningham, E.P. (1999) Recent developments in biotechnology as they relate to animal genetic resources for food and agriculture, FAO, Rome, CGRFA Background Study Paper No. 10, 23 pp. *Curr. Opin. Biotechnol.* **11**, 126–129.

Dale, P.J. (1994) The impact of hybrids between genetically modified crop plants and their related species: General considerations, *Mol. Ecol.* **3**, 31–36.

Dale, P.J. and Irwin, J.A. (1995) The release of transgenic plants from containment, and the move towards their widespread use in agriculture, *Euphytica* **85**, 425–431.

Daniell, H. and Varma, S. (1998) Chloroplast-transgenic plants: Panacea – no! Gene containment – yes! *Nature Biotechnol.* **16**, 602.

Davis, G.P. and DeNise, S.K. (1998) The impact of genetic markers on selection, *J. Anim. Sci.* **76**, 2331–2339.

De la Fuente, J.M., Ramirez-Rodriguez, V., Cabrera-Ponce, J.L. and Herrera-Estrella, L. (1997) Aluminium tolerance in transgenic plants by alteration of citrate synthesis, *Science* **276**, 1566–1568.

De Vicente, M.C. and Tanksley, S.D. (1993) QTL analysis of transgressive segregation in an interspecific tomato cross, *Genetics* **134**, 585–596.

D'Mello, J.P.F., Duffus, C.M. and Duffus, J.H. (1991) *Toxic Substances in Crop Plants*, Royal Society of Chemistry, London.

Dunham, I . (2000) Genomics – The new rock and roll? *Trends Genet.* **16**, 456–61.

Duvick, D.N. (1996) Plant breeding, an evolutionary concept, *Crop Sci.* **36**, 539–548.

Ellstrand, N.C. and Hoffmann, C.A. (1990) Hybridization as an avenue of escape for engineered genes: Strategies for risk reduction, *Bioscience* **40**, 438–442.

Estruch, J.J., Chilton, M.-D., Lotstein, R. and Beversdorf, W. (1997) Safety of transgenic corn, *Nature* **385**, 109.

FAO (1995) Recent international developments of relevance to the draft Code of Conduct for Plant Biotechnology, FAO, Rome, Document CPGR-6/95/15.

FAO (1999) Biosafety issues related to biotechnologies for sustainable agriculture and food security: FAO's role, FAO, Rome, March.

FAO (2000a) Biotechnology, biosafety and biodiversity for sustainable agriculture. FAO Position Paper for UN Commission on Sustainable Development, review of Ch14 of Agenda 21, May 2000, UN, New York.

FAO (2000b) Modern biotechnology and the management of animal genetic resources, FAO, Rome, FAO Intergovernmental Working Group on Animal Genetic Resources, Paper CGRFA/WG-AnGR-2/00/4.

Flavell, R.B. (1999) Agriculture: A path of experiment and change, *Nature Biotechnol.* Suppl. 17, BV7.

Fuchs, R.L., Ream, J.E., Hammond, B.G., Naylor, M.W., Leimgruber, R.M. and Berberich, S.A. (1993) Safety assessment of the neomycin phosphotransferase II (NPTII) protein, *Biotechnology (NY)* **11**, 1543–1547.

Gale, M.D. and Devos, K.M. (1998) Plant comparative genetics after 10 years, *Science* **282**, 656–659.

Gan, S. and Amasino, R.M. (1995) Inhibition of leaf senescence by autoregulated production of cytokinin, *Science* **270**, 1986–1988.

Goldsbrough, A. (1992) Marker gene removal: A practical necessity?, *Trends Biotechnol.* **10**, 417.

Goto, F., Yoshihara, T., Shigemoto, N., Toki, S. and Takaiwa, F. (1999) Iron fortification of rice seed by the soybean ferritin gene, *Nature Biotechnol.* **17**, 282–286.

Gould, F. (1998) Sustainability of transgenic insecticidal cultivars: Integrating pest genetics and ecology, *Annu. Rev. Entomol.* **443**, 701–726.

Gould, F. (2000) Testing Bt refuge strategies in the field, *Nature Biotechnol.* **18**, 266–267.

Gould, F., Anderson, A., Jones, A., Sumerford, D., Heckel, D.G., Lopez, J., Micinski, S., Leonard, R. and Laster, M. (1997) Initial frequency of alleles for resistance to *Bacillus thuringiensis* toxins in field populations of *Heliothis virescens*, *Proc. Natl. Acad. Sci. USA* **94**, 3519–3523.

Green, P., Keegstra, K., Ohlrogge, J., Somerville, S., Cherry, M., Sussman, M., Amasino, R. and Dellaporta, S. (1998) Plans for a functional analysis of the *Arabidopsis* genome via global gene expression analysis and gene disruption, Abstract of meeting *The Arabidopsis Genome: A Model for Crop Plants*, December 3–6, 1998, Cold Spring Harbor Laboratory, USA, p. 46.

Gressel, J. (1997a) Burgeoning resistance requires new strategies, in R. De Prado (ed.), Kluwer Academic Publishers, Dordrecht.

Gressel, J. (1997b) Genetic engineering can either exacerbate or alleviate herbicide resistance, in *Proceedings of the 50th New Zealand Plant Protection Conference*, pp. 298–306.

Gressel, J. (1999) Tandem constructs: Preventing the rise of superweeds, *Trends Biotechnol.* **17**, 361–366.

Gressel, J., Ransom, J.K. and Hassan, E.A. (1996a) Biotech-derived herbicide resistant crops for third world needs, *Ann. New York Acad. Sci.* **792**, 140.

Gressel, J., Segel, L.E. and Ransom, J.K. (1996b) Managing the delay of evolution of herbicide resistance in parasitic weeds, *Internat. J. Pest Management* **42**, 113.

Grossniklaus, U., Koltunow, A. and van Lookeren Campagne (1998) A bright future for apomixis, *Trends Plant Sci.* **3**, 415–416.

Halos, S.C. (1992) Biotechnology trends: A threat to Philippine agriculture, in A. Iftikhar (ed.), *Biotechnology: A Hope or a Threat?*, MacMillan Press, London, UK, pp. 147–165.

Halpin, C., Knight, M.E., Foxon, G.A., Campbell, M.M., Boudet, A.M., Boon, J.J., Chabbert, B., Tollier, M.-T. and Scuch, W. (1994) Manipulation of lignin quality by downregulation of cinnamyl alcohol dehydrogenase, *Plant J.* **6**, 339–350.

Hammond, B., Vicini, J., Hartnell, G., Naylor, M.W., Knight, C.D., Robinson, E., Fuchs, R.L. and Padgette, S.R., et al. (1996) The feeding value of soybeans fed to rats, chickens, catfish and dairy cattle is not altered by genetic incorporation of glyphosate tolerance, *J. Nutr.* **126**, 717–727.

Hammond-Kosack, K.E. and Jones, J.D.G. (1997) Plant disease resistance genes, *Annu. Rev. Plant Physiol. Plant Mol. Biol.* **48**, 575–607.

Harris, M. (1998) *Good to Eat*, Prospect Heights, Waveland.

Harrison, L.A., Bailey, M.R., Naylor, M.W., Ream, J.E., Hammond, B.G., Nida, D.L., Burnette, B.L., Nickson, T.E., Mitsky, T.A., et al. (1996) The expressed protein in glyphosate-tolerant soybean, 5-enolpyruvylshikimate 3-phosphate synthase from *Agrobacterium* sp. Strain CP4, is rapidly digested *in vitro* and is not toxic to acutely gavaged mice, *J. Nutr.* **126**(3), 728–740.

Hartman, G.L. and Tanimonure, O.A. (1991) Seed populations of *Striga* species in Nigeria, *Plant Disease* **75**, 494.

Hefle, S., Nordlee, J.A. and Taylor, S.L. (1996) Allergenic foods, *Crit. Rev. Food Sci. Nutr.* **36**, 69–89.

Heselmans, M. (1999) EU plant programs wither under Framework V, *Nature Biotechnol.* **17**, 1056.

Hodgson, J. (1999) Monarch Bt-corn paper questioned, *Nature Biotechnol.* **17**, 627.

Hodgson, J. (2000) Moratorium hits Danish companies, *Nature Biotechnol.* **18**, 139–140.

Humphrey, J.H., West, Jr., K.P. and Sommer, A. (1992) Vitamin A deficiency and attributable mortality among under-5-year olds, *WHO Bull.* **70**, 225–232.

Iriarte, V., Gino Aguirre, F.T. and Thiele, G. (2000) Local seed systems and PROINPA's genebank: Working to improve seed quality of traditional Andean potatoes in Bolivia and Peru, in E. Friis-Hansen and B. Sthapit (eds.), *Participatory Approaches to the Conservation and Use of Plant Genetic Resources*, IPGRI, Rome, p. 154.

James, R.R. (1997) Utilizing a social ethic toward the environment in assessing genetically engineered insect-resistance in trees, *Agricult. Human Values* **14**, 237–249.

Jayaraman, K.S. (1999) Asians link up to protect plant resources, *Nature* **398**, 360.

Jefferson, R.A. (1994) Apomixis: A social revolution for agriculture, *Biotechnol. Development Monitor* **19**, 14–16.

Jefferson, R.A. and Bicknell, R. (1996) The potential impacts of apomixis: A molecular genetics approach, in: B.W.S. Sobral (ed.), *The Impact of Plant Molecular Genetics*, Birkhauser, Boston, pp. 88–101.

Jenks, P.J. (1998) Sequencing microbial genomes – What will it do for microbiology?, *J. Med. Microbiol.* **47**, 375–382.

Joersbo, M., et al. (1998) Analysis of mannose selection used for transformation of sugar bee, *Mol. Breeding* **4**, 111–117.

Joly, P.-B. and Lemarie, S. (1998) Industry consolidation, public attitude and the future of plant biotechnology in Europe, *AgBioForum* **1**, 85–90. See http://www.agbioforum.missouri.edu.

Kaneto, T., Tanaka, A., Sato, S., Kotani, H., Sazuka, T., Miyajima, N., Sugiura, M. and Tabata, S. (1996) Sequence analysis of the genome of the unicellular cyanobacterium *Synechocystis* sp. Strain PCC6803. II. Sequence determination of the entire genome of potential protein coding regions, *DNA Res.* **3**, 109–136.

Karchi, H., Shaul, O. and Galili, G. (1993) Seed specific expression of a bacterial desensitised aspartate kinase increases the production of seed threonine and methionine in transgenic tobacco, *Plant J.* **3**, 721–727.

Khan, M.R., Ceriotti, A., Tabe, L., Aryan, A., McNabb, W., Moore, A., Craig, S., Spencer, D. and Higgins, T.J. (1996) Accumulation of a sulphur-rich seed albumin from sunflower in the leaves of transgenic subterranean clover (*Trifolium subterraneum L.*), *Transgenic Res.* **5**, 179–185.

Kiyosawa, S. (1982) Genetics and epidemiological modeling of breakdown of plant disease resistance, *Annu. Rev. Phytopathol.* **20**, 93–117.

Knauf, V.C. (1995) Transgenic approaches for obtaining new products from plants, *Curr. Opin. Plant Biotechnol.* **6**, 165–170.

Knauf, V.C. and Facciotti, D. (1995) Genetic engineering of foods to reduce the risk of heart disease and cancer, *Adv. Exp. Med. Biol.* **369**, 221–228.

Komari, T., Hiei, Y., Saito, Y., Murai, N. and Kumashiro, T. (1996) Vectors carrying two separate T-DNAs for co-transformation of higher plants mediated by *Agrobacterium tumefaciens* and segregation of transformants free from selection markers, *Plant J.* **10**, 165–174.

Kottke, M.K. (1998) Scientific and regulatory aspects of nutraceutical products in the United States, *Drug Dev. Ind. Pharm.* **24**, 1177–1195.

Krattiger, A.F. (1997) Insect resistance in crops: A case study of *Bacillus thuriengiensis* and its transfer to developing countries, International Service for the Acquisition of Agri-Biotech Applications, ISAA Briefs No. 2, ISAAA, Ithaca, NY.

Krishnan, B.R. (2000) Current status of DNA vaccines in veterinary medicine, *Adv. Drug Deliv. Rev.* **43**, 3–11.

Ku, M.S.B., Agarie, S., Nomura, M., Fukayama, H., Tsuchida, H., Ono, K., Hirose, S., Toki, S., Miyao, M. and Matsuoka, M. (1999) High-level expression of maize phosphoenolpyruvate carboxylase in transgenic rice plants, *Nature Biotechnol.* **17**, 76–80.

Lagercrantz, U. and Lydiate, D.J. (1996) Comparative genome mapping in Brassica, *Genetics* **144**, 1903–1910.

Lambais, M.R., Goldman, M.H., Camargo, L.E. and Goldman, G.H. (2000) A genomic approach to the understanding of *xylella fastidiosa* pathogenicity, *Curr. Opin. Microbiol.* **3**, 459–462.

Lappe, M. (2000) Labelling genetically engineered foods: The public right to know, Senate Publications, Sacramento, USA, 12 pp.

Lehrer, S.B. and Reese, G. (1997) Recombinant proteins in newly developed foods: Identification of allergenic activity, *Int. Arch. Allergy Immunol.* **113**, 122–124.

Leister, D., Ballvora, A., Salamini, F. and Gebhardt, C. (1996) A PCR-based approach for isolating pathoogen resistance genes from potato with a potential for wide application in plants, *Nature Genet.* **14**, 421–429.

Linder, C.R. and Schmitt, J. (1994) Assessing the risks of transgene escape through time and crop-wild hybrid persistence, *Mol. Ecol.* **3**, 23–30.

Lipton, M. (1999) Reviving the stalled momentum of global poverty reduction: What role for genetically modified plants?, Crawford Memorial Lecture, 28/10/99, CGIAR International Centers Week, Washington, D.C.

List, G.R., Orthoefer, F., Taylor, N., Nelsen, T. and Abidi, S.L. (1999) Characterization of phospholipids from glyphosate-tolerant soybeans, *J. Amer. Oil Chem. Soc.* **76**, 57–60.

Loevinshon, M. and Meijerink, G. (1998) Enhancing capacity to manage resources: Assessing the Farmer Field School approach, Paper presented at the 2nd Meeting of the Integrated Pest Management Network for the Caribbean, CARDI/CTA, Kingston, Jamaica, February 4–6, 1998.

Losey, J.E., Rayor, L.S. and Carter, M.E. (1999) Transgenic pollen harms monarch larvae, *Nature* **399**, 214.

MacKenzie, D. (1999) Red flag for green spray, *New Scientist*, 29 May 1999.

Martienssen, R.A. (1998) Functional genomics: Probing plant gene function and expression with transposons, *Proc. Natl. Acad. Sci. USA* **95**, 2021–2026.

Mason, H.S., Lam, D.M. and Arntzen, C.J. (1992) Expression of hepatitis B surface antigen in transgenic plants, *Proc. Natl. Acad. Sci. USA* **89**, 11745.

Mason, H., Ball, J., Shi, J.-J. Jiang, X., Estes, M. and Arntzen, C. (1996) Expression of Norwalk virus capsid protein in transgenic tobacco and potato and its oral immunogenicity in mice, *Proc. Natl. Acad. Sci. USA* **93**, 5335–5340.

Masood, E. (1999) Collapse of talks on safety of GMO trade, *Nature* **398**, 6.

Matthee, M. and Vermersch, D. (2000) Are the precautionary principle and the international trade of genetically modified organisms reconcilable?, *J. Agric. Environ. Ethics* **12**(1), 59–70.

McCouch, S. (1998) Toward a plant genomics initiative: Thoughts on the value of cross-species and cross-genera comparisons in the grasses, *Proc. Natl. Acad. Sci. USA* **95**, 1983–1985.

McCouch, S.R., Chen, X., Panaud, O., Temnykh, S., Xu, Y., Cho, Y.G., Huang, N., Ishii, T. and Blair, M. (1997) Microsatellite marker development, mapping and applications in rice genetics and breeding, *Plant Mol. Biol.* **35**, 89–99.

McGaughey, W.H., Gould, F. and Gelernter, W. (1998) Bt resistance management, *Nature Biotechnol.* **16**, 144–146.

Messing, J. and Llaca, V. (1998) Importance of anchor genomes for any plant genome project, *Proc. Natl. Acad. Sci. USA* **95**, 2017–2020.

Mikkelsen, T.R., Andersen, B. and Jorgensen, R.B. (1996) The risk of crop transgene spread, *Nature* **380**, 31.

Molvig, L., Tabe, L.M., Eggum, B.O., Moore, A.E., Craig, S., Spencer, D. and Higgins, T.J. (1997) Enhanced methionine levels and increased nutritive value of transgenic lupins (*Lupi-*

nus augustifolius L.) expressing a sunflower albumin gene, *Proc. Natl. Acad. Sci. USA* **94**, 8393–8398.

Mundt, C.C. (1994) Techniques to manage pathogen coevolution with host plants to prolong resistance, in P.S. Teng, K.L. Heong and K. Moody (eds.), *Rice Pest Science and Management*, International Rice Research Institute, Manila, Philippines, pp. 193–205.

Murphy, D.J. (1999) Manipulation of plant oil composition for the production of valuable chemicals. Progress, problems, and prospects, *Adv. Exp. Med. Biol.* **464**, 21–35.

Nap, J.P., Bijvoet, J. and Stiekema, W.J. (1992) Biosafety of kanamycin-resistant transgenic plants, *Transgenic Res.* **1**, 239–249.

Naylor, R. (1994) Herbicide use in Asian rice production, *World Develop.* **22**, 55–70.

Nestle, M. (1996) Allergies to transgenic foods – Questions of policy, *New Engl. J. Med.* **334**, 726–728.

Nida, D.L., Patzer, S., Harvey, P., Stipanovic, R., Wood, R. and Fuchs, R.L. (1996) Glyphosate-tolerant cotton: The composition of the cottonseed is equivalent to conventional cottonseed, *J. Agric. Food Chem.* **44**, 1967–1974.

Nordlee, J.A., Taylor, S.L., Townsend, J.A., Thomas, L.A. and Bush, R.K. (1996) Identification of a Brazil-nut allergen in transgenic soybeans, *New Engl. J. Med.* **334**, 688–692.

Nuffield Council on Bioethics (1999) http://www.nuffield.org/bioethics/publication/modifiedcrops/.

O'Laughlin, B. (1974) Mediation of contradiction: why Mbum women do not eat chicken, in M.Z. Rosaldo and L. Lamphere (eds.), *Woman, Culture and Society*, Stanford University Press, CA, pp. 301–318.

Odebiyi, A. (1989) Food taboos in maternal and child health: The views of traditional healers in ILE-IFE Nigeria, *Social Sci. Med.* **28**, 985–996.

OECD (1993a) *Safety Considerations for Biotechnology: Scale Up of Crop Plants*, OECD, Paris.

OECD (1993b) *Safety Evaluation of Foods Derived by Modern Biotechnology; Concepts and Principles*, OECD, Paris.

Padgette, S.N., Taylor, D., et al. (1996) The composition of glyphosate-tolerant soybean seed is equivalent to that of conventional soybeans, *J. Nutr.* **126**, 702–716.

Pandey, A. and Mann, M. (2000) Proteomics to study genes and genomes, *Nature* **405**, 837–46.

Persley, G.J. and Lantin, M. (2000) *Agricultural Biotechnology and the Poor*, CGIAR Secretariat World Bank, Washington D.C., 230 pp.

Phillips, P.C. (1999) From complex traits to complex alleles, *Trends Genet.* **15**, 6–8.

Phillips, R.L. and Freeling, M. (1998) Plant genomics and our food supply: An introduction, *Proc.Natl. Acad. Sci. USA* **95**, 1969–1970.

Pimentel, D.S. and Raven, P.H. (2000) Bt corn pollen impacts on nontarget lepdidoptera: Assessment of effects in nature, *Proc. Natl. Acad. Sci. USA* **97**(15), 8198–8199.

Pimentel, D., McLaughlin, L., Zepp, A., Kakitan, B., Kraus, T., Kleinman, P., Vancini, F., Roach, W.J., Graap, E., Keeton, W.S. and Selig, G. (1993), Environmental and economic effects of reducing pesticide use in agriculture, *Agric. Ecosyst. Environ.* **46**, 273–288.

Pink, D. and Puddephat, I. (1999) Deployment of disease resistance genes by plant transformation – A 'mix and match' approach, *Trends Plant Sci.* **4**, 71–75.

Pinto, Y.M. (2000) Transgenic resistance to rice yellow mottle virus disease: Highlighting some of the issues facing the potential introduction of biotechnological products into Africa, *AgBiotechNet* **2**, February, ABN 039.

Pinto, Y.M., Kok, R.A. and Baulcombe, D.C. (1999) Resistance to rice yellow mottle virus (RYMV) in cultivated African rice varieties containing RYMV transgenes, *Nature Biotechnol.* **17**, 702–707.

Pray, C.E. and Umali-Deiniger, D. (1998) The private sector in agricultural research systems: Will it fill the gap?, *World Develop.* **26**, 1127–1148.

Pusztai, A., Grant, G., Bard Ûcz, S., Alonso, R., Chrispeels, M.J., Schroeder, H.E., Tabe, L.M. and Higgins, T.J.V. (1999) Expression of insecticidal bean a-amylase inhibitor transgene has minimal detrimental effect on the nutritional value of peas in the rat at 30% of the diet, *J. Nutr.* **129**, 1597–1603.

Randell, A.W. and Whitehead, A.J. (1997) Codex Alimentarius: Food quality and safety standards for international trade, *Rev. Sci. Technol.* **16**, 313–321.

Richter, L.J., Thanavala, Y., Arntzen, C.J. and Mason, H.S. (2000) Production of hepatitis B surface antigen in transgenic plants for oral immunization, *Nature Biotechnol.* **18**, 1167–1171.

Rissler, J. and Mellon, M. (1996) *The Ecological Risks of Engineered Crops*, MIT Press, Cambridge, MA.

Rounsley, S., Lin, X. and Ketchum, K.A. (1998) Large-scale sequencing of plant genomes, *Curr. Opin. Plant Biol.* **1**, 136–141.

Rudolph, N.S. (1999) Biopharmaceutical production in transgenic livestock, *Trends Biotechnol.* **17**, 367–374.

Ruibal-Mendieta, N.L. and Lints, F.A. (1998) Novel and transgenic food crops: Overview of scientific versus public perception, *Transgenic Res.* **7**, 379–386.

Sasaki, T. (1998) The rice genome project in Japan, *Proc. Natl. Acad. Sci. USA* **95**, 2027–2028.

Sasson, A. (1998) Biotechnology and food production: Relevance to developing countries, in A. Altman (ed.), *Agricultural Biotechnology*, Marcel Dekker, New York, pp. 691–729.

Saunders, N.J. and Moxon, E.R. (1998) Implications of sequencing bacterial genomes for pathogenesis and vaccine development, *Curr. Opin. Biotechnol.* **9**, 618–623.

Senior, I.J. (1998) Uses of plant gene silencing, *Biotechnol. Genet. Engrg. Rev.* **15**, 79–119.

Shah, D.M. (1997) Genetic engineering for fungal and bacterial diseases, *Curr. Opin. Biotechnol.* **8**, 208–214.

Shelton, A.M., Tang, J.D., Roush, R.T., Metz, T.D. and Earle, E.D. (2000) Field tests on managing resistance to Bt-engineered plants, *Nature Biotechnol.* **18**(3), 339–342.

Sidhu, R.S., Hammond, B.G., Fuchs, R.L., Mutz, J.N., Holden, L.R., George, B. and Olson, T. (2000) Glyphosate-tolerant corn: The composition and feeding value of grain from glyphosate tolerant corn is equivalent to that of conventional corn (*Zea Mays L.*), *J. Agric. Food Chem.* **48**, 2305–2312.

Simoons, F. (1995) *Eat Not of This Flesh: Food Avoidance from Prehistory to the Present*, 2nd edn., University of Wisconsin Press, Madison, WI.

Smale, M. (1997) The Green Revolution and wheat genetic diversity: Some unfounded assumptions, *World Develop.* **25**, 1257–1269.

Sommer, A. (1988) New imperatives for an old vitamin (A), *J. Nutr.* **119**, 96–100.

Spillane, C. (1999) Recent developments in biotechnology as they relate to plant genetic resources for food and agriculture, FAO, Rome, CGRFA Background Study Paper No. 10, 64 pp.

Spillane, C. (2000) Can agricultural biotechnology contribute to poverty alleviation?, *CABI Ag-BiotechNet* **2**, 1–40. http://www.agbiotechnet.com.

Spillane, C. and Gepts, P. (2000) Evolutionary and genetic perspectives on the dynamics of crop genepools, in D.C. Cooper, C. Spillane and T. Hodgkin (eds.), *Broadening the Genetic Base of Crop Production*, IPGRI/FAO, CABI Publishing, Wallingford, UK, in press.

Spillane, C. and Thro, A.M. (2000) Farmer participatory research and pro-poor agricultural biotechnology, Centre for Development Research, Copenhagen, Den Ny Verden, Vol. 1, 45pp (in Danish). http://www.cdr.dk.

Tabashnik, B.E., Patin, A.L., Dennehy, T.J., Liu, Y.B., Carriere, Y., Sims, M.A. and Antilla, L. (2000) Frequency of resistance to *Bacillus thuringiensis* in field populations of pink bollworm, *Proc. Natl. Acad. Sci. USA* **297**, 12980–12984.

Tacke, E., Salamini, F. and Rohde, W. (1996) Genetic engineering of potato for broad-spectrum protection against virus infection, *Nature Biotechnol.* **14**, 1597–1601.

Tanksley, S.D. and McCouch, S.R. (1997) Seed banks and molecular maps: Unlocking genetic potential from the wild, *Science* **277**, 1063–1066.

Taylor, N.B., Fuchs, R.L., MacDonald, J., Shariff, A.R. and Padgette, S.R. (1999) Compositional analysis of glyphosate – Tolerant soybeans treated with glyphosate, *J. Agric. Food Chem.* **47**, 4469–4473.

Thro, A.M. and Spillane, C. (2000) Biotechnology assisted participatory plant breeding: Complement or contradiction?, CIAT: Cali, CGIAR Systemwide Initiative on Participatory research and Gender Analysis, Working Paper #4, 150 pp.

Timberlake, W.E. (1998) Agricultural genomics comes of age, *Nature Biotechnol.* **16**, 116–117.

Tripp, R. (2000a) Can biotechnology reach the poor? The adequacy of information and seed delivery, Paper presented at 4th International Conference on the Economics of Agricultural Biotechnology, Ravello, Italy, August 24–28.

Tripp, R. (2000b) GMOs and NGOs: Biotechnology, the policy process, and the presentation of evidence, Overseas Development Institute, London, ODI Natural Resource Perspective No. 60, 6 pp.

Van Uyen, N. and Van der Zaag, P. (1993) Potato production using tissue culture in Vietnam: The status after 10 years, *Amer. Potato J.* **70**(11), 831–832.

Voelker, T.A., Hayes, T.R., Cranmer, A.M., Turner, J.C. and Maelor Davies, H. (1996) Genetic engineering of a quantitative trait: Metabolic and genetic parameters influencing the accumulation of laurate in rapeseed, *Plant J.* **9**, 229–241.

Wagner, H.G. and Hammond, K. (1997) Animal production improvement in developing countries: Issues concerning the application of biotechnology, Paper presented at an International Workshop on Transgenic Animals and Food Production, Stockholm, Sweden, May 22–24.

Walmsley, A.M. and Arntzen, C.J. (2000) Plants for delivery of edible vaccines, *Curr. Opin. Biotechnol.* **11**, 126–129.

Waterhouse, P.M., Graham, M.W. and Wang, M.B. (1998) Virus resistance and gene silencing in plants can be induced by simultaneous expression of sense and antisense RNA, *Proc. Natl. Acad. Sci. USA* **95**(23), 13959–13964.

Weiss, G.B. (1997) Current status of functional foods in the United States, *Nippon Yakurigaku Zasshi* **110**, Suppl. 1, 23P–25P.

Whitham, S., McCormick and Baker, B. (1996) The N gene of tobacco confers resistance to tobacco mosaic virus in transgenic tomato, *Proc. Natl. Acad. Sci. USA* **93**, 8776–8781.

Wilkinson, J.Q., Lanahan, M.B., Clark, D.G., Bleecker, A.B., Chang, C., Meyerowitz, E.M. and Klee, H.J. (1997) A dominant mutant receptor from Arabidopsis confers ethylene insensitivity in heterologous plants, *Nature Biotechnol.* **15**, 444–447.

Williamson, M. (1996a) *Biological Invasions*, Chapman & Hall, London, UK.

Williamson, M. (1996b) Can the risks from transgenic crop plants be estimated?, *Trends Biotechnol.* **14**, 449–450.

Woiwod, I., Dewar, A., Hawes, C. and Heard, M. (2000) Farm-scale trials for studying the effects of GM herbicide-tolerant crops on farmland biodiversity in the UK, *Antenna (London)* **24**(2) 66–68

Wood, D. (1996) The benign effect of some agricultural specialisation on the environment, *Ecol. Econom.* **19**, 107–111.

Wood, D. (1998) Ecological principles in agricultural policy: But which principles?, *Food Policy* **23**, 371–381.

Wood, D. and Lenne, J. (1997) The conservation of agrobiodiversity on farm: Questioning the emerging paradigm, *Biodiv. Conserv.* **5**, 109–127.

Wood, D. and Lenne, J. (1999) *Agrobiodiversity: Characterisation, Utilization and Management*, CABI Publishing, Wallingford, UK.

Wraight, C.L., Zangerl, A.R., Carroll, M.J. and Berenbaum, M.R. (2000) Absence of toxicity of *Bacillus thuringiensis* pollen to black swallowtails under field conditions, *Proc. Natl. Acad. Sci. USA* **97**, 7700–7703.

WTO (1997) EC measures concerning meat and meat products (hormones) Complaint by the United States, Appelate Body Report, 18 August 1997, WT/DS26/AB/R, Geneva.

WTO (1998) EC measures concerning meat and meat products (hormones) Complaint by the United States, Appelate Body Report, 16 January 1998, WT/DS26/AB/R, Geneva.

Xiao, J., Li, J., Grandillo, S., Ahn, S.N., Yuan, L., Tanksley, S.D. and McCouch, S.R. (1998) Identification of trait-improving quantitative trait loci alleles from a wild rice relative, *Oryza rufipogon*, *Genetics* **150**, 899–909.

Yu, J. and Langridge, W.H. (2000) Novel approaches to oral vaccines: Delivery of antigens by edible plants, *Curr. Infect. Dis. Rep.* **2**, 73–77.

Zou, J., Katavic, V., Giblin, E.M., Barton, D.L., MacKenzie, S.L., Keller, W.A., Hu, X. and Taylor, D.C. (1997) Modification of seed oil content and acyl composition in the Brassicaceae by expression of a yeast sn-2 acyltransferase gene, *Plant Cell* **9**, 909–923.

2. Pest Resistance in Agriculture: An Economic Perspective

URI REGEV

1. Introduction

One of the most striking trends to have taken place in the past 100 years in modern agriculture is the decrease in its economic share in developed countries, from about 50–70 to 2–4%, concomitant with the ability to supply ever rising quantities of food to the entire world. This drastic increase in food production efficiency is largely attributable to the interaction of chemical pesticides and herbicides, and the genetic selection of new high yield but susceptible plant varieties. However, this wealth of food output has cost society dear in negative side effects (externalities) that have resulted from the tremendous increase in the use of chemical pesticides. These externalities include air, soil and water pollution, residues in the food chain, elimination of beneficial organisms, disruption of the natural ecological equilibrium, and the development of pest resistance to pesticides. Ecologists were among the first to recognize the environmental consequences of extensive chemical use in modern industry and agriculture. Carson (1962) shook the world when she warned of the environmental threats posed by pesticides, and Van den Bosch's provocative book *The Pesticide Conspiracy* (1978) alerted society to the economic consequences of environmental pollution and resistance to chemical pesticides: "Today the pesticide treadmill spins more wildly than ever. We use twice as much insecticide as in 1962, there are more insect species of pest status than ever before, insect control costs have skyrocketed, and insecticide impact on the environment is the worst in history" (Van den Bosch, 1978, p. 35).

The economic incentive problem associated with the external effects of pesticides is that our market prices do not reflect the benefits and cost entailed by these externalities. As a result, profit-maximizing criterion of competitive firms ignores external effects of pesticides, which results in resource misallocation by pesticide users involving the wrong choice of control strategy as well as the overuse of chemical pesticides. This phenomenon has not gone unnoticed, but was already observed some 25 years ago in quite a few studies (Hueth and Regev, 1974; Feder and Regev, 1975; Regev, 1976; Regev et al., 1976; Carlson, 1977). Note that resistance differs from other externalities of environmental pollution in that its impact (reducing future efficiency of currently used pesticides) affects the farmers themselves, rather than the rest of the society. This type of externality is known as

Timothy Swanson (ed.), The Economics of Managing Biotechnologies, 51–69.

the common property resource problem, which is a special case of the well-known prisoners' dilemma in game theory.

Pesticide resistance is a global phenomenon that has affected common pests including fungi, bacteria, insects, rodents, and nematodes. Since World War II, chemical pesticides have been the major weapon for suppressing human and agricultural pests. However, as early as 1950, some species began to exhibit resistance to DDT and other organochlorine compounds. After several years, resistance spread to the more expensive organophosphorus compounds. Weed resistance to herbicides developed only much later, due to the relatively slower rates of seed migration between fields, resulting in better pest control incentives (as will be explained in Section 3).

The different externalities caused by chemical pesticides have aroused conflicting reactions from individual farmers, farmers' organizations and the chemical industry. The traditional response to pesticide environmental pollution from both farmers and the pesticide industry has been to undermine the importance of the damage and emphasize how the extensive arsenal of pesticides revolutionized agricultural technology and food production. While recognizing the limitations and problems associated with their use, they maintained that no one who comprehended the food-shortage consequences to humanity could seriously recommend that chemicals be quickly replaced by alternative means of pest control. In line with this view, LeBaron and Gressel (1982, p. 1) argued that "the net effect of pesticides as with drugs has been the same; an increasing life expectancy with medicines and an increasing food production with pesticides. Chemical pesticides must certainly be a major and increasing part of agricultural technology in the decades ahead. Resistance development is an inherent result of chemical control".

Regarding resistance however, the incentives of the individual farmers differ from those of the pesticide industry. Resistance development is accelerated by excessive use of pesticides in pursuit of short-term gains, which entails long-term cost through permanent loss of, often irreplaceable, pesticides. While each individual farmer tends to ignore resistance in his pest control decisions, the chemical industry react differently, as Green et al. (1990, p. ix) contended: "The traditional response to resistance, by switching to new compounds, has become less practical due to substantial increases in the time and expense of agrochemical discovery and development". Optimal resistance management could be achieved by collective action, possibly by farmers' cooperatives who have a regional view of the problem. The economic incentives, attitudes to resistance, and the pest management policy implications of the various interest groups are further discussed in Section 3.

Any discussion of resistance would not be complete without noting the recent introduction of biotechnology that seems to present a great potential for replacing chemical pesticides in pest management. Recent breakthroughs in genetic engineering have opened the stage for dramatic changes in agriculture and, in particular, crop protection. These developments might lead to a second revolution in the control of agricultural and human pests, analogous to the discovery of chemical pesticides and antibiotics in the early 1940s. It is forecasted that genetic engineering may bring about a greater selection of high yielding varieties for food production, and will substitute chemical pesticides by developing plant varieties with pest repelling genes. However, scientists and policy makers have already begun to be concerned that conceivable external effects of biotechnology might be worse than those of pesticides, both in terms of environmental pollution and vulnerability to the rapid development of pest resistance. A possible risk of biotechnology could be the disruption of the ecological equilibrium and reduced bio-diversity engendered by the elimination of some insect groups altogether, thereby inducing the extinction of an entire pyramid of species. Biotechnology has already been introduced into several crops (e.g. corn tobacco and cotton), where it has evoked anxiety in regard to food quality, possible dispersion of the transplanted genes to other populations, and pest-resistance development to the genetically transplanted toxins. These concerns and possible tactics to encounter them are discussed in Section 4.3.

2. Resistance Development over Time

2.1. Insect Resistance

Insect resistance to pesticides has a history of nearly 90 years, but its deleterious consequences in plant pathogens was recognized in the 1940s with the introduction and extensive use of synthetic organic pesticides. The incidence of resistance continues to rise, and has been recorded to include over 500 species of insects and mites, a 13% increase since 1984 (Georghiou, 1990; see Figure 1). The initial response to the loss of efficiency induced by pesticide and herbicide-resistance development was to increase dosages and frequencies of applications. When this failed, the next move was a shift to a new toxicant, typically one that was more expensive, with no basic change in the philosophy or strategy of chemical control. This process created a vicious cycle of increasing the numbers and varieties of pests, which triggered a rise in pest control costs. A few examples can illustrate this hapless situation.

The Colorado potato beetle, a major worldwide pest of *potatoes*, was one of the first pests to exhibit resistance to DDT (1952) and has subsequently developed resistance to all synthetic insecticides in the northeastern US, apparently because of cross-resistance. For example, carbofuran was initially highly effective for controlling this pest, however, resistance developed unusually quickly, often within one growing season (Ioannidis et al., 1992). Control costs of the Colorado potato beetle were estimated in 1991 to be \$35–74/ha in resistance-free areas *versus* \$306–412/ha in high-resistance areas (Grafius, 1997). A similar experience was

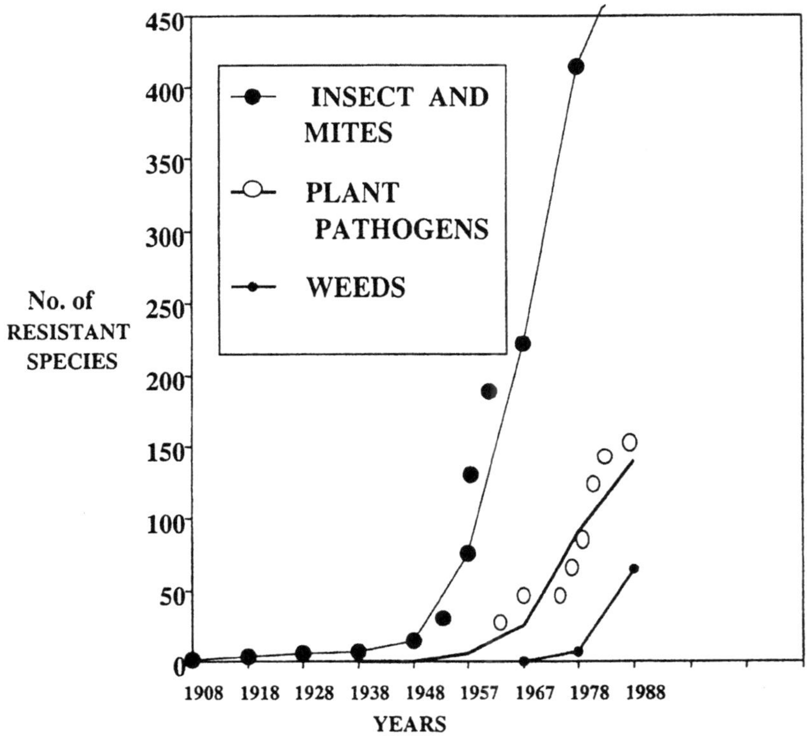

Figure 1. Magnitude of the resistance problem. Source: Georghiou (1986, p. 17; 1990, p. 19).

encountered with resistance to fungicides, which was first reported in 1940. By 1957, 76 insect species were identified as having developed resistance to chemical and organic pesticides, increasing to 228 by 1967 and 364 by 1976 (LeBaron and Gressel, 1982).

In *citrus* fruit Eckert (1990) observed that all major organic fungicides had failed to provide satisfactory control of post-harvest decay, caused by fungicide-resistant bio-types of penicillium. Fungicide resistance has become a persistent threat to the fresh citrus fruit business throughout the world. A final case in point is the continuous spraying program of pesticides for heliotis (*Helicoverpa armigera*) and whitefly (*Bemisia tabaci*) on *cotton* fields. Resistance of these pests to all major pesticides had brought by 1987 cotton production practically to a stand-still over large areas in Israel (Ishaaya, pers. comm.) and virtually everywhere else where cotton is grown (Gutierrez, pers. comm.).

These examples only touch the surface of the general situation which is presented in Figures 2 and 3, that summarize the numbers of insect and mite resistance to various insecticides.

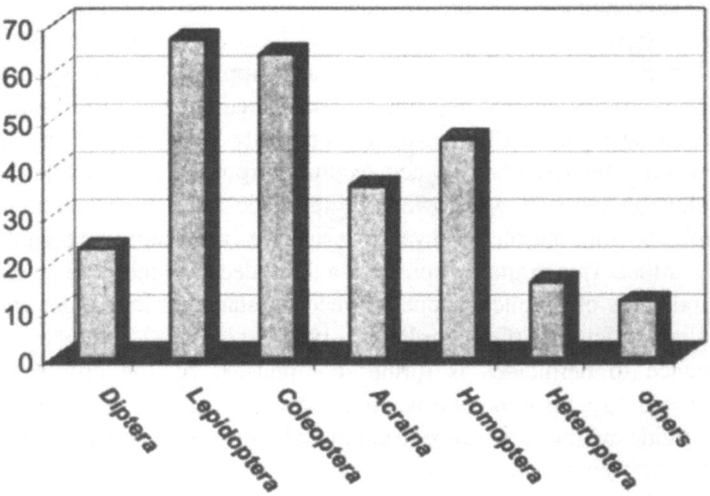

Figure 2. Number of resistant agricultural pests (total = 264). Source: Georghiou (1986, p. 19).

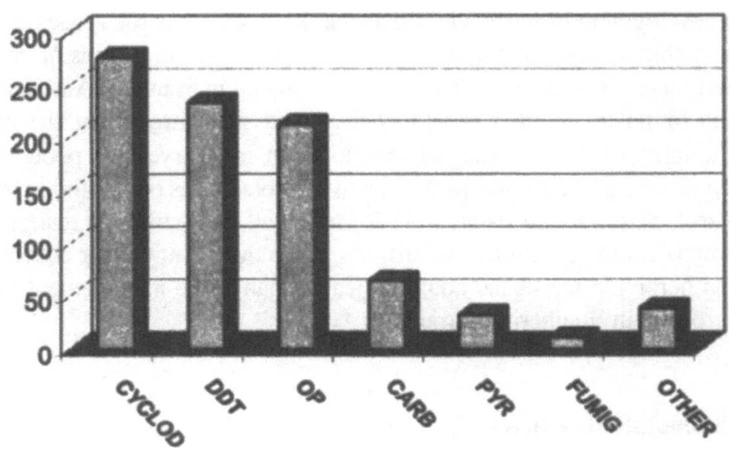

Figure 3. Number of resistance species, by insecticides-1984 (total = 866). Source: Georghiou (1986, p. 19).

2.2. Weed Resistance

Resistance of noxious weeds to herbicides is a quite recent phenomenon (see Figure 1), but has begun to be detected with increasing frequency as a consequence of the growing intensity of herbicide use. Genetic components and selection pressure are considered to be the main factors that contribute to the rate of herbicide resistance. Clearly, since total weed eradication is impossible, a higher rate of kill leaves the surviving weeds more resistant, so that faster resistance development is unavoidable (Gressel, and Segel, 1982). The first reports on herbicide resistance appeared in Ryan (1970) and Radosevich and Appleby (1973). Ryan (1970) reported the failure to control groundsel (*Senecio vulgaris L.*) in a nursery where simazine and atrazine had been used continuously since 1958. These herbicides were however effective against weed seedlings from sources outside the nursery, where selection pressure was lower. As herbicide rates were increased in 1968 in an effort to gain control, the level of resistance in the surviving seedlings have risen still further. This example represents a three-decade general trend of increased agricultural costs of production due to weed resistance to herbicides, particularly in Australian agriculture (Gorddard et al., 1995).

Resistance to herbicides is further complicated by the problem of cross-resistance. This is a phenomenon where a biotype which has developed resistance to one pesticide exhibits resistance to another that differs chemically from the original and has different modes of action (Powles and Holtum, 1990). The development of cross-resistance to commonly used herbicides is of great concern to both farmers and the pesticide industry. A case in point would be annual ryegrass (*Lolium rigidum*), a widely distributed weed, that has demonstrated a remarkable capacity for rapid development of herbicide resistance. In many cases the repeated use of a particular herbicide immediately endowed resistance to other groups of herbicides not formerly used. This problem is especially rampant in Southern Australia where ryegrass is found in pastures at widely dispersed sites throughout the cropping zones. The intensive cross-resistance development in the ryegrass population has eliminated the use of major groups of selective herbicides as predominant means of weed control. Schmidt and Pannell (1996) reasoned that herbicide resistance was a serious problem to agriculture in Australia, more than in any other country in the world, and herbicide resistance could radically change the management practices of grain growers in Southern Australia.

3. The Economic Incentives

Economic optimization in pest management involves decisions on the type of pesticides, optimal doses and number of applications, as well as timing of the initial spraying. The optimization process has often been based on short-term profit maximization for a single season only, ignoring inter-seasonal effects, as well as external effects of pesticides. Pest resistance is obviously not of concern in this type of optimization process, since it is a long-term phenomenon that does not usually develop within a single season. However, resistance should play a major role in long-term pest control optimization. This is the rationale for incorporating the effect of pest resistance into a pest management optimization model where resistance is considered as a stock variable in the context of exhaustible resources. In this context, the optimal application of pesticides implies conjunctive management of both the pest and its associated stock of susceptibility. An optimal application should be lower for pests with potential resistance development, because the cost of pesticide resistance is added to the standard marginal cost of pesticide use. This additional cost (sometimes called user cost) results from the reduced future effectiveness of chemical pesticides due to an incremental dose of pesticides at the present time. In this situation, disregarding future effects of resistance can be attributed only to ignorance or myopia (Hueth and Regev, 1974; Sarhan et al., 1979).

The long-sighted pesticide user should weigh the short-term benefits of chemical control against the long-term (user) cost, owing to pesticide-resistance build-up in the pest population. Taylor and Headley (1975) developed an illustrative population resistance model with three groups of genotypes that mate randomly according to the Hardy–Weinberg law. They suggested a dynamic programming computational procedure as an illustration of a whole class of pest control decisions. However, even this dynamic optimization approach, which considered future pest resistance in addition to the immediate short-term profits, overlooked the common property feature of pest resistance resulting from pest migration.

Many pests, especially insects, are highly mobile and when the cropping season is over, their life cycle carries them to different hosts outside the field, often to an area that is quite distant from the original field. This biological feature of many pests is the basis for resistance being a common property resource. The reason is that the degree of resistance of the incoming pests to any given field does not depend on the resistance generated by pesticide use in that field, but rather on the average resistance in the region, determined by *all* the pesticide users in the area. This common property phenomenon was observed long ago: "The individual farmer does not take resistance into his economic considerations because the level of resistance is affected not by his own pest management decisions, but as a result of the total chemical pesticides applied in the region" (Regev, 1976, p. 271).

Resistance being a common property resource implies conflicting interests between individual farmers and the farmers as a group in the region, namely a farmer who seeks to maximize profits in competitive markets will not take into consideration the effect of increasing his pesticide use on resistance development. When farmers cooperate, the cost of increasing future resistance by pesticide use in the present time is taken into consideration by the farmers' organization, since their aggregate pesticide use affects their future profits. However, farmers have to be convinced that acting collectively as long-term managers of pest susceptibility is in their own best interest, rather than to act competitively and use potent pesticides with onerous resistance consequences. But, cooperation is feasible only if transaction costs of farmers' organization are low so that the net gains of acting collectively are positive.

Naturally, the higher the pest mobility the higher will be the common property effect on farmers' pest control decisions, and this hypothesis is testable by empirical data. For a non-mobile pest species, the stock of resistance is a private (rather than a common) property, resistance is affected only by the farmer's own pesticide use, and the long sighted farmer will take future resistance risk into account when making his present pest control decisions. Thus, the hypothesis that resistance plays an important role in farmers' pest control decisions will be empirically confirmed, if data show that resistance development is slower for non-migratory pests, and that cooperative pest control decisions are positively related to pest mobility.

The main consequence of resistance being a common property (the case of migratory pests) is a market failure, which means that prices do not reflect the true costs, and farmers tend to apply pesticide dosages that are higher than the socially optimal ones. Even worse, these farmers are likely to reject alternative control methods that avoid or delay resistance development, if they even slightly diminish profits in the short run. Market failures were also observed in other common property resource situations (Regev, 1984; Regev et al., 1979, 1983; Miranowski and Carlson, 1986; Clark and Carlson, 1990), where they invariably led to resource misallocation, mostly overuse of those common property resources.

The obstacles in finding the optimal pest management strategy are further augmented by the difficulties encountered in estimating the true value of pest control decisions. Because pest control is a dynamic decision process, there is a tendency to underestimate future control measures and overestimate current effects (Carlson, 1977). Also, when the neoclassical production function is used to evaluate pest control decisions, there is an estimation bias (Lichtenberg and Zilberman, 1986). This bias can also explain why farmers respond to increased resistance by increasing usage levels, as compensation for the decreased pesticide productivity.

The pesticide industry is also concerned about pesticide-resistance development and its profit incentives go hand in hand with those of farmers' organizations. Thus, chemical companies have market incentives to restrict the usage of those pesticides with high risk of resistance development. They also will tend to invest in new compounds where resistance to currently used ones is increasing. Undoubtedly, the pesticide industry has an interest in monitoring resistance and developing new products with an underlying strategy of maintaining and expanding their market shares. However, in an effort to delay resistance development, these companies' R&D programs are mostly geared towards developing new chemicals, rather than encouraging alternative non-chemical technologies that could be of benefit to both pesticide users and the environment.

In addition to resistance, there are other negative side effects of pesticides that impose negative externalities (environmental pollution and pesticide residues in the food chain), lead to market failures, and imply overuse and wrong choice of pesticides. In these situations the negative externalities are imposed largely on all of the society, while the effects on the farmers is only negligible (except in cases of health risks to the farmers or farm workers). Without government regulation, the profit seeking pesticide industry has little motivation to relate to those externalities which do not directly affect its own or the farmers' profits. However, while public regulation is called for in the case of environmental pollution, it is not required for the case of resistant management, as the latter can be efficiently handled by cooperation of the farmers.

Given this background of the negative externalities of chemical pesticides, it is no wonder that biotechnology has excited such great interest with its potential to create genetically modified crops as a means for obtaining higher yield while avoiding the harmful pesticides. The best example of using biotechnology as a substitute for pesticides, is the *Bacillus thuringiensis* (Bt), a soil bacterium has been transplanted into corn, soybeans, tobacco, cotton and other crops. The genetically modified Bt corn does not require any pesticide application against the European corn borer, which is a major and widespread corn pest.

However, the rapid adoption of Bt corn and other genetically modified crops has caused much concern about the risks for food quality, ecological balance and the harmful effects in reducing biodiversity. Another worry is the rapid selection of pest mutations that will be highly resistance to the Bt toxin transplanted into the plant. Scientists also fear that resistance to the newly developed toxin will have high cross-resistance characteristics. In other words, a pest species that has developed resistance to the Bt toxin will be resistant to other chemical pesticides, even in the first application. The economic issues and incentives in this case are similar to those relating to chemical pesticides, i.e., common property resources and myopic decision-making by individual farmers. Therefore, any recommended policy for managing this new type of potential resistance must weigh the immediate gains of saving pesticide direct and external cost against future and as yet uncertain risks of new types of 'super resistant' pests. Furthermore, it should be anticipated that farmers may hesitate to cooperate with policies that will limit their use of this new technology, if such cooperation will reduce their immediate profits. A comparison of the current recommended policies for managing pest resistance with the newly genetically modified crops will be discussed in Section 4.3.

4. Recommended Policies

Resistance management has to be flexible, and must be determined from factors such as pest mobility, farm size, array of available chemical and non-chemical controls, and the cost of coordinating collective action. Miranowski and Carlson (1986) delineated several categories of resistance management and institutional arrangements that hinge upon pest mobility, speed of resistance development, availability of control substitutes, cost of information gathering, cost of coordination efforts, and competitiveness of the pesticide market. When pests are immobile and farmers are informed, no form of intervention in pest control decisions is required. However, when the pests are mobile, farmers' cooperation and organization becomes necessary in order to achieve optimal long-term pest and resistance management. This need for organized resistance management is even more urgent once cross-resistance is evidenced in the field, since cross-resistance can make new pesticides obsolete soon after their introduction.

4.1. PESTS

Even though some general principles for managing resistance are available, actual recommended policies vary according to the particular crop and pest. The following are only a few examples of the variety of pest control methods for managing resistance.

Rust resistance has been found to be the culprit for growing non-optimal *wheat* cultivars. Heisey et al. (1997) calculated yield losses resulting from maximizing genetic diversity and plant resistance to rust, and estimated the costs of managing resistance through genetic diversity to be around 65–70 kg/ha. Two factors, pest-resistant wheat cultivar and planting date modification, have been found to be cost-efficient alternatives for protecting winter wheat from the Hessian fly (Buntin et al. 1992).

Heliotis and whitefly, two widespread *cotton* pests, have worldwide distributions and are responsible for major economic losses over a broad range of fields and horticultural crops. In the US alone heliotis has caused more than one billion dollars of annual losses that are attributable to crop damage and control costs (Bull and Menn, 1990). In Israel, growers are encouraged to postpone spraying, and it has often been found that an initial rise in the pest population was followed by a dramatic reduction without intervention of growers. An important insect-resistance management (IRM) strategy in Israel is based on a model devised in Australia. In this model (see Figure 4), the growing season is divided into three time 'windows'. During the first period (April–July) no pesticide spraying is recommended, for as long as possible, in order to preserve the pests' natural enemies. It is also recommended that no two successive pesticides be applied with compounds that induce similar resistance mechanisms in the insect. Alternation of insecticides (mostly selective ones) is introduced in the second and third time periods in order to slow down resistance build-up (Horowitz et al., 1993). Figure 5 demonstrates the effect of this strategy on the reduction in pesticide applications on cotton in Israel. The IRM strategy also recommends that two newly introduced growth regulators (pyriproxyfen and buprofezin) be restricted to one treatment per season during peak activity of the pest. Experiments show that one treatment of these compounds appreciably altered the susceptibility of *Bemisia tabaci* to these growth regulators (Horowitz and Isaaya, 1994). In the US, Bull and Menn (1990) report on efforts to organize tri-state coordinated programs with similar features for managing pyrethroid resistance.

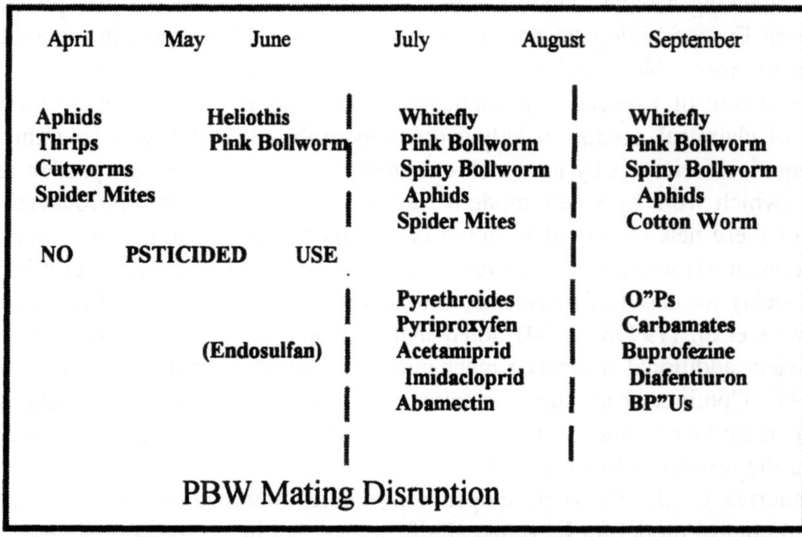

Figure 4. Scheme of IRM-IPM strategy in Israel cotton 1998. Source: Forer (1998).

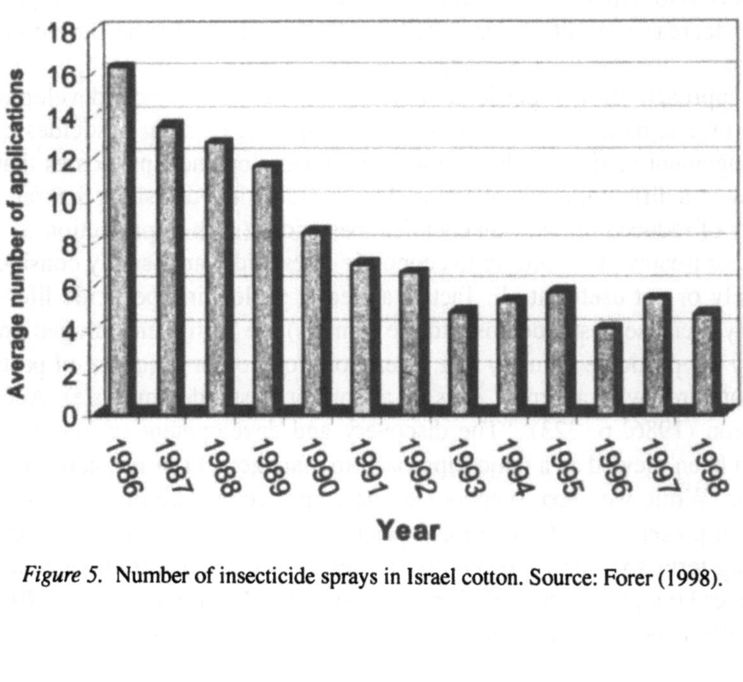

Figure 5. Number of insecticide sprays in Israel cotton. Source: Forer (1998).

The *tobacco* or cotton whitefly (*Bemisia tabaci*) is one of the more nasty pests, attacking a wide range of edible and ornamental crops. In glasshouses in Northern Europe biological control is mostly achieved by a parasitic wasp (*Encarsia formosa*). Nevertheless, for outdoor crops insecticides are still the main means of control, especially in Southern Europe. Worldwide resistance to a wide range of chemical agents, including organophosphates, carbamates, pyrethroids, and organochlorines is by now well entrenched. More recent insect growth regulators (which work by a new mode of action) have recently been introduced, but none of these new insecticides should be assumed to be immune from resistance development (Horowitz et al., 1999a). Tactics of pesticide alternations or mixtures might delay the onset of resistance, and also reduce its magnitude. For example, Horowitz et al. (1999b, p. 94) noted that "reliance on insecticide mixtures can also select additional resistance mechanisms, as demonstrated in Arizona cotton in 1995". Continuous resistance monitoring and the exploitation of non-chemical management tactics, such as natural enemies, currently provide the best model for combating resistance in these pests.

Gutierrez et al. (1979) developed a dynamic optimization (inter and intra-season) control model for Egyptian alfalfa weevil that incorporated resistance into the interaction of *alfalfa* and this pest. The empirical results presented optimal within season and long-term spraying patterns, with and without information on the development of resistance. In the long run, the quantities of pesticides were found to be high in the first few seasons, but decreased later as a result of low infestation. The results also support the common contention that overexposure to pesticides not only decreases profits more rapidly, but also speeds up the development of resistance.

The approach of the pesticide industry to pest resistance development is twofold: on one hand it tries to extend the life span of existing pesticides by various management tactics, while, on the other, it develops new pesticides that work according to a different mode of action. However, tactical decisions that fall within a strategy of reduced reliance on chemical pesticides (such as protection of natural enemies, or treatment according to economic thresholds) are usually considered as moderately or not useful at all. Tactics aimed at prolonging pesticide life (which ultimately increase pesticide costs to the farmers) are highly encouraged and promoted by the pesticide industry (e.g. alteration, rotation or sequence of pesticides, new toxophores with alternate sites of action, or pesticides mixture). According to LeBaron (1986, p. 323), "The discovery and development of new pesticides has often been viewed as a major approach to management of resistance to earlier pesticides. While we need to encourage new discoveries, we must do everything possible to preserve all of our present pesticides. This strategy is a vital and relatively long-term solution to the control of pests resistant to current pesticides, but it can never be a permanent solution". Schwinn and Morton (1990, p. 170) stated that anti-resistance strategies need to be "effective, realistic and enforceable, and should be developed by the agrochemical industry in collaboration with extramural partners . . . in order to protect the highly needed modern fungicides from becoming obsolete due to broad resistance development". These statements encapsulate the pesticide industry's approach to the problem of managing resistance. Furthermore, the authors' long-term view of anti-resistance strategies does not allude to any non-chemical control alternatives, conceivably because they do not contribute to enhancing sales and profits of the pesticide industry.

Widawsky et al. (1998) observed that host-plant resistance is a viable alternative to pest control. Under intensive *rice* production systems in eastern China, pesticide productivity is low compared to that of host-plant resistance. They found negative marginal returns to pesticide use, and host-plant resistance to be an effective substitute, achieving substantial reduction in pesticide use with no loss in rice production. Nevertheless, pesticides are overused in Chinese rice production, while host-plant resistance is under-utilized. This finding can possibly be explained by two factors: lack of information about the productivity of pesticides, and farmers' risk aversion, which leads the farmers to misperceive pesticides as a substitute for insurance.

4.2. Weeds

According to Schmidt and Pannell (1996), the main issues in herbicide resistance management include:

1. Estimates of the short-term cost of changing rotations to delay herbicide resistance.
2. The length of time before profitability collapses when continuous crop rotation is maintained.
3. The optimal time path of herbicide usage.
4. Crop rotation when weeds become totally resistant
5. Economic complexities of non-chemical control.

Schmidt and Pannell also mentioned that the best tactics to prevent or delay resistant populations are:

(a) herbicide mixtures of compounds (or rotation of herbicides) that will act at different sites of action with different modes of degradation, preferably with no or negative cross-resistance; and
(b) mechanical cultivation in the crop rotation.

Resistance can be delayed considerably by rotating with herbicides of differing groups to which the weeds are not cross-resistant (Gressel and Segel, 1982, 1990).

Ryegrass (*Lolium rigidum*) is the weed most commonly recognized for resistance development, especially in continuous cropping rotation. Gorddard et al.

(1995) suggested an optimal control model to select the optimal combination of chemical and non-chemical control measures, taking into account the tradeoffs between short-term profits and long-term costs of herbicide resistance. An optimal resistance strategy would include decreasing herbicide dosages as resistance develops, with compensatory increases in the levels of non-chemical control. Gorddard et al. evaluated three possible management strategies: non-chemical control, lower herbicide dosages, and inclusion of grazed pasture phase into the crop rotation with no selective herbicides. Each strategy entailed short-term costs due to lower yields, which are weighed against its long-term benefits. Non-chemical control can be profitable, depending on the relative costs and effectiveness of the weed control. Economic short-term losses entailed by a pasture phase in the rotation process to delay the development of herbicide resistance can be high. However, pasture may be worth growing (possibly with application of non-selective herbicides) if it causes a substantial reduction in ryegrass seed numbers, even if it generates little or no direct income (Gorddard et al., 1996).

4.3. Genetically Modified Plants for Pest and Weed Control

Genetically engineered crops containing endotoxins from Bt are an alternative pest management strategy that operate by making the plants tolerant to specific insects. Each strain of Bt produces a characteristic set of crystal proteins that are lethal to certain insects, yet does little or no harm to most other organisms, including humans, wildlife and even other insects (Tabashnik, 1997). Furthermore, cross-resistance between conventional insecticides and Bt should not be expected, since they have different modes of action and engender different mechanisms of resistance (Tabashnik, 1994).

Bt was first described in 1911 and has been available for commercial use since the 1930s, yet hardly been applied in agriculture until recently. Its emergence as a viable commercial alternative for pest control is attributable to the recent breakthroughs in biotechnology, to the increasing concern about environmental hazards of conventional pesticides, and to the rapid evolution of pest resistance to insecticides. Bt produces crystal proteins that kill insects by binding to and disrupting their midgut membranes, so that a transplanted crop (e.g. Bt corn) could provide a safe and effective means of pest control. However, their usefulness could be short-lived if insects adapt to the toxins (Mallet and Porter, 1992).

Even though resistance to Bt has been recently reported in Hawaii, US continental and Asia (Liu and Tabashnik, 1997), it has been commercially used for over two decades without reports of substantial resistance development in open field populations (Tabashnik, 1994). This could be due to the short time that Bt crops have been widely spread, or due to the inherent difficulties of conducting long-term field experiments of resistance development. Another possible explanation for lack (or slow) resistance development in the field is *fitness costs*. Genetic changes that confer resistance to Bt may compromise normal functions and thus impose fitness cost when Bt is not present, which can cause a decline in resistance in mixed populations when selection ceases.

However, there have been many theoretical arguments, simulations and laboratory experiments that demonstrate the development of resistance by various pests to a number of Bt crops (see a review in Tabashnik, 1994). Therefore, scientists consider resistance as the greatest threat to the continued success of Bt. The growing concern for development of insect resistance to Bt crops has led to the following suggested tactics to avert or slow it down by taking account of the unique features of Bt itself (Tabashnik, 1994): mixture of the toxins, synergists, mosaics (adjacent areas that are treated with different toxins), rotations, ultra-high doses and refuges. Refuges means providing temporal and/or spatial areas free from the Bt crop, and these refuges could be among tissues within plants, among plants within fields, or between fields.

Refuges is a major policy recommendation of the EPA for managing potential resistance to genetically modified plants, where a certain percentage of refuge areas is allocated for the growth of non-trangenic crops. The objective of such a policy is to preserve pest population that is susceptible to the trangenic crops in case of resistance development. This tactic has been reinforced by computer simulation results (Alstad and Andow, 1995), by theoretical projections (Liu and Tabashnik, 1997; Mallet and Porter, 1992), as well as laboratory experiments in several major pests (Tabashnik, 1994). However, the implementation of refuge policy raises many theoretical and practical problems: It increases pest levels and damages in the short run, encourages the use of conventional pesticides with detrimental environmental consequences, and may meet with non-compliance of farmers if such regulation will lower their profits.

Bt corn is a case in point for illustrating the pros and cons of the various strategies suggested for delaying resistance development. The areas planted with Bt corn have thus far exceeded 25% of the corn areas in the US. Since pest resistance to Bt corn has been found in laboratory conditions, but rarely detected in the fields, the long-term gains of such policy are questionable. Refuge area policy allows the use of conventional pesticides on the non-Bt fields, and thus hinders the possibility of a few years of pesticide-free corn fields, which may help the resurrection of beneficial insects and other organisms, establishing a new natural ecological equilibrium. Mixing Bt and non-Bt seeds is another alternative policy that may mitigate certain types of resistance development. However, Mallet and Porter (1992) showed in a theoretical model that if insects can move from plant to plant seed mixture may actually hasten insect resistance. The control of toxin concentration in genetically modified crops may also help maintain resistance development below critical levels. R&D in biotechnology may find another aspect and a corresponding gene that could be transplanted and control the pest (a sort of backstop technology). Finally, as long as resistance has not actually developed in the field (possibly because of high fitness cost of Bt-resistant pests), we should do our best in research and other preparations in case it arrives, but continue to enjoy the current gains of the new technology. If and when Bt resistance develops in the field, we could either go back to square one, or employ newly developed technologies and tactics to manage it.

Finally, one should not forget that resistance is but one of many concerns and worries regarding genetically modified crops, and adoption of these technologies should be cautious and not take them lightly.

5. Concluding Remarks

Resistance is only one problem among many negative external effects of chemical pesticides. Therefore resistance should not be treated separately from the whole complex of economic and environmental problems in pest control decisions. Farmers have incentives to consider resistance when acting collectively, but unless forced, will disregard environmental pollution such as air and water pollution and pesticide residues in the food chain. The same holds true for the pesticide industry.

To what degree conflicting economic incentives have contributed to the current pest-resistance situation is still an open question. However, the faster development of resistance by the highly mobile insects, relative to plant pathogens and weeds (Figure 1), can serve as an empirical indication for the importance of the common property role in pest control decision making. Whatever is the importance of analyzing past development of resistance, the main policy problem is how to proceed from here.

If and when biotechnology becomes a viable substitute for pesticides, society should seriously explore this option, though cautiously, given the environmental risks and benefits, and the potential for undesired selection of resistant new pest strains. The risks of biotechnology might turn out to be even more devastating than those of the conventional pesticides. In addition to reducing biodiversity, there is a growing concern that insects are smarter than humans, and will develop resistance to these new technologies faster than we can even begin to imagine. If Carson (1962) was even only partially right, perhaps we had better taken a risk-averse approach and not put all our eggs in one basket.

References

Alstad, D.N. and Andow, D.A. (1995) Managing the evolution of resistance to transgenic plants, *Science* **268**, 1894–1896.

Bull, D.L. and Menn, J.J. (1990) Strategies for managing resistance to insecticides in Heliotis pests of cotton, in M.B. Green, H.M. LeBaron and W.K. Moberg (eds.), *Managing Resistance to Agrochemicals: From Fundamental Research to Practical Strategies*, American Chemical Society, Washington, D.C., pp. 118–133.

Buntin, G.S., Ott, S.L. and Johnson, J.W. (1992) Integration of pest resistance, insecticides, and planting date for management of the Hessian Fly (*Diptera: cecidomyiidae*) in winter wheat, *J. Econ. Entom.* **85**, 530–538.

Carlson, G.A. (1977) Long run productivity of insecticides, *Amer. J. Agric. Econ.* **59**, 543–548.

Carson, R. (1962) *Silent Spring*, Houghton Mifflin, Boston, MA, 369 pp.

Clark, J.S. and Carlson, G.A. (1990) Testing for common versus private property: The case of pesticide resistance, *J. Environ. Econ. Manag.* **19**, 45–60.

Eckert, J.W. (1990) Impact of fungicide resistance on citrus fruit decay control, in M.B. Green, H.M. LeBaron and W.K. Moberg (eds.), *Pesticide Resistance: Strategies and Tactics for Management*, National Academy Press, Washington, D.C., pp. 286–302.

Feder, G. and Regev, U. (1975) Biological interactions and environmental effects in the economics of pest control, *J. Environ. Econ. Manag.* **2**, 75–91.

Forer, G. (1998) *IRM – IPM Strategy in Israeli Cotton*, The Israel Cotton Production & Marketing Board, Pest Control Division.

Georghiou, G.P. (1986) The magnitude of the resistance problem, in National Research Council, Committee on Strategies for the Management of Pesticide Resistance Pest Populations (ed.), *Pesticide Resistance: Strategies and Tactics for Management*, National Academy Press, Washington, D.C., pp. 14–43.

Georghiou, G.P. (1990) Overview of insecticide resistance, in M.B. Green, H.M. LeBaron and W.K. Moberg (eds.), *Managing Resistance to Agrochemicals: From Fundamental Research to Practical Strategies*, American Chemical Society, Washington, D.C., pp. 18–41.

Gorddard, R.J., Pannell, D.J. and Herzler, G. (1995) An optimal control model for integrated weed management under herbicide resistance, *Australian J. Agr. Econ.* **39**, 71–87.

Gorddard, R.J., Pannell, D.J. and Herzler, G. (1996) Economic evaluation of strategies for management of herbicide resistance, *Agric. Systems* **51**, 281–298.

Grafius, E. (1997) Economic impact of insecticide resistance in the Colorado Potato Beetle (*Coleoptera: Chrysomelidae*) on the Michigan potato industry, *J. Econ. Entomol.* **90**, 1144–1151.

Green, M.B., LeBaron, H.M. and Moberg, W.K. (eds.) (1990) *Managing Resistance to Agrochemicals: From Fundamental Research to Practical Strategies*, American Chemical Society, Washington, D.C., 496 pp.

Gressel, J. and Segel, L.A. (1982) Interrelating factors controlling the rate of appearance of resistance: The outlook for the future, in H.M. LeBaron and J. Segel (eds.), *Herbicide Resistance in Plants*, Wiley and Sons, New York, pp. 325–347.

Gressel, J. and Segel, L.A. (1990) Herbicide rotation and mixture. Effective strategies to delay resistance, in M.B. Green, H.M. LeBaron and W.K. Moberg (eds.), *Managing Resistance to Agrochemicals: From Fundamental Research to Practical Strategies*, American Chemical Society Washington, D.C., pp. 430–458.

Gutierrez, A.P., Regev, U. and Shalit, H. (1979) An economic optimization model of pesticide resistance: Alfalfa weevil – An example, *Environ. Entomol.* **8**, 101–107.

Heisey, P.W., Smale, M., Byerlee, D. and Souza, E. (1997) Wheat ruat and the costs of genetic diversity in the Punjab of Pakistan, *Amer. J. Agric. Econ.* **79**, 726–737.

Horowitz, A.R. and Ishaaya, I. (1994) Managing resistance to insect growth regulators in the Sweetpotato Whitefly (*Homoptera: Aleyrodidae*), *J. Econ. Entomol.* **87**, 866–871.

Horowitz, A.R., Seligman, I.M., Forer, G., Bar, D. and Ishaaya, I. (1993) Preventive insecticide resistance strategy in *Helicoverpa* (Heliothis) *armigera* (*Lepidoptera: Noctuidae*) in Israeli cotton, *J. Econ. Entomol.* **86**, 205–212.

Horowitz, A.R., Mendelson, Z., Cahill, M., Denholm, I. and Ishaaya, I. (1999a) Managing resistance to insect growth regulator, Pyriproxipen, in *Bemisia tabaci*, *Pestic. Sci.* **55**, 272–276.

Horowitz, A.R., Denholm, I., Gorman, K. and Ishaaya, I. (1999b) Insecticide resistance in whiteflies: Current status and implications for management, in I. Denholm and P. Ioannidis (eds.), *Proceeding of a Symposium on Combating Insecticide Resistance*, Thessaloniki, Greece, pp. 86–98.

Hueth, D. and Regev, U. (1974) Optimal agricultural pest management with increasing pest resistance, *Amer. J. Agric. Econ.* **56**, 543–551.

Ioannidis, P.M., Grafius, E.J., Wierenga, J.M., Whalon, M.E. and Hollingworth, R.M. (1992) Selection, inheritance and characterization of Carbofuran resistance in the Colorado potato beetle (*Colleoptera: Chrysomelidae*), *Pestic. Sci.* **35**, 215–222.

LeBaron, H.M. (leader of workshop participants) (1986) Tactics for prevention and management (of resistance), in National Research Council, National Academy of Sciences (ed.), *Pesticide Resistance: Strategies and Tactics for Management*, National Academy Press, Washington, D.C., pp. 313–326.

LeBaron, H.M. and Gressel, J. (eds.) (1982) *Herbicide Resistance in Plants*, Wiley and Sons, New York.

Lichtenberg, E. and Zilberman, D. (1986) The econometrics of damage control: Why specification matters, *Amer. J. Agric. Econ.* **68**, 261–273.

Liu, Y.B. and Tabashnik, B.E. (1997) Experimental evidence that refuges delay insect adaptation to *Bacillus thuringiensis*, *Proc. Roy. Soc. London B* **264**, 605–610.

Mallet, J. and Porter, P. (1992) Preventing insect adaptation to insect-resistant crops: Seed mixtures and refugia reconsidered, *Proc. Roy. Soc. London B* **255**, 165–169.

Miranowski, J.A. and Carlson, G.A. (1986) Economic issues in public and private approaches to preserving pest susceptibility, in National Research Council, National Academy of Sciences (ed.), *Pesticide Resistance: Strategies and Tactics for Management*, National Academy Press, Washington, D.C., pp. 436–448.

Powles, S.B. and Holtum, J.A.M. (1990) Herbicide resistant weeds in Australia, in *Proceeding of the 9th Australian Weeds Conference*, Adelaide, South Australia, August 6–10, pp. 185–193.

Radosevich, S.R. and Appleby, A.P. (1973) Studies on the s-triazines resistance in common groundsel (*Senecio vulgaris L.*), *Weed Sci.* **21**, 497–500.

Regev, U. (1976) Economic conflicts in plant protection: Individual versus social decisions, *EPPO Bull.* **6**, 265–273.

Regev, U. (1984) An economic analysis of man's addiction to pesticides, in G. Conway (ed.), *Pest and Pathogen Control: Strategy Tactics and Policy Models*, Wiley, New York, pp. 441–453.

Regev, U., Gutierrez, A.P. and Feder, G. (1976) Pest as a common property resource: A case study of alfalfa weevil control, *Amer. J. Agric. Econ.* **58**, 186–192.

Regev, U., Shalit, H. and Gutierrez, A.P. (1979) Economic conflicts in plant protection: Problems of pesticide resistance – Theory and application the Egyptian alfalfa weevil, in G. Norton and C. Holling (eds.), *Pest Management: Proceedings of an International Conference*, October 1976, IIASA Proceeding Series, Pergamon Press, Oxford, pp. 281–299.

Regev, U., Shalit, H. and Gutierrez, A.P. (1983) On the optimal allocation of pesticides: The case of pesticide resistance, *J. Environ. Econ. Manag.* **10**, 86–100.

Ryan, G.F. (1970) Resistance to common groundsel to Simazine and Atrazine, *Weed Sci.* **18**, 614–616.

Sarhan, M.E., Howitt, R.E. and Moore, C.V. (1979) Pesticide resistance externalities and optimal mosquito management, *J. Environ. Econ. Manag.* **6**, 69–84.

Schmidt, P. and Pannell, D. (1996) Economic issues in pest management of herbicide-resistance weeds, *Rev. Marketing Agric. Econ.* **64**, 301–308.

Schwinn, F.J. and Morton, H.V. (1990) Antiresistance strategies, in M.B. Green, H.M. LeBaron and W.K. Moberg (eds.), *Managing Resistance to Agrochemicals: From Fundamental Research to Practical Strategies*, American Chemical Society, Washington, D.C., pp. 170–183.

Tabashnik, B.E. (1994) Evolution of resistance to *Bacillus thuringiensis*, *Annu. Rev. Entomol.* **39**, 47–79.

Tabashnik, B.E. (1997) Seeking the root of insect resistance to transgenic plants, *Proc. Natl. Acad. Sci. USA* **94**, 3488–3490.

Taylor, C.R. and Headley, J.C. (1975) Insecticide resistance and the evaluation of control strategies for insect populations, *Can. Entomol.* **107**, 237–242.

Van den Bosch, R. (1978) *The Pesticide Conspiracy*, Doubleday & Company, New York, 226 pp.

Widawsky, D., Rozelle, S., Jin, S. and Huang, J. (1998) Pesticide productivity, host-plant resistance and productivity in China, *Agric. Econ.* **19**, 203–217.

3. Biotechnology for Planted Forests: An Assessment of Biological, Economic and Environmental Possibilities and Limitations

ROGER A. SEDJO

Forestry is currently undergoing an important transition from a wild resource, which had typically been foraged, to a planted agricultural crop, which is harvested periodically as other agricultural commodities. Only the time scale for forestry is longer. The transition of forestry from foraging to an agricultural cropping mode has been underway on a significant scale only within the past 50 years or so. Planted forests benefit from the same types of innovations that are common in other agriculture. However, economic incentives for plant domestication, breeding and improvement activities can occur when the investor can capture the benefits of the innovation. As in other types of agriculture, early improvements involved identification of trees with desired traits and attempts to capture offspring that have the desired traits through the identification of superior trees. In recent decades traditional breeding techniques were practiced in forestry. In the 1990s, however, modern biotechnology, including tissue culture and genetic modification began to be undertaken in earnest.

Forestry today is on the threshold of the widespread introduction of biotechnology into its operational practices in the form of sophisticated tissue cultures for cloned seedlings, and in the form of genetically modified organisms. As more of the world's industrial wood is being produced on planted forests, the potential to introduce genetic alterations into the germ plasma utilized in planting is obvious. In many cases the biotechnology about to be introduced in forestry is simply an extension of that being utilized in agriculture, e.g., a herbicide tolerant gene. However, biotechnology in forestry is also developing applications unique to forestry, e.g., genes for fiber modification, lignin reduction and extraction, and to promote sterility.

The environmental and ownership dimensions of biotechnology in forestry differ in some ways from agriculture. Unlike most agriculture there are no direct concerns for health or safety from the consumption of wood products using genetically modified organisms. Also, ownership and property rights issues related

Timothy Swanson (ed.), The Economics of Managing Biotechnologies, 71–95.

to biotechnological innovations appear to be more tractable in the longer harvest rotation of forestry than in typical seasonal agriculture. This is because by the time the tree is flowering and the seed is available, the technology in that seed may be obsolete. There are, however, concerns related to genetic transfers that might occur between transgenic and wild trees, and the potential implications for the natural environment. Additionally, there is the broader question of the environmental implications of producing larger portions of the world's industrial wood on rather modest amounts of land. One implication is that large areas of natural forest might be free from pressures to produce industrial wood, perhaps thereby being better able to provide biodiversity habitat. This paper addresses these issues.

The paper is organized as follows. The general introduction is followed by a discussion of the application of tradition breeding and then modern biotechnology to forests. In the next section the environmental impacts of these changes are examined in the forestry context. This is followed by a discussion of some of the perceived risks. The fifth section discusses the various types of biotechnological innovations that could be forthcoming and gives an estimate of where these innovations are currently in their development and which specific innovations are likely to become operationally available over the next decade or two. The potential economic benefits associated with selected types of biotechnology innovations are examined and the magnitude of one of the innovations, the herbicide-resistant gene, is estimated. Finally, the paper discusses the problems of economic and ecological viability of biotechnology in forestry, including the implications of forest certification for the introduction of transgenic seedlings.

1. General Introduction

The domestication of a small number of plants, particularly wheat, rice and maize, is among the most significant accomplishments in the human era. Modern civilization would be impossible without this innovation. Common features associated with plant domestication include high yields, large seeds, soft seed coats, non-shattering seed heads that prevent seed dispersal and thus facilitate harvesting, and a flowering time that is determined by planting date rather than by natural day length (Bradshaw, 1999).

Recent decades have seen continuing increases in biological productivity, especially in agriculture. This has been driven largely by technological innovations that have generated continuous improvements in the genetics of primarily domesticated plants and animals. Much of this improvement has been the result of plant improvements that have been accomplished by traditional breeding techniques through which desired characteristics of plants and animals, e.g., growth rates or disease resistance, can be incorporated into the cultivated varieties of the species in question.

Changes driven by technology, however, are not new. Hayami and Ruttan (1985) pointed out that in the U.S. most of the increased agricultural production that occurred over the past two centuries before 1930 was the result of increases in the amount of land placed in agriculture as most of the increased production reflected increased inputs in the form of labor saving technology – either animal or mechanical. In Japan, however, where land was limited, substantial improvements in rice productivity were made by careful selection of superior yield-increasing seed. Land productivity in grain production in the U.S. showed little increase until the 1930s as most of the gains in production were due to innovations that allowed more land to come into production, e.g., new equipment and mechanization. By contrast, land productivity in Japan was a function of biotechnological improvements in the form of improved seed and increased yields. In the U.S. after the 1930s, when most of the highly productive agricultural land was in use, the focus of innovation was directed at plant improvement, which increased land productivity through higher yields. Until fairly recently these improvements were achieved through the use of traditional plant breeding techniques, which gradually increased agricultural yields.

In its simplest form, biotechnology is the use of micro-organisms, plant and animal cells, to produce materials such as food, medicine, and chemicals that are useful to humankind. By this definition biotechnology comprises any technique that uses living organisms to make or modify a product, to improve plants or animals or to develop micro-organisms for a specific use (Haines, 1994a, 1994b). Such activities have been common among humans for a long time and include such activities as brewing alcoholic beverages or the use of traditional breeding techniques for improving food crops and domestic animals.

A more narrow but contemporary definition of biotechnology is 'the commercial application of living organisms or their products, which involves the deliberate manipulation of their DNA molecules'. Gradually, science has evolved its understanding of the chemical coding system, the gene, and how its message is encoded in the molecule's chemical structure. The gene is a segment of a substance called deoxyribonucleic acid; the DNA molecule and its message is encoded in the molecule's chemical structure. DNA is passed on from one generation to the next transferring a range of individual traits from parent to offspring. The science of manipulating and transferring chemical instructions from one cell to another is called genetic engineering. When the process involves the transferring of DNA from one organism to another, the result is a genetic modification that would not normally take place in nature – the production of a transgenic organism. This approach has been extremely successful in the development of new drugs, medicines and pharmaceuticals, as well as in agriculture.

A primary aim of modern biotechnology is to make living cells perform a specific useful task in a predictable and controllable way. Whether a living cell will perform these tasks is determined by its genetic make-up and by the instructions contained in a collection of chemical messages called genes.

This paper begins its investigation of biological innovations with those associated with traditional breeding techniques, but focuses primarily on those associated with genetic modification, including cloning and genetic marking, and particularly as applied to forestry.

The types of biotechnological innovations that were developed in agriculture and have applications in forestry include the two most common alterations – herbicide tolerance, e.g., Glyphosate and a natural insecticide, e.g., *Bacillus thuringiensis* (Bt), a type of bactera that infects and kills insects, which is used in cotton and corn crops.

New methods of weedy vegetation control have been developed using transgenic plants. Glyphosate (Roundup), for example, is an effective herbicide that has the desirable environmental property of rapidly decomposing into inert compounds. Thus, it can eliminate vegetation on a site without creating the traditional environmental problems such as leaving toxic residues in the soil and water. In agriculture, herbicides had been used to treat fields before planting, eliminating all existing vegetation and thus allowing the crop to begin growth with minimal weed competition. However, with ordinary crop plants, herbicides cannot be used after planting as it would damage the crop as well as the herbaceous weeds. In response to this situation a modified herbicide tolerant gene has been developed and introduced into agricultural plants, such as corn and soybean. Fields with these transgenic plants can be sprayed with the herbicide without damaging crop plants. The main advantage of this approach is lower weed control costs. Also, there is the potential to reduce total herbicide usage. In addition, increased yields are often experienced since weedy plant competition has been controlled and there is no inadvertent damage to the crop plants from the herbicide. Similarly, the introduction of a potato-bug-resistant gene, which discourages bug infestation, e.g., the Colorado potato bug, into potato plants has reduced pest control costs and increased potato yields.

Although facing resistance in some regions, overall the rate at which agribiotech applications have been commercialized has been quite rapid. In addition, there are important new applications on the horizon, e.g., increasing the protein content in milk, rice and potatoes.

2. Application of Tradition Breeding and Modern Biotechnology to Forests

2.1. Traditional Breeding

Tree improvement most often has relied on traditional breeding techniques like selection of superior (plus candidate) trees for volume and stem straightness, and grafting these into breeding orchards and producing seed orchards. When breeding orchards begin to flower, pollination of selections is artificially controlled, seeds are collected, progeny tests are established, and the best offspring are chosen for the next cycle of breeding. At the same time, selections whose offspring did not perform well in the progeny tests are removed from the production seed orchards to improve genetic quality. In the past, operational quantities of seed from production seed orchards were derived from open pollination. Today, however, more sophisticated large-scale controlled pollination techniques are in place that offer the potential of further improvement of the offspring of two superior parents.

By identifying and selecting for desired traits, breeding can select for a set of traits that can improve wood and fiber characteristics, improve the form of the tree, provide other desired characteristics, and improve growth. These traits are introduced into the genetic base that is used for a planted forest. This contributes to the more efficient production of industrial wood and to an improved quality of the wood output of the forest.

Thus far, most breeding activity has focused on increasing tree growth and disease resistance. Today, by matching superior seedlings with favorable sites and intensive management, planted forests are generating increased yields over what are commonly obtained in natural forests. Research is now focusing on improving other desired characteristics, in addition to growth rates.

The results of traditional breeding approaches to improve yields are instructive to illustrate the possibilities of traditional breeding (Table 1). For most tree species the typical approach involves the selection of superior trees for establishment in seed orchards. Experience has shown that an orchard mix of first-generation, open-pollinated seed can be expected to generate an 8% per generation improvement in the desired characteristic, e.g., yield. More sophisticated seed collection and deployment techniques, such as collecting seed from the best mothers (Family Block), can result in an 11% increase in yield, while mass-controlled pollination techniques, which control for both male and female genes (full sibling), have increased yield up to 21%.

A variant of the traditional breeding techniques is that of hybridization, which has provided robust offspring by bringing together populations that do not normally mix in nature. As in agricultural products, tree hybrids are often a means to improve growth and other desired characteristics. Hybridization crosses trees that are unlikely to breed in nature, often where parents do not occur together in sympatric populations. These crosses often exhibit growth and other characteristics

Table 1. Gains from various traditional breeding approaches: Loblolly Pine.

Technique	Effect
Orchard mix, open pollination, first generation	8% increase in yields
Family block, best mothers	11%
Mass pollination (control for both male and female)	21%

Source: Westvaco Corporation.[1]

that neither of the parent species alone can match. In the U.S., for example, several hybrid poplars have shown remarkable growth rates exceeding that found in parent populations.[2] The same is true for the *Eucalyptus grandis* and *urophylla* hybrid in many parts of the tropics and subtropics. Also, certain types of pine hybrids, pitch and loblolly hybrids, exhibit the cold hardiness found in pitch pine and the rapid growth of loblolly pine. Also, hybrids of the *P. carribaea* (Caribbean pine) and *P. elliottii* (slash pine) is are known to combine good stem form of slash with the rapid growth of the Caribbean strain.

The traits introduced can be of various types. For example, the introduction into trees of an engineered gene that confers tolerance to certain herbicides has already been accomplished and the potential for herbicide application during the establishment phase for some types of planted forests is substantial. This application is being widely used in certain agricultural crops and its adoption by farmers has been extremely rapid. Today the herbicide tolerance trait is being introduced into new plantings of certain hardwood plantations where weed problems are potentially severe. The presence of this gene, which makes the young trees tolerant to the herbicide, allows for the easy low-cost application of the herbicide early in the establishment cycle without concern that it could injure the crop. This approach not only lowers herbicide application costs, but also allows for more effective vegetative control and potential yield increases.

2.2. Molecular Biology

A second major approach to genetic manipulation of trees utilizes molecular biology. Molecular biology has two facets. The first facet is that which may aid the efficiency of traditional breeding programs. One problem with traditional approaches in tree breeding is the long growth cycles generally required by trees, which make this process very time consuming. Techniques, such as molecular biology and molecular markers, which identify areas on the chromosome where genes that control desired traits occur, can accelerate the process and enhance the productivity of the traditional approach. The second facet is where specific genes are identified and modified to affect biochemical pathways and the resulting phenotypes. For example, lignin genes can alter the amount, type and form of lignin that is produced.

In recent years molecular approaches to tree selection and breeding have shown significant promise. The molecular approach, although limited in application by its expense, involves genetic material being identified, collected, bred and tested over a wide range of sites. Rather than simply choosing specific tree phenotypes on the basis of their outward appearance, the molecular approach identifies the areas of the chromosomes that are associated with the desired traits. 'Markers' are used to identify the relative position of genes on the chromosome that control expression of a trait. This approach exploits the genetic variation, which is often abundant, found in natural populations. Molecular markers and screening techniques can be used to examine the DNA of thousands of individual trees to identify the few, perhaps less than a dozen, with the optimal mix of genes for the desired outputs. These techniques are currently being applied to the development of improved poplar in the U.S. and eucalyptus in Brazil.[3]

Recent work on hybrid poplar in the Pacific Northwest has shown a 20% increase in yields in plantations and an additional 20% on dry sites, where irrigation can be applied (east of the Cascade Mountains) (T. Bradshaw, pers. comm.). Growth rates with these plantations are impressive. Yields are about 7 tons per acre, or about 50 m^3 per ha and improvements in the yield continue (Withrow-Robinson et al., 1995, p. 13). These growth rates are approximately three times the growth rates of typical pine plantations in the South. Elsewhere in the world, for example, Aracruz in Brazil, yields of hybrid eucalyptus are reported to have more than doubled those of earlier plantings.

The second facet is where specific genes are identified and modified to affect biochemical pathways and the resulting phenotypes. For example, the promise of controlling the lignin in trees is dependent on the ability to identify and modify lignin genes, thereby altering the amount, type and form of lignin that is produced in the tree (Hu et al., 1999). As noted, the ease of gene introduction (transformation) varies with different species, generally being more difficult in conifers than hardwoods.

2.3. CLONAL APPLICATIONS

The development of cloning techniques in forestry is important for a number of reasons. First, if superior trees are available, an approach must be developed to allow for the propagation of large numbers of seedlings with the desired characteristics if these traits are to be transferred into a planted forest. Cloning provides a method that allows trees that are improved by traditional breeding techniques to be replicated on a large scale. Additionally, the clone provides the vehicle through which desired foreign or artificial genes are transferred. Thus, for genetic engineering in forestry to be viable, cloning techniques must be developed.

The ability to use inexpensive cloning techniques varies with species and genus. On the one hand, genera such as poplar tend to readily lend themselves to vegetative propagation. Eucalyptus and acacia also tend to be effective propagators. Other genera propagate less readily. In the pine family, loblolly and to a lesser extend slash pine are difficult propagators. However, much progress is being made, making the prospects for clonal production very promising in the near term (B. Goldfarb, North Carolina State University, pers. comm.). Radiata pine appears to have the best record on this account. Propagation improves when certain procedures are undertaken. For example, using the shoots emerging from newly trimmed clonal hedges increases the probability of successful regeneration. For some species, typically hardwood species, cloning can be as simple as using the vegetative propagation properties inherent in the species to accomplish the genetic replication. This might involve simply taking a portion of a small branch from a desired superior tree and putting it into the ground, where it will quickly take root (rooted cuttings). Where vegetative propagation is part of the natural process, large amounts of 'clonal' material can be propagated via rooted cuttings, the cuttings of which come from 'hedge beds'. Here the process continues until sufficient volumes of vegetative materials with the desired genes are available to meet the planting requirements.

For many species, however, the process is more difficult as simple vegetative propagation does not normally occur or occurs only infrequently. Here, 'tissue culture' techniques provide the tools to quickly produce genetically engineered plants and clones to regenerate trees with desired traits (Westvaco, 1996, pp. 8–9).

Tissue culture broadly refers to techniques of growing plant tissue or parts in a nutrient medium containing minerals, sugars, vitamins, and plant hormones under sterile conditions. It involves a set of techniques known as micropropagation, that is vegetative propagation that can produce multiple copies of an elite genotype as well as provide a means of introducing novel genes. Approaches include organogenesis and somatic embryogenesis. Typically, plant tissue is placed on a nutrient medium until new buds are initiated on the plant tissue. From these buds, shoots and, ultimately, roots are developed. Somatic embryogenesis is a method of plant tissue culture that starts with a piece of donor plant and forms new embryos. This approach has shown promise for rapidly multiplying some types of conifers and hardwoods. However, for some species[4] micropropagation cloning approaches are limited (Pullman et al., 1998).

The development of clonal approaches to propagation is important to the broad utilization and dissemination of genetically improved stock. With tree planting often involving over 500 seedlings per acre,[5] large-scale planting of improved stock requires some method of generating literally millions of seedlings, at a relatively low cost, which embody the genetic upgrading. The costs of the improved seedlings are important in a financial sense since the benefits of improved genetics are delayed until the harvest. With harvests often being 20 years or more after planting,

Type	Period
Wild forests	10,000 BC – current
Managed forests	100 BC – current
Planted forests	1800 – current
Planted, intensively managed	1960 – current
Planted, superior trees, traditional breeding techniques	1970 – current
Cloning	1990 – current
Planted, superior trees, genetic modification	2000+

Figure 1. Transitions in forest management and harvests.

large costs for improved seed may seem difficult to justify financially. However, if the costs of plantings are going to be incurred, the incremental costs associated with planting improved genetic stock are likely to be quite modest and therefore to be financially justified.

3. Environmental Impacts of Biotechnologically Induced Changes in Forestry

Currently, most of the world's industrial wood is drawn from natural forests in what is essentially, a foraging operation. In the past harvests occurred from forests created by nature as humans simply collected the bounty of nature. Figure 1 indicates how this process has changed over time as humans gradually developed silvicultural technology. Although forest management began in China as early as 100 BC (Menzies, 1985), significant areas of managed forest probably were not common until the middle ages. Planted forests began in some earnest in the 19th century in Europe and not until the middle of the 20th century in North America. The planting of genetically superior stock began about 1970, and the serious planting of genetically modified trees is just now beginning in parts of the subtropics, New Zealand and South America.

However, as Table 2 indicates, even today a large portion of the world's industrial wood supply originates in natural unmanaged forests. In recent decades, however, the widespread introduction of tree planting worldwide for industrial wood production has resulted in most of the increases in global harvests being drawn from planted forests.

Table 2. Global harvests by forest management condition circa 1999.

Forest situation harvest	Percent of global industrial wood
Old-growth	22
Second-growth, minimal management	14
Indigenous second-growth, managed	30
Industrial plantations, indigenous	24
Industrial plantations, exotic	10

Source: Sedjo (1999)*.
Notes: Old-growth includes: Canada, Indonesia/Malaysia, and Russia, * adjusted for the large harvest declines following the demise of the Soviet Union.
Second-growth, minimal management: parts of the U.S. and Canada, Russia
Indigenous second growth, managed: residual
Industrial plantations, indigenous: Nordic, most of Europe, a large but minor portion of U.S., Japan, and some from China and India.
Industrial exotic plantations
Second-growth, minimal management: the residual

The potential of the widespread introduction of genetically improved trees can have important environmental and economic effects. With increasing yields and shortened rotations, planted forests become increasingly attractive as an investment for producing future industrial wood as a substitute for the natural forest. The plantation manager can control some of the important variables, such as choosing a location for the planted forest and the species. Former agricultural sites often are desirable locations for planted forests usually being accessible and reasonably flat, thereby lending themselves to both planting and harvesting. Often, acceptable access exists via the former agricultural transport infrastructure. The planted forest can also be located in proximity to important markets.

Within limits, the manager can chose a species appropriate to the site, which may also have good market access and a reasonably short harvest rotation.

The economic advantages of planted forests have led to their widespread adoption in a number of regions throughout the globe; they are having an important influence on global timber supply. Over time, a greater share of the world's industrial wood supply has been and will be coming from planted forests. Planted forests now account for most of the increased global output and their production is replacing the timber formerly provided by native and old-growth forests, which are no longer available for harvest due to political changes, e.g., Russia, or policy changes as with the U.S. National Forest System.

4. Perceived Risks: Some Concerns

Transgenetic biotechnology has become quite controversial when applied to agriculture (e.g., see, Williams, 1998). However, in drugs, medicines, and pharmaceutical applications they are essentially without controversy.

The nature of the controversy in agriculture has developed around at least six issues:

1. *Ownership*

First is the issue of ownership of the modified genes and the question of how much ownership/control the biotechnology companies have over their transgenic products after they have been sold. Under one option the gene-altered seeds are sold under the condition that their offspring, which also contain the altered gene, will not be used as future seed in further plantings. Thus, the farmers must return to the seed developer for future seed sources. The rationale is that the company that developed the gene-altered plant has intellectual property rights to this plant throughout the patent period. This argument is buttressed by the fact that development often takes decades and costs hundreds of millions of dollars. Related concerns are tied to two considerations. The first is that of the disruptions and inconvenience associated with monitoring users for compliance with various provisions of the agreement. (This seems to be the essence in the *Washington Post* article of February 3, 1999.) The second consideration relates to the ongoing controversy regarding the broader philosophical issue of the ownership of biodiversity and improved products. For example, are wild genetic resources the property of all of humanity or of the country in which they reside? Are developed biotechnology products the property of the developer or should they be available without royalty payment to all of humanity? (For example, see, Kloppenburg, 1988; Sedjo, 1992.) This controversy continues to be manifest in the difficulties in interpreting and finalizing the 'biodiversity treaty' coming out of the UNCED 'Earth Summit' meeting in Rio in 1992. This issue recently returned to the headlines with the inability, once again, of the parties to agree to major dimensions of the proposed treaty.

The second set of issues in the overall controversy relates to the health, safety and environmental aspects of transgenic products.

2. *Health Effects of Foods*

Direct health and safety effects involve concerns about the direct health effects of foods ingested from transgenic products. Although there is little or no evidence that transgenic foods are unsafe, health concerns are raised due to the lack of long-term exposure and experience with such products. The health issue is not expressed in concerns of how the plants were produced (transgenic or traditional breeding) but rather what new proteins the plants are making and the health implications of these.

3. *Environmental Effects of GM Seeds*
A third issue is related to the question of whether transgenics can 'escape' from cultivated fields and, if so, would they have impacts on adjacent fields or the natural environment. For example, can transgenics interbreed with similar wild plants thereby changing the genetic make-up of wild plants and thereby impacting on the natural ecosystem, eventually altering that system in unanticipated ways? One method to prevent or reduce their 'escape' is to promote sterility or delayed reproduction and thereby reduce the potential for the transfer of genes from transgenics to the wild populations.

4. *Effect on Herbicide (Pesticide, etc.) Use*
GM seeds could result in the use of more (or less) herbicide, and more (or less) dangerous herbicide.

5. *Effect on Pest Population*
Another issue relates to the impact of the biotechnology on the pest population. It is well known that pests adapt through natural selection to the introduction of pest controlling chemicals in their environment. The same response would be expected to attempts at genetic pest control.

6. *Effect on Demand for Land*
Finally, there is the issue of whether biotechnology applied to agriculture will increase the demand for land thereby putting increased pressure on natural habitats. Some recent work suggests this is likely to be the case if the demand for agricultural products is elastic.

4.1. Discussion

The issue of forest products and food safety is more complex than it may initially appear. Since wood products are not usually ingested, they are usually regarded as unlikely to have any direct human health effects either in the short or long run. However, wood cellulose is often used as a filler in foods and thus is, in fact, ingested. Thus, if transgenic foods generate a health risk, it would also apply to wood in some cases.

However, the ownership issue is associated with the use of seeds from transgenic plants to create subsequent crops is likely to be less important due to the long periods required for flowering in trees. The value associated with using germ plasm from genetically altered trees after several years have passed and some of the generic alterations may be nearing economic obsolescence appears modest. Furthermore, most commercial tree planting operations obtain their seed or seedlings from an outside source. In practice, most growers do not obtained seeds from their own trees unless they are large enough to have their own breeding programs. Thus the ownership issue is reduced.

In forestry, as in agriculture, there is the problem of the pest population developing a resistance to the modified gene and thereby undermining its longer-term effectiveness. The effect of the gene on the use for chemical herbicides and pesticides is also of concern. In concept, use of herbicide-resistant GMOs could either increase or decrease the use of herbicides. Under current conditions, the use of environmentally benign herbicides, that readily decompose into inert compounds is common. The use of a herbicide-resistant seedling could either result in increased herbicide, since there are no longer concerns about damaging the crop, or less herbicide usage, since the applications are more optimally applied. In the case of pesticides, as with the Bt gene, the effect is almost certainly to reduce traditional pesticide use.

A more pressing concern, however, relates to the potential for genetic transfer from the transgenic tree to the surrounding natural environment. In cases where the tree is an exotic it is sometimes argued that the potential for transfer is unlikely since no similar species of trees are found in the natural environment. For example, since conifer species are not indigenous to South America the accidental transfer of genes from exotic conifer to indigenous trees is unlikely. Another approach to this problem is the introduction of sterility as a vehicle for preventing the release of genes that might transfer to the natural environment.

In forestry, the prevention of genes released into the wild can be facilitated by permanent sterility of the tree, a terminator gene, or by delaying the flowering of the tree to a period beyond the normal rotation period. However, the delayed flowering approach is not wholly satisfactory since, for various reasons, all trees may not be harvested in the period expected. Also questions exist regarding the harvest approach since events may occur that postpone or preclude the planned harvest. Thus, concerns arise that the probability of certain genes escaping is not zero.

If modified genes do escape, how serious are the 'expected' consequences or the 'worst case' consequences? In the case of the herbicide tolerant gene the consequences of release into the wild are probably small. This would be the case especially of the gene were not expressed in the pollen. Herbicides are unlikely to be applied to most of the natural environment. If herbicides are to be applied, types can be used to which the escaped genes do not confer tolerance. In the intermediate and longer term, the herbicide in question will almost surely be replaced periodically in the normal course of product change and development. Thus the presence of that modified gene in the natural environment appears unlikely to constitute any serious short or long-term environmental problem. Similarly for genes that affect tree form or fiber characteristics, the release of this gene into the natural environment is unlikely to provide a competitive advantage in survival and therefor unlikely to have significant or adverse consequences.

However, this situation could change if a survival gene is involved. For example, the release of a Bt gene into the wild could constitute a more serious problem if it results in the altering of the comparative competitive position of the various types of vegetation in dealing with the pest. In some sense this modified pest might be viewed as an exotic. Ultimately, the seriousness of this problem depends importantly on the probability of the transfer of a survival gene into the wild, on the scale of the transfer and on the comparative change in the competitive balance within the natural habitat. To the extent that the pest population adapts via natural selection to modified genes, however, the long-term impact of the release of the modified gene on the natural environment will be mitigated.

Although there is now a debate about the impact of genetic modification on the requirements for agriculture land use after the demand requirements have been fully considered (e.g., Angelsen and Kaimowitz, 1998),[6] the issue is unambiguous in forestry for two reasons. First, the global demand for industrial wood usually regarded to be less than unity. Secondly, the land currently being converted into plantation forests overwhelming is drawn from agricultural lands, usually marginal cropping lands or pastures. Thus, natural forest is rarely lost by the creation of a planted forest.

5. Biotechnological Innovations Forthcoming in Forestry

There is little question but that biotechnology provides lower costs for the establishment of wood plantations. Some estimates of the cost reductions are presented below. Under current conditions the vast majority of additional forest plantations will continue to be established on marginal agricultural lands. Furthermore, plantation wood has been and will continue to be substituted for harvests from natural forests thereby reducing harvesting pressure on natural forests (Sedjo and Botkin, 1997).[7]

With the planting of trees for industrial wood production there is an inherent incentive for tree improvements. Tree improvements can take many forms (Figure 2). Thus far, the most common emphasis of tree improvement programs is increased growth rates, stem form and disease resistance. Growth typically refers to wood volume growth or yields. Disease and pest resistance traits are also desired to promote or insure the growth of the tree. Resistance traits may be oriented to specific problems common in the growth of particular species or to extending the climatic range of certain species. For example, the development of frost-resistant eucalyptus would allow for a much broader planting range for this desired commercial genus. Other improvement possibilities include, as in agriculture, the introduction of a herbicide-resistant gene to allow for more efficient use of effective herbicides, especially in the establishment phases of the planted forest.

Important Attributes:

- *Growth rates.*
- *Disease and pest resistance.*
- *Climate range and adaptability.*
- *Tree form and wood fiber quality*: straightness of the trunk, the absence of large or excessive branching, the amount of taper in the trunk.
- *Desired fiber characteristics may relate to ease in processing*, e.g., the break-down of wood fibers in chemical processing.

Figure 2. Tree improvement programs.

Besides ensuring establishment, survival and rapid growth of raw wood material, tree improvement programs can also focus on wood quality. Wood quality includes a variety of characteristics including tree form, wood fiber quality, extent of lignin, and so forth. Furthermore, the desired traits vary by end product. Wood quality may involve one set of fiber characteristics for pulping and paper production and another set of characteristics for milling and carpentry. Wood desired for furniture is different from that desired for framing lumber. In addition, some characteristics are valued not for their utility in the final product, but for their ease of incorporation into the production process.

For pulp and paper production there are certain characteristics desired to facilitate wood handling in the early stages of pulp production. For example, the straightness of the trunk has value for improving the pulp and paper products in that less compression associated with straight trees generates preferred fibers. Also, straight trees are important in pulp production since it allows ease of handling and feeding into the production system. Also, paper production requires fiber with adequate strength to allow paper sheets to be produced on high-speed machines. Ease in processing includes the breakdown of wood fibers in processing and the removal of lignin, a compound found in the tree that must be removed in papermaking.

Other wood characteristics relate to utility in producing the final product. The absence of large or excessive branching, for example, influences the size and incidence of knots thereby allowing for fuller utilization of the tree's wood volume. Also, desired characteristics or properties of final paper products include paper tear strength, surface texture, brightness, and so forth. These are all properties that relate in part to the nature of the wood fiber used. Other features relate to the utility of the wood for use in final wood products, e.g., straightness facilitates production of boards or veneer in solidwood products, wood characteristics related to milling

Table 3. Forest traits that can be improved through biotechnology.

Silviculture	Adaptability	Wood quality traits
Growth rate	Drought tolerance	Wood density
Nutrient uptake	Cold tolerance	Lignin reduction
Crown/stem	Fungal resistance	Lignin extraction
Flowering control	Insect resistance	Juvenile fiber
Herbicide		Branching

Source: Context Consulting.[8]

and use in carpentry, wood color, strength and surface characteristics. In addition, wood fiber is increasingly being processed into structural products such as strand board, fiberboard, and engineered wood products, which have their own unique set of desired fiber characteristics.

In recent years pulp producers have begun to move away from simply producing standardized 'commodity' pulp into the production of specialized pulp for targeted markets. For example, Aracruz, a Brazilian pulp company, has asserted that it can customize its tree fibers to the requirements of individual customers. This requires increased control over the mix and types of wood fibers used. Customized products require customized raw materials. However, in the case of Aracruz thus far the control has been provided through cloning but not transgeneic plants.

5.1. Forestry in the Future: Traits and Examples

Gene alteration can result in unique gene combinations unachievable by traditional tree breeding. This allows species to have attributes that would not be possible through natural processes. Thus, for example, in concept, frost-resistant genes could be transferred from plants or other organisms found in cold northerly regions to tropical plants, thereby increasing their ability to survive in cooler climates.

These attributes or traits can be characterized as silvicultural, adaptability and wood quality (Table 3). Silvicultural traits would include growth rate, nutrient uptake, crown and stem form, plant reproduction (flowering) and herbicide tolerance.

Growth potential, for example, has a substantial genetic component with rates differing by 50% between families or different clonal lines. Traditional breeding approaches are steadily improving elite line yield potentials.

A subset of these traits is found in Table 4. These traits include those that are most likely to use biotechnology for further commercial development.

The first four traits of the list in Table 4 are traits that, in the judgement of many experts, could be featured prominently in biotechnological innovations in forestry over the next decade.

Table 4. Traits of interest in forestry.

Herbicide tolerance
Flowering control
Fiber/lignin modification
Insect tolerance
Disease tolerance
Wood density
Growth
Stem straightness
Nutrient uptake
Cold, wet, drought tolerance

Planted trees typically require herbicide and, in some cases pesticide applications for one or two years after planting. The introduction of a herbicide-resistant gene can reduce the costs of herbicide applications by allowing fewer but more effective applications without concern of damage to the seedlings. The use of a pest-resistant gene can eliminate the requirement to apply the pesticide altogether. Flowering control allows a delay of several years in flower initiation, non-flowering habit, or sterility. This control may be useful in preventing certain transgenic plants from transmitting genetically modified matter to other plants and/or from migrating into the wild.

As with pest resistance, disease resistance is also important and the technology for genetic modification for disease resistance is fairly well developed. In New Zealand, for example, the first applications of genetically modified pine are likely to involve the 'stacking' that is, combining of several genetically modified genes, perhaps including those of pest and disease resistance and flowering control, in the seedling.

Lignin control is viewed by the industry as an important priority. Trials with low lignin trees have already been undertaken in Aracruz Cellulose in Brazil (Hall, 2000).

6. Estimates of Some Potential Economic Benefits

Table 5 lists a number of innovations believed to be feasible within the next decade or two and suggests the possible financial gain. The innovation development costs of the innovation are not considered. All of the innovations noted in Table 5 result in a decrease in costs and/or an increase in wood volume or quality. Rates of return can be estimated from many of them. For example, the 20% increased volume due to the cloning of superior pine is estimated to provide a return of about 15–20%

Table 5. Possible financial gains from future biotech innovations.

Innovation	Benefits	Additional operating costs
Clone superior pine	20% yield increase after 20 yrs	$40/acre or 15–20%
Wood density gene	Improved lumber strength	None
Herbicide tolerance gene in eucalyptus (Brazil)	Reduce herbicide and weeding costs saving $350 or 45% per ha	None
Improve fiber characteristic	Reduce digester cost $10 per m^3	None
Reduced amount of juvenile wood	Increase value $15 per m^3 (more useable wood)	None
Reduce lignin	Reduce pulping costs $15 per m^3	None

Source: Context Consulting.

on the incremental investment cost of $40 per acre. This assumes initial yields of 15 m^3 per ha per year and a stumpage price of $20 per m^3.

The herbicide and weeding cost savings due to the herbicide tolerance trait in Brazil would generate an immediate cost reduction of $350 per ha in the establishment costs over the establishment period of 2–3 years. Obviously, this potential degree of financial benefit is substantial.

The financial impact of biotechnological innovations that reduce pulping costs can also be estimated. The value added from pulping is about $60 per m^3 or $275 per ton of pulp output. If these costs are reduced $10 per m^3, this provides a surplus (or effective cost reduction) of about $47 per ton of woodpulp, assuming wood prices are not affected.

This type of impact would be important to the forest sector. If the stumpage wood costs are $20 per m^3 and the mill experiences an increased value of $10 per m^3 due to the superior wood qualities, then the mill ought to be willing to pay an initial premium of at least $10 per m^3, or roughly 50%, for the 'improved' wood. Thus, substantial revenue improvements could be generated initially. However, if over the longer term many wood producers respond to the higher marginal wood values with increased production that would lead to falling prices for improved wood.

Furthermore, there is the issue of the cost to the wood grower of introducing technological improvement. The developer is obviously going to want compensation for the development costs. So, there is the issue of how the 'surplus' is distributed, among developer, wood producer and final wood consumer. This will be determined by the pricing policy in the context of the market structure and conditions that exist. Over the longer term in reasonably competitive conditions,

Table 6. Anticipated value increases.

Herbicide-resistance benefits:
– $35/acre ($87/ha) cost reduction for fast-growing softwoods
– $160/acre ($400/ha) cost reduction for fast-growing hardwoods

Source: Context Consulting.

the innovation can be expected to be made available at marginal cost. In this case, eventually, the full net benefits will be captured in the wood market and shared by producer and consumer. At this point the developer's rights would cease to exist either because the patent period has expired or because subsequent innovations will overtake and make of limited value the initial innovation.

7. A Crude Estimate of the Global Impact

In this section we will examine the effect of biotechnological innovations on the future timber supply, and by inference, the likely effect on harvests from natural forests. The approach used is that of a crude partial equilibrium approach,[9] which estimates the cost savings associated with the development of a specific innovation as applied to forestry – the herbicide-resistant gene. The savings in plantation establishment costs are estimated on the basis of the data presented above. These savings are translated into the lowering of the supply curve for planting activity. This results in an incremental addition to plantings. Due to the delay between planting and harvest, the direct impact on harvests is to delayed to the future timber supply.[10]

This section examines briefly the potential impact of one of the mostly likely transgenic innovations to be introduced in the near term in forestry, specifically the use of a herbicide-resistant gene.[11] Table 6 provides estimates of the cost reduction in plantation establishment for the herbicide-resistant gene and for the pest/disease-resistant gene.

7.1. Potential Cost Savings Due to Herbicide Resistance in Forestry

Forest plantation establishment involves incurring of substantial costs in an early period in order to generate larger but discounted benefits at some future time. High-yield plantation forestry involves plantations with harvest rotations from 6 to 30 years. To the extent that costs of establishment can be reduced, net benefits can be achieved. Experts estimate that herbicide resistance would reduce the costs of plantation establishment by an average of about $35/acre for fast-growing softwoods (reduced costs of 15%) and an average of $160/acre for fast-growing hardwoods (reduced costs of 30%) through the elimination of the costs of other pest mitigation activities.[12] In North America about 4 million acres are planted annually: If 98% (3.9 million) are softwood and 2.0% (0.1 million) hardwood, the potential cost reduction at current rates of planting would be $136.5 million for softwoods and $16 million for hardwoods or a total savings of $152.5 million annually.

Worldwide about 10 million acres of plantation forest are planted per year. If the plantings are roughly 50–50 conifer and hardwood *and the plantings remain unchanged*, the potential saving from the introduction of the herbicide-resistant gene is $175 million for softwoods and $800 million for hardwoods or a global potential savings of about $975 million annually.

7.2. Potential Effect on Plantation Establishment and Timber Supply of the Use of the Herbicide-Resistant Gene

Of the 10 million ha of forest planted annually, we assume that about 1 million ha represents new industrial plantations.[13] Assume that the actual costs to the industry were reduced by the full amount of the cost reduction realized through the innovation, e.g., that the innovation was priced at marginal cost. This would be an average reduction of 22.5% in plantation establishment costs. Under these circumstances what increase would be expected in the annual rate of plantation establishment? To examine this question we develop and estimate the impacts from three scenarios: the maximum impact, an intermediate impact, and a low impact.

In Figure 3 the supply of industrial wood from natural forests is given by S. At P_0 plantations become economically feasible and the effective industrial wood supply curve is given by S', with production running up S to point 'a' and then horizontally to quantity Q_0. Figure 4 can be viewed as derived from Figure 3, with the demand curve being derived from the conditions in Figure 3.

Treating Figure 4 as focusing on the annual establishment of new plantations, the introduction of the herbicide-resistant gene has the potential to reduce planter costs from Cost_0 to Cost_1, due to the herbicide-resistant gene. This is reflected in a downward shift of supply to S' and plantation starts increase from Q_0 to Q_1.

Scenario A: Maximum Impact. Given an initial total annual rate of global planting of 1.0 million ha and assuming an infinite supply elasticity and a unitary demand elasticity for forest plantation plantings (a derived demand), the estimated impact would be the establishment of an additional total planting area of 225,000 ha per year. This assumes that the additional planting would reflect current mix of planting. i.e., the additional planting would be divided evenly between conifer and hardwood. Furthermore, if we assume growth rates on plantation forests would average 20 m^3 per ha per year for softwoods and 30 m^3 per ha per year for hardwoods, the result of the additional plantings would result in a future addition

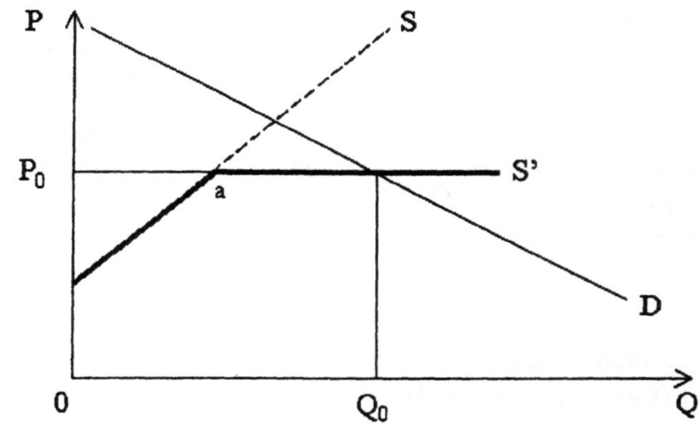

Figure 3. Industrial wood.

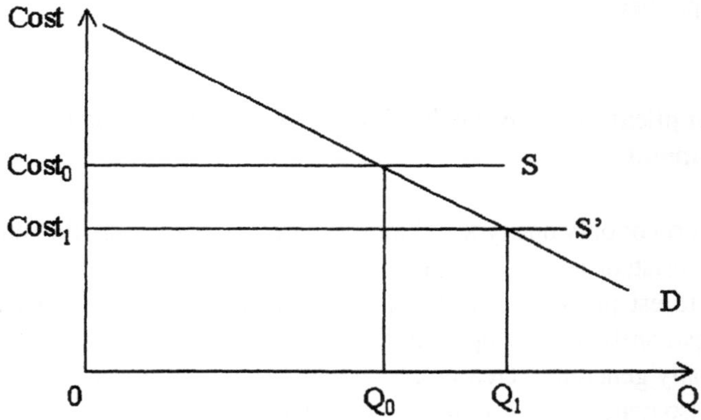

Figure 4. New plantation starts.

to total annual production at harvest of 2.5 million m^3/yr. If these increases in plantings were realized each year for a 20-year period, about 100 million m^3/yr of additional industrial wood production would be generated annually after 20 years.[14]

Scenario B: Intermediate Impact. Suppose the same conditions obtained as in Scenario A except the supply elasticity were 1.0.

In this case a total of 112,500 additional ha planted per year would result in a total increased production at future harvest of 2.5 million m^3/year. After 20 years of planting this would generate about 50 million m^3/yr of additional continuous production.

Table 7. Scenario summary

	Additional plantings	One year Additional m^3	Twenty years Additional m^3
Scenario A	225,000	5 million	100 million
Scenario B	112,500	2.5 million	50 million
Scenario C	78,750	1.97 million	39.4 million

Scenario C: Estimated Minimum Impact. The assumption is that supply elasticity remains a +1.0, as in Scenario B, but that the demand elasticity is −0.7.[15] In this case we estimate a total of 78,750 additional ha planted per year with an increase in total production at harvest of 1.969 million m^3 per year. After 20 years of planting at this rate the additional continuous wood production would be about 39.375 million m^3 per year.

8. Some Implications of Biotechnology for Forestry and for the Natural Environment

The application of only one type of biotechnological innovation, the introduction of a herbicide-resistant gene, could generate benefits estimated at $1 billion annually in reduced forest plantation establishment costs and an expansion in the rate of plantation establishment by up to 225,000 ha per year. The increased production would not only generate increased social welfare through lower commodity prices but would also generate environmental benefits in the form of decreased harvesting pressure on natural forests.

It is well documented that there has been a gradual worldwide shift in industrial wood production from natural forests to plantations. Such a trend could have advantageous effects on native forests as harvest pressures are relieved and native forests can be devoted to other purposes. The more productive are forest plantations, the more they can deflect harvesting pressures away from natural forests. Furthermore, as noted above, planted forests rarely replace natural forest but typically are in addition to natural forests. Biotechnology offers the potential of further reducing wood costs thereby contributing to human welfare. Additionally, biotechnology may promote the more rapid establishment of plantation forests thereby contributing to the maintenance of the remaining natural forest system by the reducing long-term pressures on natural forests, thus, leaving the natural forest to meet other social and biological needs.

However, the future use of genetically modified organisms (GMOs) in forestry could be influenced in the marketplace through the role being developed for forest management auditors and certifiers. Coming out of the Rio Earth Summit Meeting of 1992 have been a host of initiatives for forests to meet criteria that indicates that they are 'sustainably managed' or 'well managed' forests. One of the more active groups is the Forest Steward Council (FSC), which has a certification procedure that, at present, precludes the use of GMOs. Certified wood is expected to be used in ecolabeled products, which may have a price premium in the marketplace (Sedjo et al. 1997). This approach, if widely adopted and respected in the market, could have a major influence on the use of GMOs in forestry.

Finally, the benefits of biotechnology must be compared with the costs. Recently, biotechnology in agriculture has come under attack for its potential health, safety and environmental risks. The application of biotechnology to forestry, however, poses somewhat different considerations than biotechnology's applications elsewhere. For example, direct health and safety risks appear non-existent or negligible in most of the uses of industrial wood as ingestion occurs only in its specialized use as a food-filler. The environmental risks that exist appear to relate largely to the potential for altered genes to move out of transgenic trees into the natural environment. These risks can probably be reduced by the delay or elimination of flowering and/or by introducing the species into foreign environments where similar species are not found in the wild and gene transfer is very improbable.

Acknowledgements

The author acknowledges the assistance of a number of individuals, most of whom are specialists in the area of biotechnology. These include Barry Golbfarb, Ross Whetten, Fred Cubbage, and Dan Robinson, William Dvorak, of the College of Forest Resources at North Carolina State University, as well as Larry Tombaugh, the Dean of the College of Forestry at NCS. In addition, the author thanks David Canavera of Westvaco, Toby Bradshaw, of the University of Washington, Rex McCullough of Weyerhaeuser Company, Bob Kellison, of Champion International and, my colleague, Ramanan Laxminarayan of Resources for the Future. Finally, my thanks to David Duncan and Lisa Drake of Monsanto Company. Of course, the author is responsible for any errors.

Notes

[1] Source: Conversation with Westvaco researchers, Summerville, SC.

[2] Growth in hybrid poplar stands is 5–10 times the rate experienced in native forest growth rates (T. Bradshaw, University of Washington, pers. comm.).

[3] Personal conversation with Toby Bradshaw, Director of the Poplar Molecular Genetic Cooperative at the University of Washington, Seattle. Also see Westvaco (1997).

[4] In general, thus far there has been greater success cloning hardwoods, e.g., poplar and some species of eucalyptus, than conifers.

[5] It is estimated that 4 to 5 million trees are planted in the U.S. *every day*.

[6] It has been noted that since cattle are increasingly being placed in feedlots where they consume grains, the total demand for grain, human and animal, may be elastic. This implies that if grain prices fall, e.g., due to biotechnology, the total area of land in grains could increase. However, it should be noted that where both grain and cattle are part of society's diet, the feeding of grain to cattle has resulted in a decline in pasture area. Thus, total agricultural land, grain plus pasture, may have decreased even if the area in grains increased.

[7] The argument that plantation wood substitutes for wood from natural forests is substantially different from the issue of land involved in grain production in that forestry compares a foraging with a cropping activity. A recent FAO study (1996) estimated the global demand elasticity of industrial wood at −0.67.

[8] Context Consulting provided information on potential innovations and their likely cost implication based on the best judgements of a panel of experts.

[9] A more sophisticated modeling approach is currently being undertaken to make these estimates using the Timber Supply Model, a forest sector systems model (see Sohngen et al., 1999).

[10] It should be noted, however, that the anticipation of greater future supplies will effect current actions, including current harvests (see Sohngen et al., 1999).

[11] It should be noted that it is generally expected that the first commercial planting of transgenic seedlings will involve the insertion of a number of genes, rather than a single gene as hypothesized in this illustration.

[12] The percentages are based on an updating of plantation establishments costs as found in Sedjo (1983).

[13] Sedjo (1999) estimated this to be about 600,000 ha for the tropics and subtropics, while the model of Sohngen et al. (1999) estimated new plantations to be about 850,000 ha annually. The somewhat higher figure used in this study reflects the inclusion of new plantation establishment in the temperate regions and anecdotal evidence suggesting that the other estimates, which were made earlier, were somewhat on the modest side.

[14] At the 0.5% annual increase consumption, on a 1997 production/consumption base of 1.5 billion m^3, global industrial wood consumption would be expected to increase about 7.5 million m^3 annually.

[15] This is approximately the recent FAO estimate of −0.67 for the elasticity of demand for industrial roundwood.

References

Angelsen, A. and Kaimowitz, D. (1998) When does technological change in agricultural promote deforestation?, Paper presented at the AAEA International Conference on Agricultural Intensification, Economic Development and Environmental, Salt Lake City, 31 July–1 August.

Bradshaw, T. (1999) A blueprint for forest tree domestication, Downloaded from http://poplar2.cfr.washington.edu/toby/treedom.htm, College of Forest Resources, University of Washington, Seattle, WA (July).

FAO (1996) Long-term trends and prospects in the supply and demand for timber and implications for sustainable forest, by D. Brooks, H. Pajulja, T. Peck and B. Solberg and P. Wardle, Rome.

Haines, R.J. (1994a) Biotechnology in forest tree improvement: Providing technology for wise use, *Unasylva* **45**(177), United Nations Food and Agricultural Organization, Rome.

Haines, R.J. (1994b) Biotechnology in forest tree improvement, FAO Forestry Paper 118, United Nations Food and Agricultural Organization, Rome.

Hayami, Y. and Ruttan, V.W. (1985) *Agricultural Development: An International Perspective*, Johns Hopkins Press, Baltimore, MD (see p. 115).

Hu, W.-J., Harding, S.A., Lung, J., Popko, J.L., Ralph, J., Stokke, D.D., Tsai, C.J. and Chiang, V.L. (1999) Repression of lignin biosyntheses promotes cellulose accumulation and growth in transgenic trees, *Natural Biotechnol.* **17**, 808–812.

Kloppenburg Jr., J.R. (ed.) (1988) *Seeds and Sovereignty: The Use and Control of Plant Genetic Resources*, Duke University Press, Durham, NC.

Menzies, N. (1985) Land tenure and resource utilization in China: A historical perspective, Paper presented at a Conference on Managing Renewable Resources in Asia, Sapporo, Japan, 24–28 June.

Pullman, G.S., Cairney, J. and Peter, G. (1998) Clonal forestry and genetic engineering: Where we stand, future prospects, and potential impacts on mill operations, *TAPPI J.* **81**(2).

Sedjo, R.A. (1983) *The Comparative Economics of Plantation Forests*, Resources for the Future, Washington, D.C.

Sedjo, R.A. (1992) Property rights, genetic resources, and biotechnological change, *J. Law Econ.* **XXXV**, 199–213.

Sedjo, R.A. (1999) The potential of high-yield plantation forestry for meeting timber needs, *New Forests* **17**, 339–359.

Sedjo, R.A. and Botkin, D. (1997) Using forest plantations to spare natural forests, *Environment* **39**(10), 15–20, 30.

Sedjo, R.A., Goetzl, A. and Steverson, M. (1997) *Sustainability of Temperate Forests*, RFF, Washington, D.C.

Sohngen, B., Mendelson, R. and Sedjo, R.A. (1999) Forest management, conservation, and global timber markets, *Amer. J. Agric. Econ.* **81**(1), 1–13.

Washington Post (1999), February 3.

Westvaco (1996) Fiber farm expansion, *Forest Focus* **20**(4), Westvaco Forest Resources Division, Summerville, SC.

Westvaco (1997) *Forest Focus* **21**(1), Westvaco Forest Resources Division, Summerville, SC.

Williams, N. (1998) Agricultural biotech faces backlash in Europe, *Science* **281**, 768–771.

Withrow-Robinson, B., Hibbs, D. and Beuter, J. (1995) Poplar chip production for Willamette Valley grass seed sites, Research Contribution 11, College of Forestry, Forest Research Laboratory, Oregon State University, November.

PART B

MANAGING TECHNOLOGY: THE ECONOMICS OF MANAGING TECHNOLOGICAL DEVELOPMENTS

4. On the Economic Limits to Technological Potential: Will Industry Resolve the Resistance Problem?

TIMO GOESCHL and TIMOTHY SWANSON

1. The Economics of Pathogen Management – A Review of the Literature

The economic literature on the management of pathogens that has developed over the last 40 years does not present a unified view of the management of pathogenic biological processes. Rather it is divided along the lines of two specific problem areas: pest management in agriculture and disease management in public health. This division exists despite the fact that the general laws that govern the disease processes share the same fundamentals. While the optimal management of such processes has been discussed in the biometric literature (for instance, Wickwire, 1977), in economics little research has been done on the fundamentals of pathogen management across fields of application.

The first part of this literature survey will focus on general parallels in the development of the pathogen management literature in both areas, agriculture and public health. In the second part, we will look at the specific stages in the writings on agricultural pest management that have developed since the 1960s. This is generally the richer field where decision rules for management have been at the core of research. In the third part, we review the developments in the economics of infectious diseases in public health that have been undertaken over the course of the last ten years.

1.1. Parallels in Agricultural and Public Health Management of Pathogens

In both agricultural economics and the economics of public health, the initial interest was situated in stressing the economic efficiency perspective and in delivering improved heuristics for individuals and social planners interested in managing pathogen *populations* more effectively and efficiently. The motivation is that economic losses are directly related to the level of pathogen density and thus of immediate concern to the analyst. Pathogen control being a costly input and often requiring complementary investment in physical (machines) and human capital (training), this is an interesting problem for decision-theoretic analysis, especially

Timothy Swanson (ed.), The Economics of Managing Biotechnologies, 99–128.

once aspects of biological growth functions, i.e. dynamic models, and uncertainty are considered.

Largely, therefore, the literature on pathogen management has been a literature on the management of losses to pathogen populations whose occurrence is exogenous and stochastic, but whose underlying characteristics do not ultimately differ between time periods and are not subject to human management. This approach has justification to the extent that it informs individual small-scale decision makers about individually optimal usage of pathogen control since from this atomistic perspective, the nature of the problem is largely exogenous. The problem is that most of this literature tends to ignore the wider implications for social welfare that result from pathogen control, both along inter-personal and intertemporal dimensions. These implications relate to the spatial and evolutionary dynamics of pathogen populations that have more recently started to receive attention.

The evolution of pathogens has been taken up as a management issue rather more recently. The reason for this is – again – largely historical in that the problem of resistance development could only be observed after the large-scale introduction of biochemical control, first in agriculture and subsequently in health. The characteristic times for the appearance of resistance[1] have been empirically established to vary between 18 weeks and 40 years in the case of uniform exposure of a closed population of vectors or parasites (Anderson and May, 1991). With migrating populations and varying levels of exposure, the problem of resistance development became apparent in practice only after prolonged use of uniform inputs. Of late, however, it is being regarded as a fundamental management problem, especially in the field of antibiotics (Laxminarayan and Brown, 1998; Brown and Layton, 1996; Baumol and Oates, 1988; Cornes et al., 1995) and in high-yielding cultivars (Widawsky, 1998; Heisey and Brennan, 1991; Heisey, 1990).[2] The issue is of interest to economists predominantly because of the inherent inter-personal and intertemporal externalities that are involved and that are mediated through the transmission and evolution dynamics within pathogen populations.

1.2. The Economics of Pest Management in Agriculture

The literature on pest management in agriculture has contributed to a more sophisticated understanding regarding the optimal farm-level decision making. Historically, the initial objective was the development of static economic threshold decisions under certainty (Hillebrandt, 1960; Carlson, 1970) and uncertainty (Feder, 1979) that would enable farmers to start the application of chemical control based on the observation of a single parameter, pest density on crops. This literature has been subsequently refined to take into consideration the relative profitability of alternative thresholds (Moffitt et al., 1984; Hall and Moffitt, 1985) and different forms of risk aversion exhibited by farmers (Moffitt 1986). The biological processes that underlie crop disease dynamics have been of little importance in this part of the literature and only to the extent that they allowed further refinements of action thresholds in a one-season, one-crop model by accounting for the short-run predator-prey type relationship between the biomasses of hosts and pathogens (Moffitt and Farnsworth, 1987) in the spirit of Volterra (1931). This strand of the literature does not consider the exhaustibility of the effectiveness of control inputs in any detail.

Taylor and Headley (1973) are credited with the first consideration of intertemporal effects between the application of a pathogen control input and the development of its effectiveness over time within an economic framework, although the issue is also taken up by other authors in the same year (Hall and Norgaard, 1973). In a seminal piece, Hueth and Regev (1974) combine this aspect with the investment-theoretic perspective of 'biological capital' put forward in (Carlson and Castle, 1972) in order to develop a single-pest, single crop management model that points out the structural similarities of optimal pest management and the management of exhaustible resources. The most important similarity is the market failure in that the shadow price of the stock of 'biological capital' is not reflected through the market to the users, leading in a number of well-defined cases to over-exploitation of the resource.

The main conclusions in Hueth and Regev (1974) are twofold: (1) use is optimal if – at any point in time – marginal profits, i.e. increase in final output minus the marginal cost of the control input, equal the user cost resulting from depletion of the stock of effectiveness of the control input; and (2) for the economic threshold of application, this means the threshold varies over time, in particular such that the neglect of resistance dynamics rarely results in the overuse of chemicals.

There are a couple of specific assumptions underlying this early literature of Hueth and Regev (1974) and Hall and Norgaard (1973). In both models as well as in Moffitt and Farnsworth (1987), the emphasis is on the management within a single growing season that is subdivided into three periods. Although this approach extends the previous, static models towards considerations of optimal timing of applications during the growing period (cf. Taylor and Headley, 1973), it restricts the attention to a specific class of pathogens, namely to those that have a very high number of generations per season such that evolutionary dynamics can have an effect within a single season.[3] Secondly, none of the models considers spatial externalities arising as a result of pathogen migration. A third common assumption concerns the availability of single backstop technology to which the economy can revert when necessary such that the general analysis of the optimal allocation path is carried out against the background. The fourth limitation is that none of the models delivers an explicit account of the factors that govern the evolutionary dynamics of resistance development. Although the process is acknowledged, the interaction between pest control technologies and pathogens is treated as a 'black box'. This leaves unanswered important questions about the nature of the process, such as the possibility of reversibility of resistance development and the effect of migration on resistance. More importantly from a policy perspective, this also restricts the number of instruments available to agents to only one, namely the rate at which susceptibility of pathogens is exploited.

The subsequent literature relaxes some of these assumptions, most notably the first and the second. The latter concerns the effects of pathogen migration; the first assumption concerns the relevant time horizon and the appropriate modelling of the seasonality of agriculture.[4] This point is tackled differently by Regev et al. (1983) and by Plant et al. (1985): Regev et al. consider the case of a continuous agriculture in order to derive the optimal program for a social planner under two different regimes, one in which the planner is informed about the resistance dynamics and one where information about this process is absent. The results suggest that since the social planner unaware of the resistance development does not take into account the shadow cost of a decreasing stock of susceptible pathogens, more pest control will be used compared to the fully informed social planner.[5] Another interesting result is that – due to migration – decentralised decision makers will operate closer to the social optimum than an incompletely informed social planner as pathogen immigration reduces the marginal product of pest control to the individual decision maker. This means that non-intervention by the social planner may constitute a second-best optimum if information about the evolutionary process is prohibitively costly.

The advantage of Regev et al. (1983) over previous models is that it is amenable to the tools of optimal control theory. However, this advantage reduces the validity of the model for practical purposes as the analytical progress made regarding the level of thresholds and timing of applications is absent from the model. Plant et al. (1985) pursue these latter issues further in a discrete time model that leaves out the question of migration between fields, but explicitly models the immigration of pathogens from refugia at the start of each discrete season. Their contribution to the literature is twofold: (1) they refine the strategies for the timing of applications over several growing periods and are thus able to consider a much wider class of pathogen problems than the previous literature, and (2) the extension of the model to several periods demonstrates the crucial importance of planning horizons and discount rates in this type of setting.

As discussed in the review of the first generation of literature, the evolutionary dynamics were treated as a 'black box'. This shortcoming was subsequently rectified in two publications by Mangel and Plant (1983) and Mangel (1985).[6] These two papers bring the evolutionary dynamics to the appropriate level of analysis, namely the selection of resistance-conferring alleles in pathogens. This step not only enriches the analysis, it has profound implications for the recommended management. The reason is that in the absence of selection pressure exerted by the application of a control input, resistance against this input does not confer an evolutionary advantage. Instead individuals carrying this resistance are likely to bear an evolutionary 'cost' for this trait. This should lead to the disappearance of resistance over time and would make the process of exploitation reversible (Bishop and Cook, 1981). Consequently, models of exhaustible resources are no longer appropriate for analysing the problem as susceptibility of pathogens (or, conversely, the effectiveness of control inputs) can now be regarded as a renewable resource stock.[7]

The presence of reversibility, extensively considered in Munro (1997), implies that pesticide use will not be abandoned in the limit and the stock of effectiveness will be mined as a result of time preference. The exact amount of exploitation depends on the fitness gap between individual pathogens carrying the resistance gene and susceptible ones and naturally, the discount rate of the social planner. Although Munro does not consider this case explicitly, it also implies that if R&D in backstop technologies is sufficiently costly, no research in substitute technologies will be undertaken.

This argument in Munro (1997) points to the fact one common assumption has remained unchallenged in the agricultural literature so far. This assumption concerns the availability of a single backstop technology at a constant price to which society can revert at any given point. Although this assumption simplifies the analysis, particularly in a dynamic setting, by providing well defined transversality conditions for the programme that has to be optimised, the analysis is forced to omit the dynamics of technological change that have been a fundamental factor in shaping agriculture over the last 50 years (Huffman and Evenson, 1993). This assumption regarding R&D is therefore likely to provide little insight into the actual development of pest management in the agricultural sector over a longer time horizon. The conclusion to be drawn from this survey of the pest management literature in the agricultural context is that significant progress has been made in the analysis of the pathogen problem. Models have become increasingly refined in their treatment of population growth and the evolutionary impacts of population management through pesticides. However, the focus has largely remained on the short term and on developing practical algorithms for pest management at the farm level.

1.3. Public Health

The literature on the economics of infectious diseases is of comparatively recent origin and has initially revolved around two basic themes: the optimal management of diseases in the context of developing countries (Wiemer, 1987; Hammer, 1993), and the divergence, or lack thereof, between the social optimum and the 'market' solution (or, better, non-coordinated solution) in situations of epidemic spread of diseases (Brito et al., 1991; Geoffard and Philipson, 1996, 1997; Francis, 1997). Only recently the topic of intertemporal effects has received more attention with a focus on the development of resistance against antibiotics (Cornes et al., 1995;

Daily and Ehrlich, 1996; Brown and Layton, 1996; Laxminarayan and Brown, 1998).

The interest of the development literature in disease dynamics originates from the impacts of diseases on the earning potential of individuals in locations where infectious diseases are endemic. The incapacitation through disease is a major factor determining labour supply in developing countries and has a measurable impact on national output (World Bank, 1993). The initial interest in pathogen management in this literature is on two management issues. The first is the establishing of the economic threshold, which answers the question whether programmes should have continuous management or eradication as a target. The second issue concerns the decision about preventive *versus* curative (or reactive) measures. The results – derived in the context of schistosomiasis control in Wiemer (1987) – are strikingly similar to the ones previously arrived at in the agricultural literature: the presence of an economic threshold makes eradication of the disease uneconomical in most circumstances (unless the marginal cost of infection is extremely high even at a low pathogen population levels) and reactive measures should take precedence over preventive measures unless transmission rates are very high.[8]

A second set of literature exists around the question of divergence between the socially optimal solution and the aggregate result of private decision making in manageable epidemics. Initially, these questions were asked exclusively in terms of inter-personal externalities. Lately, inter-temporal aspects have come more to the fore.

One subset of the literature centres around the question of the optimal regulation of immunisation in a population. Conventional wisdom has it that there are positive inter-personal externalities arising from vaccination programmes that make compulsory vaccination socially desirable, but unlikely to come about without intervention. Brito et al. (1991) examine this proposition in a static framework and find that compulsory vaccination is pareto-inferior to free choice of vaccination for a wide class of utility functions. Their argument is effectively one used before in the context of compulsory social insurance schemes (Seidl, 1988): in the absence of regulation, individuals will 'self-insure' by adjusting their exposure to risk to the individual optimum. For immunisation decisions, this means that some people will vaccinate (self-insure), but many others will not, reflecting that the disutility from vaccination exceeds the private benefits. Compulsory vaccination will leave the welfare position of the first set of people unaffected, but hurt those who did not choose to vaccinate under free choice. Since inter-personal externalities exist, the optimal solution lies between these vaccination levels under free choice and compulsion and can be approximated by taxes and subsidies (Brito et al., 1991).

Static analysis has its obvious limits in that the transmission process of disease is dynamic by nature. Comparative statics may therefore misrepresent important features of the problem. The power of the dynamic perspective in problems of disease management is demonstrated in Francis (1997) and contrasted with the static result. In the latter, the same results as in Brito et al. (1991) are obtained. However, in the dynamic case, it is shown that under certain conditions, there are no externalities associated with vaccination decisions, i.e. regulation cannot improve upon the non-coordinated decision making by individuals.[9] In this model, the question that arises for decision makers is when to carry out the single investment in immunisation. Since demand for vaccination is clearly 'prevalence-elastic' with regard to the level of infections in the population, high levels of infection will lead to high rates of immunisation and vice versa. Individuals can immediately respond to the current level of prevalence. This 'closed-loop' response to the level of infections generates an aggregate behaviour in the population that coincides with the social planner's solution. The divergence between the static and the dynamic model arises from the fact the disease transmission occurs over time and has an impact on the effective demand. This shift in the underlying incentives for vaccination does not get captured in a static representation of the problem and explains the origin of the discrepancy in the results.

The concept of 'prevalence elastic' demand was first coined in Geoffard and Philipson (1996) who studied the behavioural responses of individuals to the levels of disease in a population of agents. This is used to augment purely biological models of epidemiology by taking into account that agents can adjust their exposure to the risk of infection and will thus influence the epidemiological dynamics of a disease. The most striking result is that the hazard rate of infection increases with the prevalence of the disease in standard models, but declines with prevalence in an augmented model that allows for behavioural change as demand for protection increases with prevalence. This approach shows the limits of using non-augmented biological models in the context of public health when behavioural change is relevant.[10]

In another paper, the same authors explore the problem of eradication along the lines of prevalence elastic demand for vaccination and find that there exist under none of the standard market structures, a disease will be eradicated through vaccination decisions (Geoffard and Philipson, 1997). This is because the marginal benefits to immunisation decrease with every vaccination and – for most meaningful cost schedules – eventually fall below the marginal cost of vaccination before the disease is eradicated. Eradication is only likely to be achieved with subsidies to a monopolistic vaccine producer if demand is highly price elastic and highly prevalence inelastic and when the discount rate is high. We will return to this result later in this chapter since it has some obvious similarities with the conclusions of the model developed in the following sections.

Of the most recent origin is the literature on the inter-temporal effects of pathogen management in humans, in particular the problem of increasing resistance of pathogens to treatment. Although the problem has been well known for years and discussed informally, Cornes et al. (1995) present what is probably the first model of resistance build-up with special reference to human health. The paper presents both a discrete time, three-period model of the individual decision making and its continuous time equivalent. There is a common input that is used by all decision makers and usage of this input results in a decrease in the input quality (effectiveness), thus linking all individuals into the usage of a common resource. As can be expected, non-cooperation leads to an exhaustion of the pool at too rapid a rate due to the dynamic negative externalities between individual decisions. The parallels to the fishing literature are apparent.

A second line of inquiry has been the problem of misuse of antibiotics that is discussed in Daily and Ehrlich (1996). A straightforward positive analysis of this problem is presented by Brown and Layton (1996) in the context of antibiotics in agriculture. This involves a game between farmers and consumers about the correct level of application, given that both draw from a common pool of susceptibility of pathogens to treatment. Again, the problem of the overexploitation of a commons arises in this context.

In the treatment of disease, more than one input can be used, each representing a pool of effectiveness (or resistance) exploitable through use. This has two implications: firstly, if there are substitutes available for each input, resistance may be reversible through non-use over a period of time. Secondly, substitutability of inputs makes the problem amenable to the application of results from the theory of multiple resource deposits (Weitzman, 1976; Hartwick, 1978). This would put the problem firmly into the domain of renewable resource models.

These questions are addressed in a paper by Laxminarayan and Brown (1998). They analyse the problem when reversibility of resistance build-up is negligible.[11] Then the theory of multiple resource pools developed in the context of the extraction of ore can be successfully applied. The contribution of their paper is twofold: it presents one of the first attempts to bring the literature on the evolutionary ecology of pathogens to bear on the problem of management. These enriched model dynamics lend support to the economic model and lead to conclusions on optimal management that differ significantly from the epidemiological literature by including cost considerations. Its second contribution is the application of an pathogen management model to an empirical dataset on antibiotics usage.

1.4. The Contribution of This Chapter

This chapter aims to make a number of small contributions to the literature on pathogen management at a level that is sufficiently encompassing to view the results as applicable to both agriculture and public health. The overarching purpose is to provide a positive analysis of the interaction between pathogen dynamics and the industrial structure that is in place to deal with pathogens. Specifically, this chapter asks whether this industrial structure is likely to eradicate the pathogen problem if it is technologically feasible to do so.

There are a number of smaller points this chapter also provides. Firstly, the chapter introduces the lag load concept of evolutionary dynamics into the economic literature that has originated with Maynard Smith (Stenseth and Maynard Smith, 1984). This concept offers a simpler and more flexible form of modelling evolutionary dynamics over time, thus fulfilling Mangel's dictum that little gain is derived from building more complex evolutionary models (Mangel and Plant, 1983). Secondly, it presents a model of decision making about treatments that accommodates migration and evolution within a threshold model. This model in turn is able to generate a prevalence-elastic demand curve in the manner of (Geoffard and Philipson, 1997) at the aggregate level and thus reconciles the threshold literature in agriculture with the single decision models in public health. The third point is that it captures the dynamics of the R&D process by no longer assuming the existence of a single backstop technology, but by explicitly modelling the R&D behaviour by industry in the face of evolution.[12] With these three components, a positive analysis of the interaction between pathogen dynamics and R&D provides new insights into the nature of the inherent problems of the current industrial structure that exists to manage pathogens in both agriculture and public health.

2. The Problem

The underlying assumption of a substantial part of this literature is that humanity is engaged in an ever-lasting battle against its biological competitors, a battle which cannot be won with any certainty and finality (Laxminarayan and Brown, 1998).[13] According to this view, humanity can contain the rate of loss to pathogens at some specific level for every host specie through the application of pharmaceuticals and pesticides combined with R&D in these inputs, but attempts to lower this loss rate through increases in dosage speed up pathogen evolution by more than can be compensated efficiently through R&D.[14] The conclusion is there exists what could be termed a 'non-accelerating evolution rate of loss'. If loss falls below this level, evolution increases rapidly such that deviation from this natural rate only incurs a short-term benefit which is outweighed by its long-run costs (Morse and Schluederberg, 1990; Krause, 1981).[15]

On the other hand, many scientists regard a future in which losses to pathogens are reduced to zero as a realistic possibility. More specifically, this scenario is considered attainable within commercially meaningful periods of 20 to 40 years (Ulph and Sianesi, 1998). A sufficiently low discount rate would render the present-value of such a state of nature high enough to justify both R&D costs and the potential expenses involved in lost evolutionary battles in the meantime. This implies policy recommendations exactly opposite to the ones laid out above. If sufficient evolutionary pressure was applied to pathogens and a sufficient level of R&D accomplished, the costs of defending a high-input/high-output type of interaction with surrounding biological systems would vanish in the long run.

This evidence poses to a social planner two mutually exclusive options: one is to continue to apply substances which inhibit pathogen growth at ever increasing levels and to develop simultaneously more effective substitutes for these substances through R&D, in the hope that the technology which allows the decisive strike to be delivered will be produced in the process. The second is to target the last remaining areas where gains can be made without inducing strong evolutionary responses and then cease to innovate. As long as the respective likelihood of the outcome of the evolutionary competition is unknown, this decision cannot be made on firm grounds.[16] The question to be asked, however, is whether the final technology, if available, will in fact result in the evolutionary race between humans and their biological competitors to be decided in favour of society.

This chapter provides a positive analysis of the forces that inhibit disease eradication in the absence of intervention. The main proposition is that even if the premises of the first line of research were wrong and the battle between humans and pathogens can be decided in favour of humans in the end, there are important grounds for suggesting that in the absence of intervention, this possibility will not realised by uncoordinated decision makers responding to market signals. The reason for this lies in the specific incentives that these ecological processes pose to a decentralised economic system, specifically in the form of prevalence-elastic demand functions for treatment (cf. Geoffard and Philipson, 1997).

The conclusions of this chapter will be developed in the context of two interacting sectors of the economy. The one sector is characterised by a large number of decision makers that are individually concerned with the production of a final good. Pathogens inflict some damage on the final output from this sector and their growth can only be contained by using some additional input in the production. Specific examples of such a sector would be the agricultural sector with farmers as decision makers on the level of pesticides applied or the health sector where patients ultimately determine the dosage of, for instance, antibiotics.[17] The other sector produces this specific additional intermediate input used in the final sector. The effectiveness of this input declines, however, as a function of the use rate of the input in the final sector. This decline can only be overcome through R&D. Specific examples here would be the agrochemical or the pharmaceutical industry where development of resistance reduces the effectiveness of chemicals for crop protection or of drugs. The analysis of the spill-over between the industry providing the intermediate input and the final users is similar to Geoffard and Philipson (1997). The difference is that their paper deals with the incentives for eradication for a monopolist when there is no natural product life-cycle as in our example.[18] In the present case, R&D is continuously necessary in order to keep up with pathogen evolution.

In contrast to Laxminarayan and Brown (1998), the model presented here allows for the battle between humans and their biological competitors to be won by society. The question we are concerned with here is: is the 'winnability' of this contest a sufficient condition for this two-sector economy to effect 'victory' by itself? Will the market edge its way towards this resolution? Despite the importance that 'winnability' has from the vantage point of a social planner, this paper claims that this feature does not alter the outcome of an uncoordinated decision making process over pathogen management in a systematic way. Independent of the 'winnability' property, the two-sector economy will keep pathogen populations at positive, but stable levels and will exhibit cyclical R&D expenditure levels without addressing the option of 'winning', if it is available. The stochastic nature of the innovation process, however, can result in successful eradication as a random event.

This leads to two conclusions: society cannot expect uncoordinated pathogen management to result in pathogen eradication, and the *ex ante* answer to the question whether the system is winnable or not cannot come from observing the behaviour of the managed system. The reason for this pessimistic outlook is the difference between the way ecological processes are tied together and the way in which decision makers that influence these processes relate.

In the following model, we construct a system incorporating two dynamic processes as a stylised model of the ecological processes in nature as proposed by some ecologists: (1) pathogen population changes driven by a modified version of the classical exponential growth function, and (2) changes in the susceptibility of pathogens to treatment which derive from a simple lag load formulation customary in evolutionary biology (Stenseth and Maynard Smith, 1984). Activities in the final goods sector and in the intermediate goods industry affect both processes, the final sector by applying inputs which depress pathogen growth, and the industry by producing inputs which decrease the fitness of pathogens. We then examine the implications of this arrangement for the question of 'winnability'. Finally, we establish the conditions which result in the economic system resolving the evolutionary problem.

3. The Interaction between Pathogens Dynamics and Human Choice

3.1. Agriculture as a Co-Evolutionary System

This model of agriculture as a co-evolutionary system closely follows the ideas of Maynard Smith about the co-evolution of species (Stenseth and Maynard Smith, 1984). Elsewhere I have shown that models of this kind have a natural equilibrium of stable pathogen population levels, degrees of resistance and population density by hosts (Goeschl, 1998). At these equilibria, rates of genetic innovation are positive, i.e. we observe what evolutionary ecologists have termed a 'Red Queen race': a permanent and continuing interchange between co-evolving species where each specie changes in non-deterministic ways in order to counter moves on the part of the other – in order for the species to remain in a steady state of ecological fitness (Frank, 1996; Rosenzweig, 1996; Schaffer and Rosenzweig, 1978; Van Valen, 1977). The equilibrium is guaranteed by the way nature allocates resources to species of different ecological fitness.

Man has replaced this natural mechanism for resource allocation by processes of human choice, leading for instance in agriculture to a higher degree of uniformity among crop plants and to a system which runs at higher evolutionary speeds than one would observe in nature (Scheffer, 1997; Swanson, 1995; Barrett, 1981). In this sector, a continuous flow of inputs is necessary to maintain the human share of agricultural products. For an equilibrium to be brought about, the societal process of conversion of lands to intensive agriculture would have to be modified, an issue which has been taken up elsewhere (Goeschl and Swanson, 1998).

3.2. Population Dynamics

This section follows closely Plant et al. (1985) with the exception of considering migration of pathogens. We assume that pathogens grow on a population of uniform hosts of mass 1. One could think of this mass as a population of humans or farmlands of identical plot size. Now suppose – such as Mangel (1985) – that treatment is the application of some optimal amount L of the intermediate input per unit of time in order to contain pathogen growth. At those host sites where there is no treatment, pathogens grow according to the standard exponential growth function,[19] minus emigration of pathogens from the site to others plus immigration of pathogens from other sites, given by

$$\dot{N}_j = N_j r - eN_j + m \sum N_{-j}, \tag{1a}$$

where N_j denotes the pathogen population level at site j, N_{-j} the pathogen population level outside j, r the natural growth rate of pathogens, e the rate of emigration, and m the rate of immigration.

At sites where pathogens are being fought, we have to take into account the effect of treatment. As customary in entomology, we will view treatment here as a substance that increases the death rate among pathogens as a linear function of dosage and independent of the population level (cf. Hueth and Regev, 1974). This results in

$$\dot{N}_j = N_j(r - L(1 - v_j)) - eN_j + m \sum N_{-j}, \tag{1b}$$

where v is the share of ecologically successful pathogen in the population and L is the optimal level of treatment per time unit.

This growth function enhances the standard version by introducing a resistance component into the equation of motion. Pathogens are either resistant, i.e. able to produce offspring under current ecological conditions, or susceptible to the intermediate input. The population share of resistant pathogens is denoted by v. The ecological conditions are to a large extent set by humans[20] and involve in our specific case the use of a treatment load, L. The parameter v hence denotes the share of pathogens which are able to function under these given conditions.

Assuming a closed system, immigration and emigration have to cancel each other out at the aggregate level of the total pathogen population. This means that for the number of sites sufficiently large, m, the rate of immigration, is linearly related to e, the rate of emigration.[21] Let the share of hosts where the intermediate input is applied be denoted by f, then the average population dynamics of the pathogen (viruses, fungi, etc.) are given by this version of the standard exponential growth function:

$$\dot{\bar{N}} = \bar{N}(r - fL(1 - v)), \tag{2}$$

where f is the share of hosts undergoing treatment.

It is easy to establish the rest point for the differential equation (2): pathogen growth is zero either when population levels are zero, i.e. $N = 0$, or when the dosage neutralises any growth produced by bringing the death rate into line with the birth rate, i.e. when the share of treated hosts is $f = r/(1 - v)L$.

3.3. Evolution of Resistance

Evolution enters into the model via changes in resistance over time, i.e. resistance to treatment. The engine for this change lies in the lag load of the species, i.e. the average distance of the pathogen population from its local fitness maximum. Positive lag loads merely state that the chance for mutant or recombinant members of the pathogen population to pass on fitness-enhancing traits is positive, and that this chance increases with the distance from the local maximum (see Felsenstein, 1971). This basic concept underlies many models that explain evolutionary change over time. Most other models of evolution of resistance adopt some formulation of the Hardy–Weinberg Law based on a one-locus, two-allele model (cf. Mangel and Plant, 1983; Plant et al., 1985; Munro, 1997). The lag load concept captures a wider class of combinatorial effects and time scales and makes it easier to integrate migration and technological innovation. Also, more complex genetics-based models do not yield qualitatively different outcomes as Mangel and Plant (1983) point out.

Generally, the relative change in resistance is the change induced by the lag load of pathogens plus that caused by the host lag load. Formally,

Evolution of Resistance (after Maynard Smith, 1976)

$$\frac{\dot{v}}{v} = \frac{\partial v}{\partial P} \cdot \dot{P} + \frac{\partial v}{\partial H} \cdot \dot{H}, \tag{3}$$

where P denotes the pathogen lag load and H the host lag load.

In natural systems, these changes operate in opposite directions, the host reducing resistance through genetic innovation, the parasite increasing it. In managed populations, it is technological innovations in agricultural inputs that takes over the role of host innovation.

We derive the lag loads from the fitness function in the manner of Roughgarden (1979) such that the fitness of the individual pathogen is 1 plus its contribution to growth. For a pathogen population with a randomly distributed share of resistant individuals v and of avirulent individuals $(1 - v)$, the average population fitness is

$$\begin{aligned} \bar{F} &= v(1+r) + (1-v)[f(1+r-L) + (1-f)r] \\ &= 1 + r - fL(1-v). \end{aligned} \tag{4}$$

The lag load is the relative distance of the average population fitness from the local adaptive peak (Stenseth and Maynard Smith, 1984).[22] Applying (4) to this definition results in

$$\text{Lag} = \frac{\hat{F} - \bar{F}}{\hat{F}} = \frac{fL(1-v)}{1+r}. \tag{5}$$

Assuming quadratic adaptive speed, the change of resistance per time unit effected by pathogens due to exposure to treatment is $-\partial v/\partial \,\text{Lag} \cdot \text{Lag}^2 i$.

An underlying assumption of this lag load model is that there is some form of interaction between these pathogens such that fitness differentials can have a real effect. So it is suited to small-scale environments where individual pathogens compete directly with each other. Migration of pathogens in and out of host sites therefore not only affects local population growth (as discussed above), but also widens the area over which fitness differentials have an effect by bringing pathogens in contact with each other. Comins (1977) shows in an influential paper that migration indeed has evolutionary effects in that the speed of genetic response to selection pressure can be mitigated or amplified by the effects of migration within the population. The gist of the argument is that if resistance genes are sufficiently recessive and migration is high, the aggregate pathogen population can absorb a good deal of treatment without producing a large share of resistant individuals. In other words, if the fitness cost associated with carrying resistance genes is not substantial, then migration can retard the speed at which the effectiveness of the intermediate input declines. The rate of absorption is therefore directly linked to e, the rate of migration. We denote this absorption[23] by $a = a(e)$ with $a'(e) > 0$ and let R be a random variable that stands for the realisation of R&D at this particular point in time, in other words, an innovation in the protection of the host population. This innovation will be embodied in the intermediate good.

The net impact of these two changes on resistance results in a relative change in resistance of

$$\frac{\dot{v}}{v} = \left[(fL - a(e)) \frac{(1-v)^2}{1+r} i - R \right], \tag{6}$$

where a denotes the rate of absorption, i the adaptive speed of pathogens, and R the realisation of industrial R&D process.

Equation (6) shows the relationship between the development of the pathogen population through growth rate r, its control through treatment at rate fL and the impact of an innovation of size R on the RHS and the development in resistance over time on the LHS. The first term of the RHS summarises the factors which govern the rate at which members of the pathogen population that are ecologically more successful than the average invade the population if the benefits of resistance to the current type of treatment outweigh the costs. This leads to a rise in overall resistance determined by the adaptive speed of the pathogen population, i. The second term summarises the reduction in resistance brought about by changes in the 'host' which are a probabilistic function of the R&D investment in the intermediate sector. Pathogen evolution rests when $v = 0$, i.e. if the pool of mutants with beneficial traits is exhausted; when $v = 1$, i.e. when all pathogens have reached the local adaptive peak and are able to produce offspring; and when $(1 + r)R(x) = i(fL - a)(1 - v)^2$, i.e. when R&D exactly compensates for the evolutionary pressure exerted by treatment.

3.4. Winnability: An Operational Definition

Equations (2) and (6) clearly show the basic problem inherent in pathogen management: if pathogen growth is sufficiently high and the damage associated with pathogen populations significant, then more of the pathogens should be exposed to treatment. This, however, decreases the effectiveness of the treatment at the rate of pathogen evolution which – in return – increases with the proportion of pathogens treated. It is therefore necessary to invest in R&D to devise new forms of treatment in order to contain pathogen growth.

This system of population and evolution dynamics is termed 'winnable' if the system is capable of rendering unnecessary any further investment in R&D while successfully containing pathogen populations. This means that for X denoting the total R&D expenditure and $R(X)$ the expected increase in the effectiveness of the intermediate input given X, there exists an $f \geq 0$ such that $\dot{N} \leq 0$ and $\dot{v} \leq 0$ for $X = 0$. Inspection of the population dynamics and ecological evolution shows that for the system to be winnable, the rate of absorption $a(e)$ has to equal or exceed $f(e)L$, i.e. the evolutionary pressure which will be exerted by decision makers given the migration rate e. In other words, either migration is substantial and absorption gains more from this than the level of treatment or the costs to pathogens of being resistant to treatment must be significant. The higher a, the lower R&D has to be to compensate for treatment-induced resistance and, conversely, the higher levels of treatment can be without triggering an erosion of its effectiveness.

Trivial as this may appear in terms of the arithmetics, there is considerable uncertainty in the biological literature regarding the question whether a differs from zero for different pathogens. In particular, although the theoretical argument that high levels of migration will dampen the evolutionary response of pathogens is sound, this relationship is not easy to quantify. Much importance is attached to answering this puzzle for the purpose of eradication programs (cf. Anderson and May, 1991). After all, if absorption is a significant factor in pathogen evolution, the way seems to be clear for a future without substantial damages due to pathogen activity. The question addressed in the following section is what the nature of the incentives is that this system poses to decision makers, and whether results differ when absorption is a factor.

The next section will show that the externalities present in pathogen treatment at the level of decision makers in the final goods sector support an optimistic conclusion: high migration rates are associated not only with high absorption, but also with lower levels of treatment, resulting in very low net drift towards resistance by pathogen evolution. 'Winnability' is therefore sets the correct incentives at the level of the final sector decision making from an evolutionary point of view. If we assume that levels of R&D are not affected by the migration rate of pathogen and remain positive, then 'winnability' would indeed be a sufficient condition for this two sector economy to move towards this state.

4. Decision Making in Pathogen Management

4.1. Final Goods Sector

4.1.1. *Individual Decision Making*

We will first look into the behaviour of decision makers in the final goods sector and determine the aggregate demand for treatment. The aim is to derive analytically the so-called 'threshold decision rule' which is popular both in crop protection (Moffitt and Farnsworth, 1987; Hall and Moffitt, 1986) and in the treatment of common infectious diseases (Laxminarayan and Brown, 1998). The individual 'manager' is faced with a growing pathogen population unless she decides to start treatment.[24] Consequently, at any point in time, she has to consider two alternative pay-offs, one from continuing to do nothing and the other from starting treatment which incurs a single fixed cost of F.[25] In the spirit of Plant et al. (1985), we think of treatment as involving the application of a technically optimal amount L to the host.[26] Managers do not take into account pathogen evolution as the application time is arguably small relative to the time scale over which evolution occurs. Assuming conveniently that damage from pathogen is linear in population levels (Plant et al., 1985), the pay-off from treatment, POT, equals the value of damage avoided minus the cost of treatment minus the fixed cost of commencing treatment. Therefore, dropping the host site index j,

$$\mathrm{POT} = -\int_{t=s}^{s+AT} D \cdot N(t)[r - L(1-v) - e]\,\mathrm{d}t - \int_{t=s}^{s+AT} Dm \sum N_{-j}\,\mathrm{d}t - \int_{t=s}^{s+AT} c \cdot L\,\mathrm{d}t - F, \tag{7}$$

with s being the starting time, AT the application time, D the damage inflicted per unit of pathogens, c the unit cost of treatment, and F the fixed cost.

Optimising (7) with respect to application time, we find that given population levels $N(s)$ and resistance $v(s)$ at starting point s, the optimal length of treatment[27] is

$$AT^* = \frac{1}{r - L(1-v(s)) - e} \cdot \ln \frac{cL + Dm \sum N_{-j}}{-DN(s)[r - L(1-v(s)) - e]}. \tag{8}$$

After this period, treatment stops and the host faces a reduced pathogen population. We will call this level of pathogens the 'discontinuation threshold', N_d. The solution to (8) enables the manager to compare POT with the pay-off of doing nothing and suffer the associated damage of pathogen growth over this period, denoted by PDN. We find that

$$\mathrm{PDN} = -\int_{t=s}^{s+AT^*} D \cdot N(t)(r-e)\,\mathrm{d}t - \int_{t=s}^{s+AT^*} Dm \sum N_{-j}\,\mathrm{d}t. \tag{9}$$

Consequently, the manager will choose to start the application when population levels increase to the point where POT $\geq$ PDN. Hence, the action threshold is

$$N_a = \frac{AT^*cL + F}{-D[\exp\{[r - L(1-v(s)) - e]AT^*\} - \exp\{(r-e)AT^*\}]}. \tag{10}$$

This threshold is a well-known concept in pathogen control, especially in the area of crop protection (cf. Oerke et al., 1994): managers, in this case farmers, will not engage in pest control until population levels reach such a level that the expected damage from refraining from pesticide use exceeds the costs of pesticide application. Likewise, in the absence of a vaccine, patients may decide not to take action against a pathogen until it has reached a sufficient level to warrant intervention. Unsurprisingly, this threshold increases with the costs of treatment, both fixed and variable, with the optimal duration of application and with the resistance of pathogens.

4.1.2. *Aggregate Demand for Treatment*

The start of pathogen growth does not occur at the same moment for each individual host. Agricultural seasons start at different points in time at different locations; humans become exposed to pathogens as they spread through the population over time. At any given point therefore, there will be a distribution of pathogen levels across hosts. We know, however, that the levels will lie somewhere between the action threshold N_a and the discontinuation threshold N_d. For the starting points of treatment sufficiently smoothly distributed along the time axis, the aggregate demand for treatment is a linear function of the share of managers currently applying the intermediate input. This share, again denoted by f, is then

$$f = \frac{AT^*}{AT^* + WT^*}, \tag{11}$$

where WT^* is the optimal waiting time between having reached the discontinuation threshold and reaching the current action threshold, which is

$$WT^*(t) = \frac{1}{r} \ln \left[\frac{N_a(s + AT^*(t))}{N_a(s) \exp\{[r - L(1 - v(s + AT^*(t))) - e]AT^*(t)\}} - 1 \right]. \tag{12}$$

The intuition behind Equation (11) is that managers can find themselves in one of two states, either treating the pathogen population or waiting until it becomes economical to do so. Hence the relative share of managers in either of these states will conform to the relative length of either of the periods. Consequently, the monopolist in possession of the patent for the currently used technology will sell fL of the input per unit of time.

The aggregate demand for treatment reacts to changes in the price of the intermediate input and to changes in resistance as we would expect. Examining the derivatives of the share of managers involved in treatment at any given time, we find that $\partial f/\partial c < 0$ and $\partial f/\partial v > 0$, such that the demand of treatment decreases with its unit cost and increases with resistance. Interestingly, a social planner would reach the opposite conclusion on that last point: as resistance increases, the marginal social productivity of the intermediate input declines such that less of it will be allocated (Munro, 1997).

Not only does the individually optimal level of crop damage increase both with the price of the intermediate input and resistance, there is also a direct effect of migration on the aggregate demand for treatment. Taking the derivative of (11) with respect to e, we find that application time decreases with increases in migration for two reasons. First, emigration decreases local population growth and thus requires less treatment the higher the dissipation of pathogens into the environment of the host site. Second, immigration will increase with higher migration rates which renders the treatment less competitive as there is a source of pathogen growth that the treatment is ineffective against. This means that treatment will be discontinued at an earlier point in time the higher the rate of migration. Although this also decreases the optimal waiting time WT^*, the impact on WT is less than on AT. We know therefore that the partial derivative of aggregate demand with respect to migration is $\partial f/\partial e < 0$. This means that although the rate of migration has no effect on the aggregate pathogen population in the economy, the average population will increase when migration increases since farmers reduce their crop protection measures.

The conclusion from examining the final goods sector is therefore that because absorption increases with migration, i.e. $\partial a/\partial e > 0$, increasing rates of migration, and therefore, increased 'winnability' reduce the evolutionary drift in the pathogen population both from the side of increased absorption and decreased selection pressure through treatment. This means that decision making in the final sector is compatible with the hypothesis that winnability is sufficient for an uncoordinated system to effect pathogen eradication. The following section will analyse whether the same holds for the industry providing the intermediate good that embodies R&D.

4.2. INDUSTRY

Our model of the intermediate goods industry assumes that there is a sequence of firms that – by virtue of making a patentable drastic innovation – become for a limited time monopolist in the market, namely until they are being superseded by a competitor that has created the next innovation. This is a standard construction in models of infinite sequences of innovation as we are considering here (Aghion and Howitt, 1992; Grossman and Helpman, 1991; O'Donoghue, 1998). The drastic innovation in our case is the discovery of compounds that are able to negatively affect the growth of those pathogens that are resistant against the currently available technology.[28] We suppose constant and identical unit production costs for the intermediate input, denoted by k. To simplify the analysis, drastic innovations in this model are of constant size I (i.e. a fixed share of resistant pathogens is becomes susceptible as a result) and innovations are generated by a Poisson process with probability $\lambda x(t)$ where λ is the arrival rate and $x(t)$ the R&D effort of some individual firm. We will assume for simplicity that any new product is also able to contain those currently susceptible, although the results are not believed to depend on this point.

4.2.1. *The Incumbent Monopolist*

The incumbent monopolist, i.e. the one holding the patent for the current technology, faces the problem of maximising intertemporal profits by choosing the price of the intermediate good. Hence the problem can be written as

$$\max_{c(t)} \pi(t) = L[f(t)c(t) - k]. \tag{13}$$

Solving (13) for the optimal price, c^*, we arrive at

$$c^*(t) = -\frac{f(t)}{\partial f(t)/\partial c(t)}. \tag{14}$$

Since the optimal price is a function of current state variables only, this pricing policy is Markov-perfect, but not necessarily sub-game perfect. Solving (13) with (14) in hand, we obtain

$$\pi^*(t) = L\left[-\frac{f(t)^2}{f_{c^*}(t)} - k\right]. \tag{15}$$

What is the likely development of these intertemporal profits over time? Most importantly, $f(t)$ depends on the level of resistance in the economy, i.e. $v(t)$. The change in profits over time is hence

$$\dot{\pi}^*(t) = \frac{\partial \pi}{\partial v} \cdot \dot{v} = L\left[-\frac{2f(t)}{f_{c^*}(t)}\frac{\partial f(t)}{\partial f_{c^*}(t)} + \frac{f(t)^2}{f_{c^*}(t)^2}\frac{\partial f_{c^*}(t)}{\partial v(t)}\right] \cdot \dot{v}. \tag{16}$$

The first-order effects on the RHS of (16) – the first term in the brackets representing the increase in demand and, hence, revenues – are clearly positive while the second-order effects – the effect of changes in v on marginal revenue – are negative. This means that profits will increase if evolution favours resistant strains of pathogens ($\dot{v} > 0$), although at a decreasing rate as marginal revenues decrease with resistance rising. The intuition is clear: as we learned from the analysis of the final sector, aggregate demand moves in the same direction as resistance. This enables the monopolist to increase the price, although the optimal marginal increase declines with increasing v. The movement of intertemporal monopolistic rents is therefore positively related to pathogen evolution. This need not be a problem, however, as rents generate incentives for competitors to enter the market: given sufficiently high rents, R&D may well be high, thus driving resistance down over time. The final part of the analysis is therefore to examine the incentive structure for R&D.

4.2.2. *The R&D Sector*

As in other n-firm patent race models of this type, firm i, the incumbent monopolist, is not undertaking any research because his expected value of innovation is diminished by his ownership of the current technology. This is known as the 'replacement effect'. At the same time, all other $(n - 1)$ firms may be involved in R&D such that there is a cumulative probability of $(n - 1)\lambda x_{-j}$ of the current technology being superseded. According to Reinganum (1985), the current value $V(t)$ of the next innovation to firm j is then

$$\rho V(t) = \pi(t) - (n - 1)\lambda x_{-j}(t)V(t), \tag{17}$$

where V denotes the current value of next innovation, ρ the discount rate, n the number of firms in the industry, and λ the arrival rate of innovations per unit of research expenditure.

R&D investment of a risk-neutral firm j is optimal where the marginal cost of R&D equals the marginal expected benefits such that

$$1 = \lambda \frac{\pi(t)}{\rho + (n-1)\lambda x_{-j}(t)} + \lambda x_j(t) \frac{\partial \pi(t)/\partial x_j(t)}{\rho + (n-1)\lambda x_{-j}(t)}. \tag{18}$$

With constant marginal costs of R&D, firm j must choose $x_j(t)$ such that the marginal benefits equal 1, given the (necessarily identical) choices of all the other competitors in the race, $(n-1)x_{-j}(t)$. The marginal benefits can be divided up into two terms, the first capturing the increase in the probability of successful innovation, the second capturing the effect on expected profits. This second term is unusual in patent races, but captures an important effect present in this model; namely that innovating firms have to take into consideration the decrease in intertemporal profits that accompanies any innovation: innovations – by their very nature and due to the demands of the patent system – must be more effective against pathogens than their predecessors and hence invariably affect resistance in a negative way. This undercuts the profitability of innovations as the final goods sector responds to a drop in resistance with a reduction in aggregate demand.

The pay-off function for firm j derived from (18) is continuous and concave in $x_j(t)$. This makes the situation ideal to be considered under the Nash conjecture. A continuous and concave pay-off function in the strategy space of player j is sufficient for a pure-strategy Nash equilibrium of identical players. This is known as the Glicksberg theorem (Fudenberg and Tirole, 1998). Under the Nash conjecture, $x_j(t) = x_{-j}(t)$. The optimal research intensity for each player is then $x^*(t)$,

$$(n-1)x^*(t) = \frac{L\left[-\frac{f(t)^2}{f_{c^*}(t)} - k\right] - \frac{\rho}{\lambda}}{\left[1 + \lambda I L\left[-\frac{2f(t)}{f_{c^*}(t)}\frac{\partial f(t)}{\partial v(t)} + \frac{f(t)^2}{f_{c^*}(t)^2}\frac{\partial f_{c^*}(t)}{\partial v(t)}\right]\right]}. \tag{19}$$

Equation (19) states that the optimal research expenditure of the industry is expected monopolistic profit minus the real opportunity cost of capital in the R&D industry[29] discounted by the cumulative externality of innovation on the value of the patent as demand falls with higher effectiveness of treatment. This discounting of the patent value naturally increases with the size of innovations I, and the probability of discovery λ. What this tells us is that there exists what one could call an 'industry-wide Arrow effect' in the intermediate goods sector: not only does the owner of the patent that protects the currently used technology refrain from innovation because her pay-off would be strictly lower than that of any of the competitors (Arrow, 1962); but the whole R&D sector suffers from the expected loss of profits associated with R&D activity.

It is easy to check that $\partial x^*(t)/\partial v(t) > 0$ and $\partial^2 x^*(t)/\partial v(t)^2$, i.e. that x is concave in v. The intuition is that the aggregate demand for treatment drops more sharply with an increase in price when v is higher which reduces the marginal profitability of owning the patent when resistance is high.

This means that the optimal R&D expenditure policy closely tracks the development of resistance: expenditure is high when present value is high; the present value is only high if expected profits are high and the probability of innovation low; profits again are high only when resistance is high. Solving (19) for an individual firm's R&D expenditure and taking the limit as resistance goes to zero, we get

$$\lim_{v \to 0} x^*(t) \leq 0. \tag{20}$$

This means that innovation ceases in the limit as resistance disappears.

4.3. Discussion

4.3.1. *The Overall Behaviour of the System*

The problem of non-cooperative management is one that involves two distinct dynamic processes at the same time: one is the development over time of pathogen population levels, the other the evolution of strains in the population that are resistant to current forms of treatment. A manager in the final goods sector will start treatment as soon as pathogens pass a certain population level. This action threshold that positively depends on the price of the intermediate good and the resistance of the pathogens ensures that overall, population levels are stable and that the loss to pathogens is contained.[30]

Demand for treatment is satisfied by an intermediate industry in which the current treatment technology is produced by a patent-holding monopolist. The monopolist initially faces a low profit level as a result of having provided an effective input that decreases resistance by the fixed amount I: the share of managers engaged in treatment has dropped sharply as the optimal duration of treatment shortened in response to the innovation. Over time, however, and since aggregate demand is concave in resistance, profits rise as technological effectiveness declines in response to evolutionary pressure on pathogens. This increasing availability of rents present in the intermediate goods market induces R&D spending among competitors in the industry. Initially, R&D spending reacts very slowly to rising resistance as the relative (negative) impact of an innovation on resistance (and hence profits) is substantial, but spending must increase over-proportionately over time. With R&D spending rising, the probability of an innovation rises as well and finally leads to some competitor displacing the incumbent with a new technology. The overall behaviour of the system is therefore cyclical, but stable. Figure 1 depicts a typical period of resistance evolution over time when a series of firms replace each other after having innovated and thus decreased resistance at t_1, t_2, t_3, and t_4.

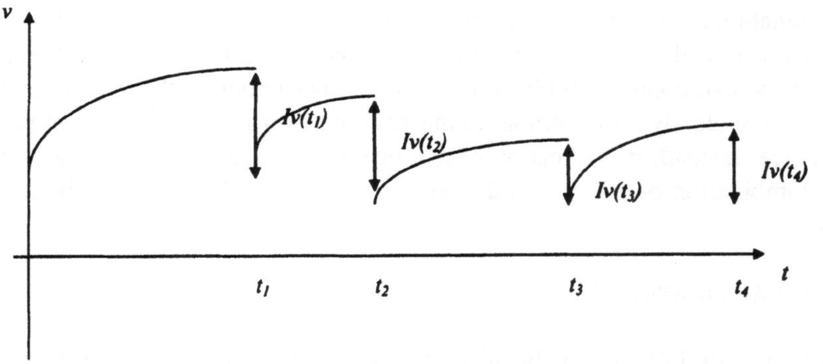

Figure 1. Development of resistance over time.

4.3.2. *The Impact of Winnability*

As the analysis in Section 1 shows, the behaviour of the final goods sector is influenced by pathogen migration. Specifically, high migration rates are associated with decreased intensity of treatment. The response in the final sector is therefore conducive to the idea that an uncoordinated economy can realise the potential for pathogen eradication as the evolutionary response in pathogen populations is low when migration is high. This means that a small level of R&D may be sufficient to move pathogens into a situation where resistance is so low and irresponsive to treatment that containment is possible without continuing R&D. The question must therefore hinge on the behaviour of the intermediate goods sector.

To examine the likely outcome, we have to look at the specific components of the evolution function. Turning to the incentives for innovation first, it is apparent from Equation (19) that low levels of resistance or even decreasing resistance exacerbate the negative externalities from innovation for industry. This is because 'winnability' effectively means that the industry is competing against a 'competitor' that is able to innovate without costs. The 'competitor' in this case is nature. In the same way that any innovation within the industry inflicts a negative externality on all the competitors, nature punishes industry spending on R&D. This causes R&D spending to collapse.

The answer to winnability must therefore come from the aggregate demand for treatment. Only if the share of managers is below a/L, does the pathogen population enter into a phase of resistance decline. For this to occur, however, industry must have previously provided a technology with an associated $v(t)$ such that $f^*(v(t)) < a/L$. Unless a is very substantial, this means that winning the evolutionary battle depends on the fortunate conjunction of at least two innovations over a very short stretch of time and on sufficiently low production costs which enable the firm that innovated last to generate a profit even at very low aggregate demand.[31]

Winnability therefore has no structural effect on the dynamics of the system. Moreover, it will not be possible to answer the question whether pathogen damage can be overcome from observing industry behaviour as the incentives for the industry are clearly unfavourable to the goal of a consistent reduction of pathogen resistance. Instead, if eradication of pathogens is possible, it requires a serendipitous combination of events and the question of winnability can only be answered *ex post*.

4.4. Qualifications

On the way to deriving our results, we have made a number of assumptions, the most important of which is the premise that innovations in the industry are drastic and that they affect a fixed share of pathogens. Without re-iterating the justifications given for these assumptions, it is interesting to consider what the implication of endogenously determined innovation size would be. Since pay-offs to innovation are clearly decreasing in the size of innovations, we would expect a tendency for innovations to become smaller. On the other hand, as Aghion and Howitt (1992) point out, innovators have an incentive to choose an excessive innovation size to increase profit margins. Since this effect is related to the elasticity of demand, however, it is unlikely to balance the negative effects of innovation size on profits as demand becomes more elastic for highly effective inputs. Overall, therefore, endogenising the innovation size will accelerate the replacement process at the technological frontier of the industry, but will not create incentives to decrease resistance more than before.

A second point concerns the fact that we have chosen the migration rate to be exogenously given as a 'natural' parameter. The transmission mechanisms in reality are a joint outcome of biological and topographical factors as well as human choice: land use patterns affect the possible linkage between pathogen populations by putting hosts close together or far apart. Migration of the vector of the pathogen, such as human migration, affects the geographical spread of pathogens among human hosts. This is a natural instrument available for regulation to social planners and there may be incentives for decision makers both in the final and intermediate goods industry to affect these rates.

Further possible extensions of the model could be the inclusion of stochasticity in pathogen evolution and growth which would enrich the analysis of decision making processes of managers in the final sector and of firms in the intermediate industry. Further work is also necessary on endogenising the industry structure that gives rise to the innovations.

5. Conclusion

This chapter has looked at the potential for a two sector economic system to effect pathogen eradication if this was technologically possible. The motivation has been the idea that there may be allocative problems in the way ecological processes are managed in a two sector economy which may render successful pathogen management unlikely even if the premises of the 'eternal battle'-hypothesis were not correct. The latter predicts that the evolutionary battle can never be won with finality.

Our results show that the incentive structure created by the combined working of population growth and evolution is not conducive to an eradication of pathogens. In particular, the question whether the race between humans and society is winnable or not has no bearing on the basic structure of the incentives. Although the final sector responds 'correctly' to the presence of high migration rates, the incentives for industry to maintain a sufficient rate of innovation during the transitional period to the point when R&D becomes unnecessary are exactly opposite The intuition behind this negative result is that investment into resistance decrease made by a firm through technological innovation inflicts a negative externality on the industry. The reason for this 'sector-wide Arrow effect' is that high levels of resistance enhance the scope for acquiring monopolistic rents from managers in the final goods sector. Winnability, however, can have an effect due to the stochastic nature of innovation as the outcome can be achieved as a random event.

A final observation is that there is also a social trade-off between the desirability of winnability and its associated costs. high migration rates not only allow possible eradication in the future, they also create current costs by increasing optimum level of epidemiological damage. The aggregate population level increases with the migration rate. This means that in the absence of intervention and somewhat counter-intuitively, society may be worse off in a situation where pathogens can be eradicated. Whether or not this is the case depends of course on the relative cost of R&D. This paper therefore predicts that the potential to manage society's biological competitors is seriously constrained by economic factors even if technologies are available, and that society is worse off if the technologies are available, but are not realised.

Acknowledgement

This research is supported by a Marie Curie Research Fellowship from the European Commission, DG XII, Contract No. ERBFMBICT972571.

Notes

[1] This is customarily defined as the time it takes for 50% of the pathogen population to exhibit resistance against the control input (Anderson and May, 1991).

[2] The most notable case is that of DDT in the course of Malaria eradication programs in India (Cornes et al., 1995).

[3] In one of the best researched parasite species, avian coccidia (*Eimeria tenella*), the time to resistance is between 6 and 65 generations, corresponding to between 1 and 27 years. This is a common time scale for this process (cf. Anderson and May, 1991, p. 612). As the growing season for crops is usually not longer than one year, resistance development does not appear to be of great importance within a single planting period.

[4] There is an obvious trade-off between analytical clarity and the realism of the model in that a dynamic programming approach is unlikely to yield interpretable closed form solutions at meaningful time horizons ('the curse of dimensionality'), but represents a more realistic depiction of the problem.

[5] The result is not firm since it depends on the costs of developing new types of pest control inputs and thus on the characteristics of the relevant innovation function.

[6] An almost identical paper to Mangel (1985) was later published by Munro (1997).

[7] The possibility of reversibility is mentioned in Regev et al. (1983), but not explored.

[8] The reason for the last conclusion in Wiemer's model stems from the fact that the marginal gains from reducing the rate of transmission are lower than those from curing already infected patients. Different results arise in the context of vector-borne diseases such as Malaria (Hammer, 1993).

[9] These specific conditions are: the disease is transmitted directly, person-to-person. Individuals do not recover from illness, once infected. The effectiveness of a vaccination has infinite duration. Individuals are homogeneous with regard to the cost of vaccination and susceptibility to infection. There is no behavioural response in individuals.

[10] The authors explore the theme in the context of AIDS where behavioural choices clearly have an impact. With other infectious diseases, this element may carry significantly less weight, e.g. in the case of water-borne diseases in developing countries where adjustment to disease prevalence is prohibitively costly.

[11] The conditions that make reversibility unlikely are explained in Anderson and May (1991).

[12] It is especially in this context that the lag load model proves helpful as its general form that does not depend on allele frequencies helps to accommodate the shifts in resistance in a straightforward manner.

[13] For instance, cf. "Ingenuity, knowledge, and organisation alter but cannot cancel humanity's vulnerability to invasion by parasitic forms of life. Infectious disease which antedated the emergence of humankind will last as long as humanity itself, and will surely remain, as it has been hitherto, one of the fundamental parameters and determinants of human history" (McNeill, 1977).

[14] The industries which exists to manage these externalities, the agrochemical and plant-breeding industry for agriculture and the pharmaceutical industry for human and animal health, are thus engaged in intertemporal rent-seeking activities at the expense of the welfare of future generations.

[15] To take an analogous concept from macroeconomics, in the same way that the concept of the NAIRU ('non-accelerating inflation rate of unemployment') relates productivity growth to unemployment, the NAERL relates the productivity growth of crop or health protection to the equilibrium level of losses to pathogens.

[16] The concept of the expected value of information would suggest, however, to go on investing in R&D and to gather information about the likelihood of the two scenarios through the observation of usage and success.

[17] We are not concerned with the specific problems that may arise in some of these situations such as principal-agent problems between patients and doctors.

[18] Biologically, the difference is one between viral and other (bacterial, fungal, etc.) types of diseases. For the first type, the question of optimal vaccination is the prime management issue.

[19] The reason for choosing the exponential model rather than the logistic one is that we assume that society is sufficiently concerned about pathogen loss such that at the point where there is a crowding cost in the pathogen population, society has already taken action.

[20] Agriculture is to the greatest extent a man-modified environment and manufactured drugs play a significant role in human health.

[21] Strictly speaking, $e/m = S(1 - 1/S)$ where S is the number of sites.

[22] Here this is an individual that is resistant under current conditions and thus has fitness $\hat{F} = (1 + r)$.

[23] If resistance has a negative cost to the pathogen, then there is of course 'negative absorption'.

[24] There are many ways to derive the threshold level rule. With a discrete control variable, it could also be formalised as an optimal control problem. This does not yield qualitatively different results (cf. Moffitt and Farnsworth, 1987).

[25] The cost is not essential for the argument, but makes the decision situation crisper. We could think of F as a cost of going to a doctor for the purpose of diagnosis or the fixed cost involved with purchasing agrochemicals.

[26] We could think of L as the dosage of treatment where the marginal benefits equal the disutility from side effects.

[27] Equation (8) imposes two conditions on the current intermediate input. The first condition is that $r < L(1 - v)$, the second that $cL + Dm \sum N_{-j} < -DN(s)[r - L(1 - v)]$. The first condition means that in order to be used, the current technology has to be effective, i.e. it may not merely dampen pathogen growth, but has to have a negative impact on population levels. The second states that current technology has to be competitive, i.e. the intertemporal cost of treatment has to be lower than the intertemporal damage avoided.

[28] This attributes the innovation to improvements in the productivity of the input in the final goods sector rather than improvements in the marginal costs of production as in most models (cf. Tirole, 1988).

[29] This is the relative productivity of a unit of money in the economy overall, r, over the productivity of a unit of money in research, λ.

[30] As has been pointed out in other models (Cornes et al., 1995; Hueth and Regev, 1983), the use of treatment may be excessive, however, since the managers do not take into account the evolutionary impact of their choice. Since the effects of evolution diffuse over many managers and since the threshold decision rule used only takes into consideration the optimal time horizon of the treatment during which changes in resistance will be hard to observe, managers may be entirely rational to be unconcerned about the long-term effects of their choices.

[31] Since withholding the product from the market if positive profits can be earned achieves less pay-off then marketing the product (since the owner of the patent suffers from the chance of being replaced by a new entrant), it is unlikely that patent shelving will be excessive.

References

Aghion, P. and Howitt, P. (1992) A model of growth through creative destruction, *Econometrica* **60**, 325–351.

Anderson, R.M. and May, R.M. (1991) *Infectious Diseases of Humans. Dynamics and Control*, Oxford Science Publications, Oxford University Press, Oxford.

Arrow, K. (1962) Economic welfare and the allocation of resources for invention, in R. Nelson (ed.), *The Rate and Direction of Inventive Activity*, Harvard University Press, Cambridge, MA.

Barrett, J.J. (1981) The evolutionary consequences of monoculture, in J.A. Bishop and L.M. Cook (eds.), *Genetic Consequences of Man Made Change*, Academic Press, London, pp. 209–249.

Baumol, W.J. and Oates, W.E. (1988), *The Theory of Environmental Policy*, 2nd edn., Cambridge University Press, Cambridge.

Bishop, J.A. and Cook, L.M. (eds.) (1981) *Genetic Consequences of Man-Made Change*, Academic Press, London.

Brito, D., Sheshinski, E. and Intriligator, M. (1991) Externalities and compulsory vaccinations, *J. Public Econ.* **45**(1), 69–90.

Brown, G. and Layton, D.F. (1996) Resistance economics: Social cost and the evolution of antibiotic resistance. *Environ. Developm. Econ.* **1**(3), 349–355.

Carlson, G.A. (1997) A decision theoretic approach to crop disease prediction and control, *Amer. J. Agricult. Econ.* **52**, 216–222.

Carlson, G.A. and Castle, E.N. (1972) Economics of pest control, in National Academy of Science (ed.), *Control Strategies for the Future*, Washington, D.C., pp. 79–99.

Comins, H.N. (1977) The development of insecticide resistance in the presence of migration, *J. Theor. Biol.* **65**, 399-420.

Conrad, J. and Clark, C. (1987) *Natural Resource Economics. Notes and Problems*, Cambridge University Press, Cambridge.

Cornes, R., Long, N.V. and Shigemura, S. (1995) Drugs, pest and resistance, mimeo, Department of Economics. McGill University.

Daily, G.C. and Ehrlich, P.R. (1996) Impacts of development and global change on the epidemiological environment, *Environ. Developm. Econ.* **1**(3), 311–346.

Feder, G. (1979) Pesticides, information and pest management under uncertainty, *Amer. J. Agricult. Econ.* **61**, 97–103.

Felsenstein, J. (1971) On the biological significance of the cost of gene substitution, *Amer. Naturalist* **105**, 1–11.

Francis, P.J. (1997) Dynamic epidemiology and the market for vaccinations, *J. Public Econ.* **63**(3), 383–406.

Frank, S. (1996) Statistical properties of polymorphism in host-parasite genetics, *Evolut. Ecol.* **10**, 307–317.

Fudenberg, D. and Tirole, J. (1991) *Game Theory*, MIT Press, Cambridge, MA.

Geoffard, P.Y. and Philipson, T. (1996) Rational epidemics and their public control, *Internat. Econ. Rev.* **37**(3), 603–624.

Geoffard, P.-Y. and Philipson, T. (1997) Disease eradication: Private versus public vaccination, *Amer. Econ. Rev.* **87**(1), 222–230.

Goeschl, T. (1998) The economics of the Red Queen: Biodiversity in agriculture and long-term food supply, Paper presented at the 2nd International Conference of the European Society for Ecological Economics, Geneva.

Goeschl, T. and Swanson, T. (1998) Optimal management of agricultural biodiversity: Genetic resources, gene banks and land use, Paper presented at the World Congress of Environmental and Resource Economists, Venice, Italy, June 1998.

Grossman, G.M. and Helpman, E. (1991) Quality ladders in the theory of growth, *Rev. Econ. Studies* **58**, 43–61.

Hall, D.C. and Moffitt, L.J. (1985) Application of the economic threshold for interseasonal pest control, *Western J. Agric. Econ.* **10**, 223–229.

Hall, D.C. and Norgaard, R.B. (1973) On the timing and application of pesticides, *Amer. J. Agricult. Econ.* **55**, 198–201.

Hammer, J.S. (1993) The economics of malaria control, *World Bank Research Observer* **8**(1), 1–22.

Hartwick, J. (1978) Exploitation of many deposits of an exhaustible resource, *Econometrica* **46**(1), 201–217.

Heisey, P.W. (ed.) (1990) Accelerating the transfer of wheat breeding gains to farmers: A study of the dynamics of varietal replacement in Pakistan, CIMMYT Research Report No. 1, CIMMYT, Mexico, DF.

Heisey, P.W. and Brennan, J.P. (1991) An analytical model of farmers' demand for replacement seed, *Amer. J. Agricult. Econ.* **73**(4), 1044–1052.

Hillebrandt, P.A. (1960) The economic theory of the use of pesticides, *J. Agricult. Econ.* **13**, 464–472.

Holmberg, S.D., Solomon, S.L. and Blake, P.A. (1987) Health and economic impacts of antimicrobial resistance, *Rev. Infect. Diseases* **9**(6) 1065–1077.

Hueth, D. and Regev, U. (1974) Optimal agricultural pest management with increasing pest resistance, *Amer J. Agric. Econ.* **56**, 543–552.

Huffman, W.E. and Evenson, R.E. (1993) *Science for Agriculture: A Long-Term Perspective*, Iowa State University Press, Ames, IA.

Krause, R.M. (1981) *The Restless Tide: The Persistent Challenge of the Microbial World*, The National Foundation for Infectious Diseases. Washington, D.C.

Laxminarayan, R. and Brown, G. (1998) Economics of antibiotic resistance. A theory of optimal use,. mimeo, Department of Economics, University of Washington.

Mangel, R. (1985) *Decision and Control in Uncertain Resource Systems*, Academic Press, New York.

Mangel, M. and Plant, R.E. (1983) Multiseasonal management of an agricultrual pest I: Development of the theory, *Ecol. Modeling* **20**, 1–19.

Maynard Smith, J. (1976) A comment on the Red Queen, *Amer. Naturalist* **110**, 325–330.

McNeill, W. (1977) *Plagues and Peoples*, Anchor, New York.

Moffitt, L.J. (1986) Risk-efficient thresholds for pest control decisions, *J. Agricult. Econ.* **27**, 69–75.

Moffitt, L.J. and Farnsworth, R. (1987) Thresholds for chemical control of agricultural pests in a dynamic ecosystem, *Can. J. Agric. Econ.* **35**, 627–637.

Moffitt, L.J., Hall, D.C. and Osteen, C.D. (1984) Economic thresholds under uncertainty with application to corn nematode management, *Southern J. Agricult. Econ.* **16**, 151–157.

Morse, S.S. and Schluederberg, A. (1990) Emerging viruses: The evolution of viruses and viral diseases, *J. Infect. Diseases* **162**, 1–7.

Munro, A. (1997) Economics and biological evolution, *Environ. Resource Econ.* **9**, 429–449.

Nutter, F.W., Teng, P.S. and Royer, M.H. (1993) Terms and concepts for yield, crop loss and disease thresholds, *Plant Disease* **77**(2), 211–215.

O'Donoghue, T. (1998) A patentability requirement for sequential innovation, *Rand J. Econ.* **29**(4), 654–679.

Oerke, E.-C., Dehne, H.-W., Schönbeck, F. and Weber, A. (1994) *Crop Production and Crop Protection. Estimated Losses in Major Food and Cash Crops*, Elsevier, Amsterdam.

Palti, J. and Ansher, R. (1986) Crop value, economic damage threshold and treatment thresholds, in *Economic Advisory Work in Crop Pest and Disease Management*, Springer-Verlag, Berlin, pp. 45–56.

Plant, R.E., Mangel, M. and Flynn, L.E. (1985) Multiseasonal management of an agricultrual pest II: The economic optimization problem, *J. Environ. Econ. Manag.* **12**, 45–61.

Regev, U., Shailt, H. and Gutierrez, A.P. (1983) On the optimal allocation of pesticides with increasing resistance: The case of Alfalfa weevil, *J. Environ. Econ. Manag.* **10**, 86–100.

Reinganum, J.F. (1985) Innovation and industry equilibrium, *Quart. J. Econ.* **100**, 81–99.

Rosenzweig, M.L. (1973) Evolution of the predator isocline, *Evolution* **27**, 84–94.

Rosenzweig, M.L. (1996) And now for something completely different: Genetic games and Red Queens, *Evolutionary Ecol.* **10**, 327.

Roughgarden, J.R. (1979) *Theory of Population Genetics and Evolutionary Ecology: An Introduction*, MacMillan, London.

Schaffer, W.M. and Rosenzweig, M.L. (1978) Hommage to the Red Queen I. Coevolution of predators and their victims, *Theor. Population Biol.* **14**, 135–157.

Scheffer, R. (1997) *The Nature of Disease in Plants*, Cambridge University Press, Cambridge.

Seidl, C. (1988) Das Sozialversicherungsparadoxon, *Wirtschaftwissenschaftliches Studium* **17**, 305–308.

Stenseth, N.C. and Maynard Smith, J. (1984) Coevolution in ecosystems: Red Queen evolution or stasis?, *Evolution* **38**(4), 870–880.

Swanson, T. (1995) Uniformity in development and the decline of biodiversity, in T. Swanson (ed.), *The Economics and Ecology of Biodiversity Decline: The Forces Driving Global Change*, Cambridge University Press, Cambridge.

Taylor, R.C. and Headley, J.C. (1973) Resistance and the optimal control of pest populations, Working Paper, Department of Agricultural Economics, University of Missouri, Columbia.

Tirole, J. (1988) *The Theory of Industrial Organization*, MIT Press, Cambridge, MA.

Ulph, D. and Sianesi, B. (1998) Species loss through the genetic modification of crops – A policy framework, Paper presented at the World Congress of Environmental and Resource Economists, Venice, Italy, June 1998.

Van Valen, L. (1977) The Red Queen, *Amer. Naturalist* **111**, 809–810.

Volterra, V. (1931) *Leçons sur la théorie mathématique de la lutte pour la vie*, Gauthier-Villar, Paris.

Weitzman, M. (1976) The optimal development of resource pools, *J. Econ. Theor.* **12**, 351–364.

Wickwire, K. (1977) Mathematical models for the control of pests and infectious diseases: A survey, *Theoret. Population Biol.* **11**, 182–238.

Widawsky, D. (1998) Pesticide productivity, host-plant resistance and productivity in China, *Agricult. Econ.* **19**(1–2), 203–217.

Wiemer, C. (1987) Optimal disease control through combined use of preventive and curative measures, *J. Developm. Econ.* **25**(2), 301–319.

World Bank (1993) *World Development Report 1993: Investing in Health*, published for the World Bank by Oxford University Press, New York.

5. Providing the Correct Incentives for Genetic Modification

LUCY O'SHEA and ALISTAIR ULPH

1. Introduction

Despite a steadily decreasing population growth rate, the absolute level of population is rising and thus ever-increasing supplies of food are required – especially in regions where environmental degradation compromises the ability to produce food. The prospect that genetically modified crops might greatly enhance yields is probably the single greatest argument in favour of genetic engineering of plants (Reiss and Straughan, 1996). One major cause of reduced crop yields is due to crop pests. Worldwide, around a third of all potential crop production is lost through pests (Reiss and Straughan, 1996). The potential for increased crop yields arises from the reduction of crop losses due to pests. Indeed, according to Reiss and Straughan (1996) the most immediate source of increased crop yields will be through enhanced pest resistance. Enthusiasts argue that genetic pest resistance will lead to enhanced yields, reduce the use of pesticides and result in reduced consumer prices. However, there is a concern that GM (genetically modified) crops may present a threat to the environment. In this chapter we limit our analysis to the question of possible impacts on biodiversity caused by a reduction in the availability of food for organisms due to the incorporation of a pest-resistant gene into agricultural crops. The impact on biodiversity will obviously depend on how biodiversity is measured. We utilise the most simple representation of biodiversity – the number of species per unit area.[1]

The central question we shall address in this chapter is whether policy on GM crops should be directed towards the growth of the crops themselves or the R&D which leads to the technology for growing GM crops.[2] Obviously, this is not the only issue to be addressed, but we focus on this issue because it is not an issue that has been much addressed in the environmental economics literature. Perhaps this reflects the fact that most economists' intuition would be that if, as we shall assume, it is the growth of GM crops rather than the undertaking of GM R&D *per se* which damages the environment,[3] then one should regulate the growing of GM crops. Such regulation may indirectly affect the rate of R&D that is done, but that is a consequence of the policy not a direct target of the policy. In a closely related paper by Sianesi and Ulph (1999) which addresses this question, that is indeed precisely the conclusion that is reached, namely that the only intervention that is

Timothy Swanson (ed.), The Economics of Managing Biotechnologies, 129–143.

warranted is the decision about whether to grow GM or non-GM varieties of crops, and there is no need to alter the normal incentives to do R&D.[4]

However, we will argue that these conclusions follow from a very specific model of how it is believed that GM crops affect the environment. Like us, Sianesi and Ulph (1999) are concerned solely with the potential threat that growing GM crops will pose to the diversity of species. In their model, species diversity is directly related to the total supply of non-GM crops that is grown. The externality in their model is thus that the growing of non-GM crops generates an external benefit in terms of species variety. Thus if left to market forces, a suboptimal level of non-GM crops results. Thus, the policy required is to subsidise the growing of non-GM crops. Therefore, in their model irrespective of whether GM technology is available or not, a subsidy on non-GM crops is necessary. However, there is no link in their model between the number of crops grown as either GM or non-GM varieties and the number of species – in the limit if a sufficient quantity of only one crop in the non-GM variety is grown, with all other crops grown as GM varieties, then one could sustain any level of diversity desired. In other words, their model assumes that species are polyphagous, i.e. that they can feed on many sources of food and if one source disappears they can easily switch to another. This seems to us an implausible characterisation of the feeding habits of different species. Assuming greater specialisation in feeding habits demands that a degree of variety of crops is maintained. Thus, we follow most of Sianesi and Ulph's (1999) model but modify the assumption about the link between GM crops and species diversity. Thus, we adopt an equally simple, but very different view of the relationship between species diversity and crops. In our approach the variety of animal and insect species that survives is directly linked to the variety of crops for which at least some minimum threshold quantities of non-GM varieties is grown. With this characterisation of the link between GM crops and biodiversity we get rather different policy conclusions. There will be outcomes where there is no need to directly intervene in the decisions about which crops to grow, although in other possible steady states of our model it will be necessary to subsidise the growing of the minimum threshold levels of non-GM varieties for some, but not all crops. The need to intervene in the decision about whether to grow crops as GM or non-GM varieties arises only because of the existence of GM technology. This is in direct contrast to Sianesi and Ulph (1999) and arises because of the different way in which species dependence on food supplies is modelled. More importantly, we will show that it is necessary to intervene in the R&D decision, essentially by taxing the profits that would be earned from R&D in order to reflect the social costs imposed on society by making available GM technology to a wider range of crops.

The format of the chapter is as follows. In Section 2 we set out the model and assumptions. Within Section 2 the problem at hand is set out and the results are analysed in three parts. Firstly, the optimal amounts of crops grown are determined and explained. Secondly, we determine the optimal proportion of GM-only crops.

Three possible optimal levels of GM-only crops are found. Thirdly, the optimal proportion of crops for which GM technology is made available is determined as well as the approach path to its equilibrium level. The policy implications of the findings of this section are also presented. Finally, in Section 3 we conclude.

2. Model

We shall stick as closely as possible to the model by Sianesi and Ulph (1999) and modify it only with respect to the link between GM crops and biodiversity. We assume that there is a continuum of possible crops that can be grown denoted by the interval [0,1]. Although crops grown are inherently different from each other, we shall assume that the model is symmetric with respect to different crops, in a sense that will become clear as we proceed. For each crop it is possible to grow a non-GM variety at a unit cost k. If and only if the GM technology has been discovered for a particular crop, it will be possible to grow a GM variety of that crop at a unit cost c, which is the same for all crops. To make the problem interesting, we assume $c < k$. The advantage of GM technology therefore is simply to lower the unit cost of growing crops. However, crops are sufficiently different from each other, that developing the technology to grow a GM variety of one crop does not give one the technology to grow GM varieties of other crops, and so one has to continuously invest in GM R&D if one wishes to expand the range of crops for which GM technology is available. Specifically we assume that at time t, for fraction p_t of all crops, GM technology has been developed, and that if one carries out GM R&D at the rate g_t, then the proportion of crops for which GM technology is available can be expanded at the rate $\dot{p} = g_t$.

As shown by Sianesi and Ulph (1999), without loss of generality, we shall assume that $p_0 = 0$. The cost of carrying out the R&D is given by $\gamma(g)$, where $\gamma' > 0$, $\gamma'' > 0$ for all g; in particular $\gamma'(0) > 0$, so that there are strictly positive and increasing marginal costs of doing GM R&D, even for the very first unit of R&D.

Now obviously on the proportion $(1 - p_t)$ of crops for which there is no GM technology available, it is only possible to grow non-GM varieties of those crops, and we denote by z_t the amount of the non-GM variety grown for each of these crops (although crops are different, everything in the model is symmetrical so if it pays to grow z_t of one crop with no GM technology it will pay to grow z_t for all crops for which there is no GM technology). For the proportion of crops p_t for which GM technology has been discovered at time t we do not assume that only GM varieties need be grown. We assume that on a proportion $q_t \leq p_t$ of crops only GM varieties are grown, and the amount of GM variety of each crop grown we denote by w_t. On the remaining proportion $(p_t - q_t) \geq 0$ of crops we assume that both GM and non-GM varieties of each crop are grown, and we denote by

x_t the amount of GM variety of each crop grown, and y_t the amount of non-GM variety of each crop grown. Now in Sianesi and Ulph (1999) it was just assumed that $\forall\, t,\, q_t = p_t$, so that once GM technology is available for a crop only the GM variety would be grown. It turns out that for their model, such an assumption is justified, at least in steady state, but because of the different way we model the link between GM crops and biodiversity it is essential that we maintain the possibility of growing both GM and non-GM varieties of the same crop. We now turn to the different models of the GM-biodiversity link.

In Sianesi and Ulph (1999) the number of bird/insect species, S_t is related to the total food supply of non-GM crops, denoted by

$$f_t \equiv (1 - p_t)z_t$$

through an equation

$$\dot{S}_t = \phi(f_t/S_t)S_t.$$

But this implies that even if $q_t = p_t \approx 1$, provided z_t is made large enough one can achieve, over time, whatever level of variety of species one desires. So species of insect (and hence birds) which may have fed on the non-GM variety of one crop, which is subsequently made insect resistant through GM technology, are assumed to be able to switch to feeding on the non-GM varieties of other crops. Insects are assumed to be completely non-discriminatory in their feeding habits. This seems to us biologically a very strong assumption and inconsistent with the recognition of specialised feeding habits among species.

So in our paper we adopt an equally simplistic but radically different assumption about the link between species diversity and crops. We assume that there is a direct link between the variety of species of insects and birds and the variety of crops which are not GM, provided that for insect species that feed on a particular crop at least $\bar{y} > 0$ of the non-GM variety of that crop has to be grown, otherwise the species is driven to extinction.

Now since the only rationale in our model for growing non-GM varieties of crops for which GM technology is available will be to ensure the survival of the insects/birds which depend on that crop, it will turn out that for the proportion of crops $(1 - q_t)$, $y_t \geq \bar{y}$. We shall therefore simply identify the variety of species of insects/birds dependent on crops by $S_t = (1 - q_t)$. Since it will also turn out in our model that $\dot{q}_t \geq 0$, this implies that once a crop is grown only in its GM variety, the associated species of insects and birds which depend on that crop are irreversibly lost.

Society's instantaneous preferences for variety of species is captured in a value function $V(1 - q_t)$ with the properties $V' > 0$, $V'' \leq 0$, $V'(0) = \infty$, so that there is a strictly positive but diminishing marginal benefit to a given variety of species, and an infinite marginal disutility associated with the loss of the last species.

Finally we consider the benefits obtained from consuming crops. We assume that consumers are completely indifferent to whether they consume the GM or non-GM variety of a crop,[5] but that they have a preference for consuming a variety of crops. This is captured by the usual Dixit-Stiglitz C.E.S. preferences, so that at any moment of time, consumer benefits are given by:

$$U(p, q, w, x, y, z) \equiv q\frac{w^{1-\beta}}{1-\beta} + (p-q)\frac{(x+y)^{1-\beta}}{1-\beta} + (1-p)\frac{z^{1-\beta}}{1-\beta}; \quad \beta > 0, \beta \neq 1.$$

Let $\varepsilon = 1/\beta$ be the elasticity of substitution between crops and obviously with $\beta > 0$ crops are imperfectly substitutable. As in Sianesi and Ulph (1999) denote by

$$B(\xi) \equiv \max_x \frac{1}{1-\beta}x^{1-\beta} - \xi x$$

the maximum social surplus (ignoring environmental effects) from a having a crop produced at unit cost ξ. Carrying out the maximisation yields

$$x = \xi^{-\varepsilon}, \quad B(\xi) = -\frac{1}{1-\varepsilon}\xi^{1-\varepsilon}$$

so that $B'(\xi) < 0$.

We assume that $\bar{y} < k^{-\varepsilon}$; as we will see this implies that if one grew only the non-GM variety of a crop at unit cost k and chose to grow the amount of that crop which maximises social surplus, then one would grow enough of it to meet the minimum threshold amount of the crop to allow the insects and birds that depend on that crop to survive. So, in the absence of any GM technology, there would be no reason to intervene in the growing of crops.[6]

2.1. Optimal Paths of GM R&D and Growth of GM Crops

The problem facing the social planner is to choose w_t, x_t, y_t, z_t, g_t and q_t to maximise:

$$\int_0^\infty e^{-\delta t}\Bigg\{q_t\frac{w_t^{1-\beta}}{1-\beta} + (p_t - q_t)\frac{(x_t+y_t)^{1-\beta}}{1-\beta} + (1-p_t)\frac{z_t^{1-\beta}}{1-\beta} - q_t w_t c - (p_t - q_t)(cx_t + ky_t) - (1-p_t)kz_t - \gamma(g_t) + V(1-q_t)\Bigg\}\,dt$$

such that

$$\dot{p}_t = g_t; \quad y_t \geq \bar{y}; \quad q_t \leq p_t; \quad p_t \leq 1; \quad p_0 = 0.$$

The first-order conditions are:

$$w_t^{-\beta} \leq c; \quad w_t \geq 0, \tag{1}$$

$$(x_t + y_t)^{-\beta} \leq c; \quad x_t \geq 0, \tag{2}$$

$$(x_t + y_t)^{-\beta} \leq k - \mu_t; \quad y_t \geq 0, \tag{3}$$

$$(y_t - \bar{y})\mu_t = 0, \tag{4}$$

$$z_t^{-\beta} \leq k; \quad z_t \geq 0, \tag{5}$$

$$\pi_t \leq \gamma'(g_t); \quad g_t \geq 0, \tag{6}$$

$$[B(c) - (B(c) - (k - c)\bar{y}) - V'(1 - q_t)] \leq \lambda_t; \quad q_t \geq 0, \tag{7}$$

$$\lambda_t(p_t - q_t) = 0, \tag{8}$$

$$\dot{\pi}_t = \delta\pi_t - [B(c) - B(k) - (k - c)\bar{y} + \lambda_t - \omega_t], \tag{9}$$

$$\omega_t(1 - p_t) = 0. \tag{10}$$

We analyse these conditions in three stages.

2.1.1. *Optimal Amounts of Crops Grown*

In this subsection we take as given p_t and q_t and determine w_t, x_t, y_t and z_t.

From (1) and (5) it is straightforward to see that, $\forall\, t$,

$$w_t = c^{-\varepsilon}; \quad z_t = k^{-\varepsilon}; \quad w_t > z_t. \tag{11}$$

Although we have omitted finding the private market solution explicitly, it is trivial to see what it would be. When only GM or non-GM varieties are grown the amount chosen is what would be demanded if the crop was sold at its marginal (unit) cost of production, and so is the amount that would be grown in competitive markets with no need for policy intervention. In the absence of a constraint on y, this would be zero in the private solution as farmers would grow all crops for which GM technology existed as GM-only. In the social optimisation the level of y attained is the constraint level. To see this using (3) and (2) and $y_t \geq \bar{y} > 0$, then $(x_t + y_t)^{-\beta} = k - \mu_t \leq c$. Since $k > c$, that implies $\mu_t > 0$, which from (4) implies that $y_t = \bar{y}$, $\forall\, t$. Since $\bar{y} < k^{-\varepsilon} < c^{-\varepsilon}$, from (2) we must have $x_t = c^{-\varepsilon} - \bar{y} > 0$, $\forall\, t$. What this means is that where both GM and non-GM varieties of the same crop are grown, the total amount of the crop grown is the same as would be the case if only the GM variety had been grown, so consumers will get the same benefit from the 'mixed' case of GM and non-GM varieties as from the pure GM-only case. But out of the total, the minimum amount of non-GM variety is grown ($\bar{y}$) to sustain the existence of the species of insect and birds which depend on these crops. To induce the 'mixed' case which includes the required levels of non-GM crops, all crops in the mixed category are sold at a price equal to the unit cost of producing GM crops and the shortfall in profits from producing non-GM crops is made good by the government which grants a subsidy equal to the difference in the costs of production, $(k - c)$ for each unit produced of non-GM crops up to the threshold, $\bar{y}$.

We summarise the above as:

RESULT 1. For all time periods t, the following pattern of crops is grown:

(i) on a proportion of crops $q_t \leq p_t$, only GM varieties are grown: an amount $c^{-\varepsilon}$ is grown of each crop and sold at price c;
(ii) on a proportion $(1 - p_t)$ of crops, only non-GM varieties are grown: for each such crop an amount $k^{-\varepsilon}$ is grown and sold at price k;
(iii) on a proportion $(p_t - q_t)$ of crops both the GM and the non-GM varieties are grown: for each of these crops, $\bar{y}$ of the non-GM variety is grown and $(c^{-\varepsilon} - \bar{y})$ of the GM variety is grown: both varieties are sold at price c; and
(iv) the only policy required is that on the proportion $(p_t - q_t)$ of crops for which both GM and non-GM varieties are grown, the amount of $\bar{y}$ of the non-GM variety should be subsidised at the rate $(k - c)$ per unit.

2.1.2. *Optimal Proportion of GM-only Crops Grown*

We now determine the optimal value of q_t taking as given the value of p_t. It is easier to begin with the case where $0 \leq q_t < p_t$, so that from (8), $\lambda_t = 0$ and (7) becomes:

$$\begin{aligned}&[B(c) - (B(c) - (k - c)\bar{y}) - V'(1 - q_t)]\\&\quad = [(k - c)\bar{y} - V'(1 - q_t)] \leq 0; \quad q_t \geq 0. \qquad (7')\end{aligned}$$

The interpretation of (7′) is straightforward. An increase in q_t by 'one crop' has three effects. First for this crop, only the GM variety is now grown, which has a social benefit $B(c)$. Second, the crop will no longer be grown as a mix of GM and non-GM varieties: since this crop was sold at a common price of c, but an amount $\bar{y}$ was subsidised at a rate of $(k - c)$, the net social cost of no longer growing this crop in a mix of GM and non-GM varieties is $B(c) - (k - c)\bar{y}$. These two effects together mean that the net social benefit of increasing q_t by 'one crop' is $(k - c)\bar{y}$, i.e. it eliminates the need for society to subsidise the growing of $\bar{y}$ level of the non-

GM variety of a crop for which the GM technology is available. The third effect is that increasing q_t reduces the variety of species and the marginal social cost of this is $V'(1-q_t)$. So, assuming that $q_t < p_t$, the net social gain from an increase in q_t is $(k-c)\bar{y} - V'(1-q_t)$. At an optimum, it must not be possible to vary q_t and obtain a net gain. There are therefore two cases. (A) $V'(1) \geq (k-c)\bar{y}$. This says that the marginal social cost of the loss of even one species is at least as great as the cost of subsidising the growth of non-GM varieties. If that is the case, then $q^* = 0$, $\forall\, t$, i.e. one never grows only GM varieties of crops, so all species are retained. (B) $V'(1) < (k-c)\bar{y}$. Since $V'(0) = \infty$, there must exist a q^*, $0 < q^* < 1$ such that $V'(1-q^*) = (k-c)\bar{y}$.

In case (B) the level of q_t depends on the relationship of the availability of GM technology and the optimal implementation of that technology or growing of GM crops. If $p_t > q^*$ then $q_t = q^*$. In this case, the proportion q^* of crops will be grown as GM-only, whereas the proportion $(p_t - q^*) > 0$ of crops will be grown as both GM and non-GM.

Suppose that $p_t \leq q^*$, then $V'(1-p_t) \leq V'(1-q^*)$, and from (7), $\lambda_t = (k-c)\bar{y} - V'(1-p_t) \geq 0$; $q_t = p_t$. In this case, all the crops for which GM technology is available will be grown as GM varieties. We summarise these results as follows:

RESULT 2. For any value of p_t, $0 \leq p_t \leq 1$, there are three possible values of q_t:

(i) if $V'(1) \geq (k-c)\bar{y}$, $q_t = 0$, $\lambda_t = 0$, $\forall\, t$;
(ii) if $V'(1) < (k-c)\bar{y}$, and $p_t > q^*$, where $V'(1-q^*) = (k-c)\bar{y}$, then $q_t = q^*$ and $\lambda_t = 0$; and
(iii) if $V'(1) < (k-c)\bar{y}$ and $p_t \leq q^*$, then $q_t = p_t$, and $\lambda_t = (k-c)\bar{y} - V'(1-p_t) \geq 0$.

These three possibilities are illustrated in Figure 1.

2.1.3. *Optimal Path for GM R&D*

Steady State

To find a solution to the optimal path for p_t or equivalently, g_t we begin by considering the steady-state solution. Steady-state values are denoted by an asterisk. In steady state $\dot{\pi}_t = 0$, or $\pi_t = \pi^*$. Equation (9) becomes:

$$\pi_t = \pi^* \equiv \frac{B(c) - B(k) - (k-c)\bar{y} + \lambda^* - \omega^*}{\delta}. \tag{12}$$

In addition $\dot{p}_t = 0$, or $p_t = p^*$, so that using (6) and (12) steady state requires that

$$\frac{B(c) - B(k) - (k-c)\bar{y} + \lambda^* - \omega^*}{\delta} \leq \gamma'(0), \tag{13}$$

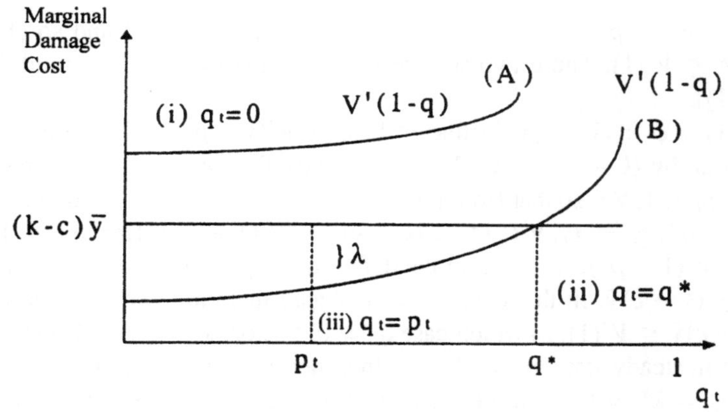

Figure 1. Determination of optimal GM-only crops.

where $\gamma'(0)$ is the marginal cost of carrying out the first unit of R&D. It will be useful to define

$$\phi = B(c) - B(k) - \delta\gamma'(0),$$

where ϕ is the instantaneous marginal private net return to carrying out the first unit of research in GM technology, and would be the instantaneous marginal social net return to the first unit of GM R&D in the absence of any concern about the environmental consequence of GM crops (by net return, we mean the marginal gross return net of the marginal cost of doing R&D). The definition of ϕ implicitly reflects the result that in the absence of any concern about environmental consequences of GM, once the GM technology exists for a crop, only the GM variety would be grown. To make the problem interesting we assume $\phi > 0$, otherwise it would never pay to carry out GM R&D even if there was no environmental impact. We can rewrite the condition for steady state as

$$\phi \leq (k - c)\bar{y} - \lambda^* + \omega^*. \tag{14}$$

Now we shall follow Sianesi and Ulph (1999) in arguing that if the steady-state value of p^* is strictly positive, then along the approach path to steady state we must have had $g_t > 0$ and hence (6) would have held with equality; by continuity (14), which is the steady-state version of (6), must also hold with equality. However, if $p^* = 0$, so GM R&D is never carried out, then (14) could hold as a strict inequality.

There are then three possible outcomes for steady state:

(i) $\phi \leq \min[(k-c)\bar{y}, V'(1)]$. In this case we argue that $p^* = 0$, and hence $p_t = 0$, $\forall\, t$. From (10) $\omega_t = \omega^* = 0$. There are two sub-cases. If $(k - c)\bar{y} \leq V'(1)$ then from Section 2.1.2 $q^* = 0$, $q_t = 0$, $\forall\, t$, $\lambda_t = \lambda^* = 0$, $\forall\, t$. Hence (14) holds as $\phi \leq (k - c)\bar{y}$. If $(k - c)\bar{y} > V'(1)$ then from Section 2.1.2 $q^* > 0$,

$q_t = p_t = p^* = 0, \forall\, t, \lambda_t = \lambda^* = (k - c)\bar{y} - V'(1)$ and so (14) holds as $\phi \leq V'(1)$. These two sub-cases are summarised by $\phi \leq \min[(k - c)\bar{y}, V'(1)]$.

(ii) $V'(1) < \phi < (k - c)\bar{y}$. Define p^* by $\phi = V'(1 - p^*)$ and recalling that q^* is defined by $(k - c)\bar{y} = V'(1 - q^*)$ we have $0 < p^* < q^* < 1$. This implies that $p_t < 1, \forall\, t$, so that from (10) $\omega_t = \omega^* = 0, \forall\, t$. From Section 2.1.2 since $p^* < q^*, q_t = p_t, \forall\, t, \lambda^* = (k - c)\bar{y} - V'(1 - p^*)$. Hence (14) becomes $\phi \leq V'(1 - p^*)$, which from the definition of p^* holds with strict equality.

(iii) $\phi \geq (k - c)\bar{y}$. In this case $p^* = 1$. Again there are two sub-cases. Firstly, $(k - c)\bar{y} \leq V'(1)$, in which case $q_t = 0, \forall\, t$, or $(k - c)\bar{y} > V'(1)$, in which case in steady state, $q_t = q^* < 1$. In either case, in steady state $q_t < p_t$, and so $\lambda_t = \lambda^* = 0$. From (10) we must have $\omega^* \geq 0$, and from (14) we must have $\phi = (k - c)\bar{y} + \omega^*$, so then $\omega^* = \phi - (k - c)\bar{y} \geq 0$.

These results are summarised in:

RESULT 3. There are three possible steady states for the proportion of crops p^* for which GM technology will be developed.

(i) If $\phi \leq \min[(k - c)\bar{y}, V'(1)]$, $p^* = 0$.
(ii) If $V'(1) < \phi < (k - c)\bar{y}$, p^* is defined by $\phi = V'(1 - p^*)$ and $p^* < q^*$.
(iii) If $\phi \geq (k - c)\bar{y}$, $p^* = 1$.

The three different levels of returns to R&D relative to V' and $(k - c)\bar{y}$ are illustrated in Figure 2. The interpretation of this result is straightforward. If the marginal private return is less than the minimum social cost of doing GM R&D, defined as the minimum of the subsidy required to ensure the survival of species or the marginal cost of the loss of just one species, then it is never worth doing GM R&D. On the other hand, if the marginal private returns to R&D are higher than the cost of subsidising the growing of non-GM varieties, then it is worth undertaking GM R&D so that all crops have GM varieties.

Finally, if the marginal private return is less than the cost of the subsidy to grow non-GM varieties, but higher than the marginal social cost of losing a species, then it will be desirable to do GM R&D on some fraction of crops, and for those crops grow only GM varieties, where the steady-state proportion is determined by the rule that the private marginal return to the last unit of GM R&D just equals the marginal social cost of the last species lost.

2.1.4. *Approach to Steady State and Policy Implications*

The approach to steady state in terms of p_t and q_t is straightforward to describe in the three cases, identified in Result 3.

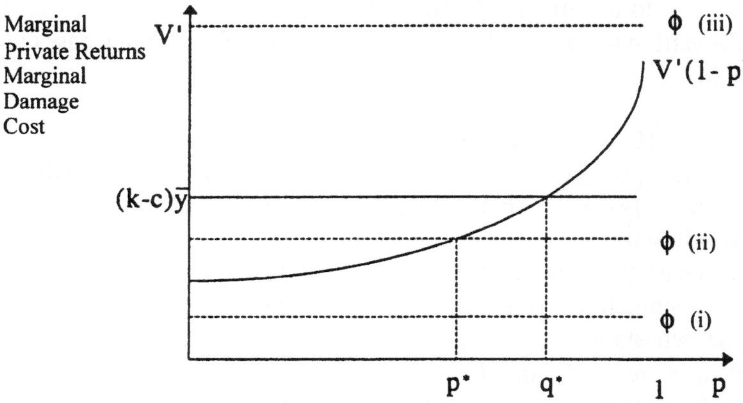

Figure 2. Comparison of R&D returns to: non-GM subsidy and social cost of GM crops.

(i) In this case, $p_t = q_t = 0, \forall\, t$.
(ii) In this case $q_t = p_t$, $\forall\, t$ and p_t rises as described below until it reaches $p^* < q^*$.
(iii) In this case there are two phases; in the first phase $q_t = p_t$ and both rise at a rate as described below until $p_t = q_t = q^*$; in the second phase, p_t continues to rise until $p_t = 1$, but q_t remains constant at q^*.

The way that p_t evolves in cases (ii) and (iii) is described by $p_0 = 0$, $\dot{p}_t = g_t$ where, integrating (9) we get

$$\gamma'(g_t) = \pi_t = \int_t^\infty \mathrm{e}^{-\delta(\tau-t)}[B(c) - B(k) - (k-c)\bar{y} + \lambda_\tau - \omega_\tau]\,\mathrm{d}\tau. \tag{15}$$

Equation (15) is just the condition that ensures that R&D is carried out until the marginal cost of doing R&D equals the present value marginal gross social return from R&D. The instantaneous marginal gross social return from GM R&D can be written as $B(c) - B(k) - \theta_\tau - \omega_\tau$ and $\theta_\tau = (k-c)\bar{y} - \lambda_\tau$. $B(c) - B(k)$ is the marginal gross private return to GM R&D, and represents the gain to consumer surplus from a reduction in the cost of growing crops. We follow Sianesi and Ulph (1999) in assuming that this private return can be fully captured by those engaged in GM R&D. Of course there are standard reasons why that is unlikely to be the case, but the arguments are not specific to GM R&D, and so we ignore them for the purpose of this chapter.

What we are concerned with are the variables θ_τ and ω_τ which are the policy variables designed to bring the private rate of return on GM R&D in line with the social rate of return. We shall interpret θ_τ as a tax that needs to be levied on the return from GM R&D to reflect the social cost of GM R&D. We interpret ω_τ as a 'licence fee' to undertake GM R&D on crops. From our previous analysis its straightforward to see what value θ_τ and ω_τ must take for the three cases identified in Result 3.

(i) $\theta_\tau = \min[(k-c)\bar{y}, V'(1)], \omega_\tau = 0, \forall\, \tau$
(ii) In this case, we are always in the region where $p_\tau = q_\tau < 1$, so that $\omega_\tau = 0$, $\lambda_\tau = (k-c)\bar{y} - V'(1-p_\tau)$, so that $\forall\, \tau, \theta_\tau = V'(1-p_\tau)$, with $\theta^* = V'(1-p^*)$ in steady state.
(iii) In this case there will be three phases. The first phase corresponds to (ii), so that throughout this phase $p_\tau = q_\tau < 1$, and $\omega_\tau = 0, \theta_\tau = V'(1-p_\tau)$. This phase persists until $p_\tau = q^*$. In the second phase, $q_\tau = q^* < p_\tau < 1$, so that $\lambda_\tau = \omega_\tau = 0$, and $\theta_\tau = (k-c)\bar{y}$. Finally, $p_\tau = p^* = 1$, and in this case $\lambda_\tau = 0$, $\omega_\tau = B(c) - B(k) - (k-c)\bar{y}$, and $\theta_\tau = (k-c)\bar{y}$. Clearly, $\theta_\tau + \omega_\tau = B(c) - B(k)$. ω_τ can be thought of as a 'licence fee' which is used to signal to biotech firms that the limit $p_\tau = 1$ has been reached and there are no more crops left to modify, by effectively taxing away any remaining surplus to doing R&D.

We summarise the discussion in the following:

RESULT 4. To secure the optimal investment in R&D the social planner imposes a licence fee to develop crops for GM R&D, ω_τ, and a tax to reflect the social cost of GM R&D, θ_τ. θ_τ and ω_τ are determined as follows from each of the three cases in Result 3.

(i) $\theta_\tau = \min[(k-c)\bar{y}, V'(1)]; \omega_\tau = 0.$
(ii) $\theta_\tau = V'(1-p_\tau); \omega_\tau = 0.$
(iii) Phase 1: $(0 \le p_\tau \le q^*); \theta_\tau = V'(1-p_\tau); \omega_\tau = 0.$
Phase 2: $(q^* < p_\tau < 1); \theta_\tau = (k-c)\bar{y}; \omega_\tau = 0.$
Phase 3: $(p_\tau = 1); \theta_\tau = (k-c)\bar{y}; \omega_\tau = B(c) - B(k) - (k-c)\bar{y}.$

Figure 3 illustrates the optimal path of R&D. As stated above, between zero and q^*, $p_t = q_t$ both of which continue to rise until q^* is reached, at which point q_t stops rising and remains equal to q^*. p_t keeps rising until it reaches 1.

In terms of policy outcomes, the key point to note is that unlike Sianesi and Ulph (1999), it will now be necessary to intervene in the market for R&D to reflect the social costs of GM R&D. These social costs arise to the extent that the availability of GM technology translates into crops grown as GM-only. The optimal proportion of crops grown as GM-only is determined by the relative costs on the environment in terms of reduced species numbers and the subsidy which has to be paid to sustain the threshold level of non-GM varieties of crops, but the current value of q_t depends on the availability of GM technology. A single instrument is now no

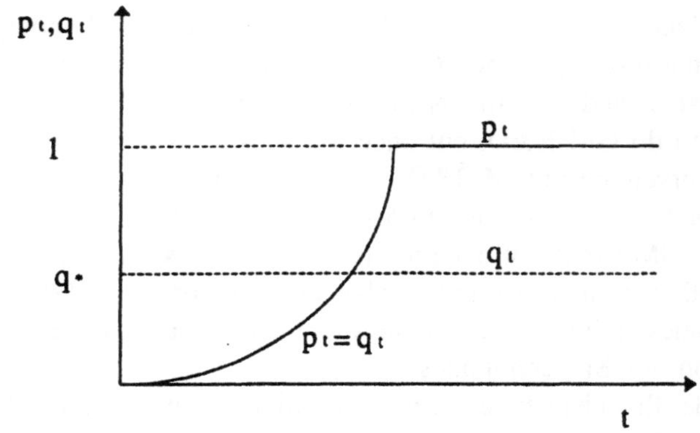

Figure 3. Dynamic path of R&D.

longer sufficient to ensure both the optimal levels of GM-only crops and R&D. In addition to the subsidy granted to the cultivation of non-GM crops, the returns to R&D should be taxed. Three scenarios depending on the size of R&D returns are outlined. Obviously, the socially optimal level of R&D is zero when its marginal net private returns fall short of its marginal social cost. To ensure this outcome a tax is imposed. In this case there is no need to subsidise non-GM crops since the critical level of non-GM crops is not binding. Once returns exceed the marginal loss of species but are lower than the subsidy required to ensure an appropriate amount of non-GM it is optimal to tax R&D at the rate of the marginal loss of species. The tax is increased up to the point that the availability of GM technology reaches the optimal proportion of GM-only crops. At this point no further crops are grown as GM-only and the tax on R&D returns is capped at the level of the subsidy required to maintained the threshold level of non-GM variety of crops. In the final phase where the last crop is modified an additional charge in addition to the R&D tax is imposed. The purpose of this additional charge or 'licence fee' is to choke off any further demand for GM R&D.

3. Conclusion

In this chapter we have attempted to illuminate the debate on whether GM crops should be grown and if so, in what quantity. We have also examined the issue of whether intervention in the GM R&D market is warranted. Unlike Sianesi and Ulph (1999) we found that intervention was indeed necessary. The reason for this difference in policy conclusion stems from the different assumption about how species diversity is related to the growing of GM and non-GM varieties of crops. In Sianesi and Ulph (1999) species diversity is related only to the total availability of non-GM varieties of crops, where, in terms of environmental impact, the non-

GM variety of one crop is a perfect substitute for the non-GM variety of another. A subsidy to the growing of non-GM crops is necessary to address the environmental externality, and this affects the profitability of GM crops and indirectly the incentive to do GM R&D, but once the subsidy is set there is no need for any further intervention in GM R&D. By contrast, we assume that the number of species is directly related to the number of non-GM crops grown at some initial level, so a non-GM variety of one crop is not a perfect substitute for the non-GM variety of another. In this case the number of crops for which GM technology is available becomes an important variable and it is necessary to intervene directly in the introduction of GM technologies.

To conclude, this chapter represents an initial step in analysing the impact of GM on the environment. The environmental aspect focused upon in this chapter is biodiversity. We have chosen an extremely simple model which has produced some policy conclusions. There is no doubt that the model of species reliance on resources does not capture all the nuances in the relationship between species and their food sources. However, from our point of view, the pertinent question is whether introducing a greater degree of complexity in this relationship adds anything in terms of policy. We are not sure that it does. In addition, in an attempt to concentrate on the potential impact of GM on species diversity we have omitted the possibility of intertemporal spillovers in the R&D process. However, there appears to be alot of similarity in the genetic make-up of species. This similarity referred to as synteny enables specific useful genes identified in a well-studied species to be localised to the corresponding genomic site in other species that have not been studied in the same detail (de Vicente and Hodgkin, 2000). Thus, as the sequence of incorporating the insect-resistant gene in successive crops proceeds the process becomes progressively easier. At first glance it would appear that returns to R&D may increase and the steady-state value of p_t will move closer to 1. Such R&D spillovers may indicate a need to intervene in the R&D market for GM. However, the existence of such spillovers are not unique to GM R&D and we argue that their inclusion does not detract from the basic message of this chapter – the need for intervention in the R&D market in GM technologies when biodiversity depends on a variety of non-GM crops rather than on a large quantity of a single non-GM crop.

Notes

[1] An example of another measure of biodiversity is distance or degree of dissimilarity between two species (Weitzman, 1992). Instead of focusing on species, measures could also relate to different levels of aggregation such as habitat, ecosystem, subspecies or gene as well as others. Choice of the level of aggregation will depend on the focus of concern.

[2] Environmental economists have been concerned with issues of regulating pollution or possible measures to encourage R&D to develop green technologies but has not addressed the issue of the

balance of policy between regulating a new polluting technology once that has been developed or slowing down the development of a new polluting technology.

[3] Of course this is a strong assumption, and we make it on the basis that even if GM R&D involves growing some GM crops (perhaps as trials) this is unlikely to be on such a scale to be a real threat to the environment, and that the real threat to the environment comes only if full-scale growing of GM crops takes place.

[4] As we will see, this is based on the assumption that there are none of the usual failures in the R&D market. Obviously, these would call for intervention, but these arguments are not specific to GM R&D.

[5] Again this is a strong assumption designed to focus on the central issue of this paper. Effectively we are saying that consumers have no reason to dislike GM crops (e.g. for possible health reason) and their concern about the environmental effect of GM crops is captured in a separable fashion through the value of species diversity function V.

[6] The reason for such an assumption is to omit externalities bar the one we are interested in which is the reduction in species diversity caused by growing GM crops.

References

de Vicente, C. and Hodgkin, T. (2000) Key issues in molecular techniques to improve conservation and use of plant genetic resources, Paper presented at the Biotechnology Environmental Policy and Agriculture Workshop, Rome, May 2000.

Goeschl, T. and Swanson, T. (1999) Endogenous growth and biodiversity: The social value of genetic resources for R&D, CSERGE GEC99-11.

Reiss, M.J. and Straughan, R. (1996) *Improving Nature? The Science and Ethics of Genetic Engineering*, Cambridge University Press, Cambridge.

Sianesi, B. and Ulph, D. (1999) Species loss through the genetic modification of crops – A policy framework, unpublished.

Weitzman, M.L. (1992) On diversity, *Quart. J. Econ.* **107**, 363–405.

6. Decision Making under Temporal Uncertainty and Irreversibility: Benchmark Values for the Release of Transgenic Crops in the EU

JUSTUS WESSELER

1. Introduction

The scientific revolution in the biological sciences with its rapid advances in molecular biology offers great potentials for productivity gains in agriculture. Food crops that have higher yields and better nutrition content, plants that are resistant to drought and pests, livestock that are immune to disease, and fisheries that are sustainable, are possible developments which can result from the application of biotechnology (Krimsky and Wrubel, 1996).

However, opposing the expected gains, there are risks related specifically to the widespread use of transgenic crops. Gene flow in plants can enable domesticated plants to become pernicious weeds, or enhance the fitness of wild plants which might be serious weeds, thus shifting the ecological balance in a natural plant community. New viruses could develop from virus-containing transgenic crops. Plant-produced insecticides might have harmful effects on unintended targets. While some of these scenarios are highly unlikely, little is known about the overall impact that transgenic crops can have on biodiversity, ecosystem balance and the environment (Kendall et al., 1997).

Proponents of genetic engineering press for the rapid release of transgenic crops while opponents either reject the use of transgenic crops in general or want to postpone their release until further information on the related risks is available. An immediate release of a transgenic crop will provide immediate and future benefits through the expected positive effects on yields, product quality, production costs, and/or other characteristics of the crop (e.g. Nelson et al., 1999).[1] On the other hand, an immediate release will expose society to potential environmental risk. Therefore, a decision to delay or reject a release delays or avoids those risks, but also the benefits of an immediate release. Any such decision includes, implicitly or explicitly, a comparison of costs and benefits. Even a decision which is based on the assumption that the risk cannot be estimated and therefore transgenic crops should not be released implicitly assumes that the expected risks are higher than the expected benefits. As decisions have to be made, most developed countries

Timothy Swanson (ed.), The Economics of Managing Biotechnologies, 145–156.

have established regulating agencies which approve the release of transgenic crops. The problem these agencies face is that if they decide to release the new crop and discover later that the transgenic crop has a negative impact on health and/or the environment, they may be able to prevent consumption and thus to reduce the impact on health, but they cannot retrieve the genes released into the environment. They may regret that they have allowed the release of the transgenic crop and did not wait until further information on the impact of this transgenic crop on health and the environment was available. On the other hand, every delay in release is a loss in immediate expected benefits. Therefore, the agency has not only to weigh the benefits of an immediate release against the expected risk but also against the option to delay the decision into the future.

This decision-making problem can be described as one under temporal uncertainty and irreversibility (Sianesi and Ulph, 1998; Wesseler and Weichert, 1998). Temporal uncertainty exists because future prices, yields and other benefits as well as environmental risks of transgenic crops are uncertain; irreversibility exists as once transgenic crops are released, their genetic information cannot be gathered again. Therefore, the value of the option to delay the release of transgenic crops is most likely positive and, hence, the benefits have to exceed the costs by a certain factor (Arrow and Fisher, 1974; Henry, 1974; Dixit and Pindyck, 1994).

A simple numerical example will be used to illustrate the approach: a company has developed a herbicide-tolerant crop. The immediate increase in gross margin, π_0, through the introduction of the new variety is about US$ 1 Mio. For the next year, t_1, it is expected that the additional gross margin will change. Two situations are possible. Either the additional gross margin, π_{1H}, will increase to US$ 1.5 Mio., if more farmers adopt the technology and other factors are also in favor of the technology or, if the factors are less favorable, the additional gross margin, π_{1L}, decreases to US$ 0.5 Mio. To keep it simple, the additional gross margin remains the same from t_1 onwards. The probabilities of both scenarios (favorable/non-favorable) are assumed to be the same, $p = 1 - q = 0.5$. Further, it will be assumed that the gross margins are independent of the development of the economy and the riskless rate of return of 10% can be used as the discount rate. The release of the new transgenic crop into the environment is assumed to have a negative impact on bio-diversity. A contingent valuation study shall have revealed that the negative impact will be in the order of US$ 8 Mio. There are no additional costs and benefits that have to be considered. In this case the net-present-value of the immediate release (NPV_I) of the transgenic crop into the environment would provide the following solution:

$$\begin{aligned} \text{NPV}_I &= -I + \pi_0 + \frac{(p \cdot \pi_{1H})}{r} + \frac{((1-p) \cdot \pi_{1L})}{r} \\ &= -8 + 1 + 7.5 + 2.5 = 3.0. \end{aligned} \tag{1}$$

The NPV_I of Equation (1) is positive and hence, under the traditional cost-benefit approach, the new variety should be released into the environment. However, this evaluation did not consider the possibility of delaying the release into the environment until more information becomes available on factors that may have an impact on the result. If the release was postponed by one year and it is assumed that the new state will be known with certainty, it is obvious that the transgenic crops should not be planted if the additional gross margin decreases to US$ 0.5 Mio. On the other hand, if the other state occurs, the NPV will be the following:

$$\mathrm{NPV}_D = 0.5\left(-\frac{I}{1+r}+\sum_{t=1}^{\infty}\frac{\pi_{1H}}{(1+r)^t}\right)=\frac{4.25}{1.1}=3.86. \tag{2}$$

The result of Equation (2) shows that in the case of delaying the decision, a higher NPV would be achieved, $\mathrm{NPV}_D > \mathrm{NPV}_I$, and hence this strategy should be adopted. Ignoring the option to delay the release into the environment would result in a sub-optimal decision under this scenario.

In the following, the model to describe the comparison of costs and benefits under uncertainty and irreversibility which has been presented earlier (Wesseler, 1998) will be reviewed. The results of the model describe the factor by which the benefits have to exceed the costs. Based on different scenarios for the European Union benchmark values for the irreversible costs will be derived, that can be used as a guideline for decision making.

2. The Model

The following assumptions were made about social benefits and costs. A hypothetical agency has to decide on the release of transgenic crops. The agencies' decision are based only on the benefits and costs of the release as explained below. Hence, the political economy of the decision-making process is not considered in the model.

The agency considers as social benefits V only the additional benefits that result from the use of transgenic crops compared to non-transgenic crops (in the following called conventional crops) and as social costs I only the additional irreversible costs related to the release of transgenic crops. Strategic costs and benefits of the company requesting the release of the transgenic crop are ignored. Further, the agency considers only domestic costs and benefits. Across-border effects are ignored.[2]

The additional social benefits of transgenic crops as compared to conventional crops are assumed to originate from changes in yields, prices and/or variable production costs under the assumption of perfect elasticity of demand and perfect non-elastic supply. Overhead costs are assumed to be the same for transgenic and conventional crops. Therefore, the additional benefits can be described by the difference in gross margin between transgenic and conventional crops. Positive environmental effects of transgenic crops and possible health effects due to the consumption of transgenic crops are assumed to be reflected in yields, prices and variable production costs. If, for example, soil erosion is reduced due to the practice of zero tillage in combination with a herbicide like *Round-up®* and a *Round-up®*-resistant crop, positive on-site effects would result in a higher yield of the crop and/or less use of fertilizer. Also, possible health effects of transgenic plants are assumed to result in price adjustments.

Additional welfare benefits arising from the application of the new technology through "peace of mind" (Monsanto, 1998, p. 4) are assumed to be balanced by concerns about the new technology and are therefore ignored.[3]

In summary, all benefits resulting from the release of transgenic crops are assumed to be internalized.

The irreversible costs of the release of transgenic crops are assumed to be the loss in biodiversity (Mooney and Bernardi, 1990; ACRE, 1997; Tiedje et al., 1989). Those costs are not known with certainty and it is also unknown whether an increase in information will result in an increase or decrease of the expected value of those costs. The contribution by Wesseler and Weichert (1998) analyses the situation where benefits and irreversible costs both follow a Brown–Wiener stochastic process. This allows to identify the ratio by which the benefits have to exceed the irreversible costs. An application of the approach not only requires information about the specific stochastic process of the irreversible costs but also about the parameter values. Also, the justification of contingent claim analysis for loss in biodiversity will depend on the identification of a relevant spanning asset. The alternative use of dynamic programming leads to the problem of identifying the correct discount rate (Wesseler and Weichert, 1998). Therefore, here it is assumed that the irreversible costs I are known with certainty at the time when the decision is made.

Bearing in mind the assumptions described above, the objective of the regulatory agency can simply be described as maximizing the value $F(V)$ of the decision to release transgenic crops under a given level of information:

$$F(V) = \max E[V_T - I)\,\mathrm{e}^{-\mu T} \mid Q], \tag{3}$$

with E the expectation operator, V_T the present value of the incremental benefits at the time of release T, I the irreversible costs, μ the discount rate and Q the level of information.

As the benefits are uncertain, it will be assumed that V follows a stochastic process. As there are different views about the benefits of transgenic crops, two main views will be modeled using two different continuous time stochastic processes. One assumption is the benefits V follow a stochastic process with a positive trend using a geometric Brownian motion. More specifically:

$$dV = \alpha V\,dt + \sigma V\,dz, \tag{4}$$

where α is the trend variable, σ is the standard deviation and dz is a Brown–Wiener process. It has been shown elsewhere (McDonald and Siegel, 1986) that by assuming $F(V) = AV^{\beta}$ the optimal value of V, will be of the form:

$$V^* = \frac{\beta}{\beta - 1} \cdot I, \tag{5}$$

with

$$\beta = \frac{1}{2} - \frac{r-\delta}{\sigma^2} + \sqrt{\left[\frac{r-\delta}{\sigma^2} - \frac{1}{2}\right]^2 + 2r/\sigma^2} > 1, \quad \beta > 1,$$

where r is the risk-free interest rate, and δ the difference between the discount rate μ, which is the risk adjusted market rate of return and the trend α.

If the irreversible costs I are set $I = 1$, Equation (3) shows that the benefits from the release of transgenic crops have to be higher by the factor $\beta/(\beta - 1) > 1$ to justify an immediate release from the economic point of view, whereas the traditional cost-benefit analysis would suggest an immediate release if $V^* \geq I$ (Abel et al., 1996).

A continuous increase in benefits through transgenic crops is not necessarily a valid assumption. Critics argue that benefits, if at all, will be only available for a short period of time. Weeds and pests become resistant to the herbicides and crop-produced pesticides and this much faster than previously expected (Bergelson et al., 1998; Huang et al., 1999). This view about transgenic crops can be modeled by assuming a mean-reverting process with respect to benefits, where initial additional benefits V from transgenic crops decrease over time until they become zero:

$$dV = \eta(\bar{V} - V)V\,dt + \sigma V\,dz, \tag{6}$$

where η is the speed of mean-reversion, $\bar{V}$ the value to which V tends to return, in the following set to zero assuming no additional benefits after some years, and V is the value of the initial additional benefits through the introduction of transgenic crops.

An approach to find the optimal hurdle is provided by Dixit and Pindyck (1994, pp. 161–167). Defining the option function $F(V)$ as:

$$F(V) = AV^{\theta} H\left(\frac{2\eta}{\sigma^2}V; \theta, b\right) \tag{7}$$

with

$$\theta = \frac{1}{2} + (\mu - r - \eta\bar{V})/\sigma^2 + \sqrt{\left[(r - \mu + \eta\bar{V}/\sigma^2 - \frac{1}{2}\right]^2 + 2r/\sigma^2},$$

Table 1. Hurdle rate V^* for given parameter values.

Parameter	Geometric Brownian motion	Mean-reverting process
discount rate, μ	0.08	0.08
risk-free RoR, r	0.04	0.04
standard deviation σ	0.20	0.20
trend α	0.04	
mean-reverting, η		0.76
hurdle rate V^*	*2.00*	*1.07*

where A is a constant and $H(\ldots)$ a hypergeometric function. Analytical solutions for V^* do not exist but can be found numerically.

If reasonable parameter values are used the numerical value of the hurdle rate can be identified. Following common practice, the discount rate μ is assumed to be 8%, the risk-free rate of return r to be about 4% and the standard deviation σ to be 20% (Dixit and Pindyck, 1994). The average growth rate α is expected to be 4%, a rather low value, whereas the immediate benefits from transgenic crops are assumed to be in the order of 20%. Further, it is assumed that the speed of mean-reversion will be approximately 7 years.

Using these *guesstimates* provides the results shown in Table 1. The critical value V^*, the factor by which the benefits have to exceed the irreversible costs, is in the order of two under a geometric Brownian motion. The benefits V^* have to be two times the irreversible costs to justify an immediate release of transgenic crops. Surprisingly, the hurdle rate assuming a mean-reverting process is much lower. The benefits only have to exceed the irreversible costs by a factor of 1.07.[4] Tables A1 and A2 of the Appendix show the different hurdle rates for a range of parameter values. The values can be translated into monetary terms calculating the expected benefits from transgenic crops for a country or region like the European Union.

The present value of benefits $\bar{V}_0$ can be described as a function of adoption rate, λ, land allocated to crops where transgenic varieties exist, L, increase in gross margin per ha, dGM, and discount rate, μ:

$$\bar{V}_0 = \frac{\lambda \cdot \mathrm{dGM} \cdot L}{\mu}. \tag{8}$$

In the EU cotton, maize, potatoes, rape, soybeans, and sugar beet are grown on almost 15.5% of the arable land or almost 12 Mio. ha (FAO, 2000). Assuming adoption rates of 25, 50, and 100% and an increases in the gross margin per ha of

Table 2. Range of possible hurdle rates and maximal tolerable irreversible costs of first generation transgenic crops (Mio. Euro).*

V^*	Maximal tolerable irreversible costs (Mio. Euro)				
	Scenarios				
	a	b	c	d	e
1.03	14018.20	7009.10	3504.55	1752.28	876.14
1.06	13748.18	6874.09	3437.05	1718.52	859.26
1.10	13211.67	6605.83	3302.92	1651.46	825.73
1.16	12528.50	6264.25	3132.12	1566.06	783.03
1.20	12052.30	6026.15	3013.07	1506.54	753.27
1.28	11312.27	5656.13	2828.07	1414.03	707.02
1.33	10880.25	5440.13	2720.06	1360.03	680.02
1.42	10205.16	5102.58	2551.29	1275.65	637.82
1.47	9883.19	4941.60	2470.80	1235.40	617.70
1.51	9593.32	4796.66	2398.33	1199.17	599.58
1.62	8976.17	4488.09	2244.04	1122.02	561.01
1.74	8335.86	4167.93	2083.96	1041.98	520.99
1.81	8022.91	4011.46	2005.73	1002.86	501.43
1.90	7622.41	3811.20	1905.60	952.80	476.40
2.00	7253.50	3626.75	1813.38	906.69	453.34
2.19	6632.85	3316.43	1658.21	829.11	414.55
2.46	5903.33	2951.66	1475.83	737.92	368.96
2.81	5165.43	2582.71	1291.36	645.68	322.84
3.41	4249.00	2124.50	1062.25	531.13	265.56
3.61	4015.09	2007.55	1003.77	501.89	250.94
3.73	3887.14	1943.57	971.78	485.89	242.95
4.60	3153.42	1576.71	788.36	394.18	197.09
6.05	2398.49	1199.25	599.62	299.81	149.91
6.90	2101.42	1050.71	525.35	262.68	131.34
9.90	1465.50	732.75	366.38	183.19	91.59
11.81	1228.50	614.25	307.12	153.56	76.78
13.62	1065.30	532.65	266.33	133.16	66.58
18.89	767.80	383.90	191.95	95.98	47.99
19.95	727.17	363.59	181.79	90.90	45.45
38.95	372.46	186.23	93.12	46.56	23.28

*Crops include cotton, maize, potato, rape, soybean, and sugar beet. Source: own calculations.

about 25, 50 and 100 Euro and a discount rate of 8% results in the values shown in Table A3. The results can be summarized as followed:

scenario a: $(\lambda = 1 \wedge \text{dGM} = 100\ \text{ECU})$,

scenario b: $(\lambda = 1 \wedge \text{dGM} = 50\ \text{ECU}) \vee (\lambda = 0.5 \wedge \text{dGM} = 100\ \text{ECU})$

scenario c: $(\lambda = 1.00 \wedge \text{dGM} = 25\ \text{ECU}) \vee (\lambda = 0.50 \wedge \text{dGM} = 50\ \text{ECU}) \vee (\lambda = 0.25 \wedge \text{dGM} = 100\ \text{ECU})$,

scenario d: $(\lambda = 0.50 \wedge \text{dGM} = 25\ \text{ECU}) \vee (\lambda = 0.25 \wedge \text{dGM} = 50\ \text{ECU})$

scenario e: $(\lambda = 0.25 \wedge \text{dGM} = 25\ \text{ECU})$.

If the monetary values are divided by the hurdle rate V^*, the critical values for the irreversible costs, I^*, the costs associated with a release of transgenic crops into the environment are obtained.

$$I^* = \frac{\bar{V}_0}{V^*}. \tag{9}$$

They range for the different scenarios from about 23.28 Mio. Euro on the lower side up to 14018.20 Mio. Euro on the upper side (Table 2). Converted on a ha basis the range is about 2 to 1208 Euro/ha. The extremely wide range indicates that the identification of the stochastic process and the parameter values are of utmost importance.

3. Conclusions

Temporary uncertainty and irreversibility are two important characteristics of the benefits and costs related to the release of transgenic crops into the environment. The economic literature on real option pricing theory has shown that under temporary uncertainty and irreversibility an additional value, the value of the option to delay the decision, has to be included as an additional cost into the traditional cost-benefit framework. Therefore, decisions on the release of transgenic crops that are based on the traditional cost-benefit framework may be wrong.

The two stochastic processes used to model the benefits of transgenic crops reveal important results. Under the assumption that benefits follow a geometric Brownian process, the benefits have to be much higher to justify an immediate release than under the assumption of a mean-reverting process. The difference in the results shows that it is not only important to include the option of delaying the release of transgenic crops into the cost-benefit analysis, but also that the result will depend to a large extent on the assumptions about the benefits from transgenic crops in the longer run.

The scenarios show that the range of the tolerable irreversible costs is extremely wide, ranging from 23.28 Mio. Euro up to 14018 Mio. Euro or 2 to 1208 Euro/ha.

The wide range indicates that is important not only to get information on the benefits side but also on the cost side to be able to improve the decision making about the release of transgenic crops. The results at the lower side also indicate that the release of transgenic crops not necessarily will result in net economic benefits.

Appendix

Table A1. Hurdle rates V^* for different parameter settings, geometric Brownian process.

trend	discount rate μ^a				standard deviation σ^b				
α	0.10	0.08	0.06	0.04	0.10	0.2	0.4	0.8	1.2
0.01	1.3333	1.4678	1.7403	2.4574	1.1429	1.4678	2.4843	6.0484	11.8088
0.02	1.3904	1.5774	2.0000	3.4142	1.1896	1.5774	2.7583	6.9034	13.6177
0.04	1.5774	2.0000	3.4142	****	1.4215	2.0000	3.7321	9.8990	19.9499
0.06	2.0000	3.4142	****	****	2.4254	3.4142	6.7016	18.8941	38.9487
0.08	3.4142	****	****	****					

[a]The standard deviation σ is set to 0.2 and the risk-free rate of return r to 0.04.
[b]The expected rate of return μ is set to 0.08 and the risk-free rate of return r to 0.04.
Source: own calculations.

Table A2. Hurdle rates V^* for different parameter settings, mean-reverting process.

Mean-reverting	discount rate μ^a				standard deviation μ^b				
speed η	0.10	0.08	0.06	0.04	0.10	0.2	0.4	0.8	1.2
0.05	1.1710	1.2037	1.2500	1.3173	1.1710	1.2037	1.2500	1.3173	4.6004
0.10	1.1211	1.1371	1.1579	1.1856	1.1211	1.1371	1.1579	1.1856	3.6131
0.20	1.0760	1.0822	1.0894	1.0980	1.0760	1.0822	1.0894	1.0980	2.8085
0.30	1.0552	1.0586	1.0620	1.0661	1.0552	1.0586	1.0620	1.0661	2.4233
0.40	1.0433	1.0453	1.0474	1.0498	1.0433	1.0453	1.0474	1.0498	2.1871
0.50	1.0363	1.0376	1.0390	1.0405	1.0363	1.0376	1.0390	1.0405	2.0235
0.60	1.0349	1.0360	1.0370	1.0380	1.0349	1.0360	1.0370	1.0380	1.9032
0.70	1.0452	1.0463	1.0474	1.9557	1.0452	1.0463	1.0474	1.9557	1.8082
0.80	1.1010	1.1184	1.5122	1.9944	1.1010	1.1184	1.5122	1.9944	1.7319
0.90	1.2540	1.3650	1.5691	1.9990	1.2540	1.3650	1.5691	1.9990	1.6690
1.00	1.2824	1.3852	1.5753	1.9997	1.2824	1.3852	1.5753	1.9997	1.6162

[a]The standard deviation σ is set to 0.2 and the risk-free rate of return r to 0.04.
[b]The expected rate of return μ is set to 0.08 and the risk-free rate of return r to 0.04.
Source: own calculations.

Table A3. Total benefits from transgenic crops under different changes in gross margins/ha and rates of adoption (Mio. Euro).

Country	Arable land ('000 ha)	Pot. TG crops* ('000 ha)	%	Increase 25 Euro/ha Adoption (%)			Increase 50 Euro/ha Adoption (%)			Increase 100 Euro/ha Adoption (%)		
				100	50	25	100	50	25	100	50	25
EU (15)	**75818**	**11606**	**15.31**	3626.75	1813.38	906.69	7253.50	3626.75	1813.38	14507.00	7253.50	3626.75
Austria	1397	333	23.87	104.21	52.10	26.05	208.42	104.21	52.10	416.84	208.42	104.21
Belgium-Luxembourg	768	166	21.58	51.78	25.89	12.95	103.56	51.78	25.89	207.12	103.56	51.78
Denmark	2365	212	8.96	66.25	33.13	16.56	132.50	66.25	33.13	265.00	132.50	66.25
Finland	2126	129	6.05	40.22	20.11	10.05	80.44	40.22	20.11	160.88	80.44	40.22
France	18305	3576	19.54	1117.50	558.75	279.38	2235.00	1117.50	558.75	4470.00	2235.00	1117.50
Germany	11832	2092	17.68	653.69	326.84	163.42	1307.38	653.69	326.84	2614.75	1307.38	653.69
Greece	2823	716	25.37	223.82	111.91	55.95	447.63	223.82	111.91	895.26	447.63	223.82
Ireland	1343	55	4.09	17.16	8.58	4.29	34.31	17.16	8.58	68.63	34.31	17.16
Italy	8283	1788	21.59	558.89	279.44	139.72	1117.77	558.89	279.44	2235.54	1117.77	558.89
Netherlands	900	303	33.67	94.69	47.34	23.67	189.38	94.69	47.34	378.75	189.38	94.69
Portugal	2153	271	12.57	84.56	42.28	21.14	169.12	84.56	42.28	338.23	169.12	84.56
Spain	14344	972	6.77	303.69	151.84	75.92	607.38	303.69	151.84	1214.75	607.38	303.69
Sweden	2799	158	5.65	49.41	24.70	12.35	98.81	49.41	24.70	197.63	98.81	49.41
United Kingdom	6380	835	13.09	260.91	130.45	65.23	521.81	260.91	130.45	1043.63	521.81	260.91

*Area harvested in the EU in 1997 (cotton, maize, potatoes, rape, soybean, sugar beet).
Source: FAO 2000 and own calculations.

Notes

[1] If there were no direct benefits, there would be no incentive for farmers to buy the seeds of transgenic crops.

[2] The last two conditions were included to keep the model simple. Had they been omitted, the analysis would have been complicated by the need to allocate cost and benefits correctly, as the benefits and costs of a multinational company are not necessarily equivalent to those at the domestic market.

[3] Monsanto (1999) cites as one positive benefit from transgenic crops the positive mental effect on users, because of the positive impact of transgenic crops on the environment. They call this kind of benefits "peace of mind".

[4] This observation can be explained by the fact that under increasing stochastic benefits a later release reduces the risk of negative net benefits because of the positive trend, whereas the mean-reverting process has no positive trend effect to counterbalance downside risk.

References

Abel, A.B., Dixit, A.K., Eberly, J.C. and Pindyck, R.S. (1996) Options, the value of capital, and investment, *Quart. J. Econ.* **61**, 753–777.

ACRE (Advisory Committee on Release to the Environment) (1997) Insect resistance genes, *ACRE Newsletter*, June 8.

Arrow, K. and Fisher, A.C. (1974) Environmental preservation, uncertainty, and irreversibility, *Quart. J. Econ.* **88**, 312–319.

Bergelson, J., Purrington, C.B. and Wichmann, G. (1998) Promiscuity in transgenic plants, *Nature* **395**, 25.

Crawley, M.J. (1991) Die Ökologie genetisch veränderter Organismen: Eine Einschätzung der Umweltrisiken, Arbeitsmaterialien zur Technikfolgen-Abschätzung und -Bewertung der modernen Biotechnologie Nr. 2, University of Hamburg.

Dixit, A.K. and Pindyck, R.S. (1994) *Investment under Uncertainty*, Princeton University Press, Princeton, NJ.

Dixit, A.K. and Pindyck, R.S. (1995) The new option view of investment, MIT working paper WP#3794-95-EFA.

FAO (2000) Statistical Database http://apps.fao.org/cgi-bin/nph-db.pl.

Huang, F., Buschman, L.L., Higgins, R.A. and McGaughey, W.H. (1999) Inheritance of resistance to bacillus thuringensis toxin (Dipel ES) in the European corn borer, *Science* **284**, 965–967.

Henry, C. (1974) Investment decisions under uncertainty: The irreversibility effect, *Amer. Econ. Rev.* **64**, 1006–1012.

Kendall, H.W., Beachy, R., Eisner, T., Gould, F., Herdt, R., Raven, P.H., Schell, J.S. and Swaminathan, M.S. (1997) *Bioengineering of Crops*, Environmental and Socially Sustainable Development Studies and Monograph Series 23, The World Bank Press, Washington, DC.

Krimsky, S. and Wrubel, R. (1996) *Agricultural Biotechnology and the Environment: Science, Policy, and Social Issues*, University of Illinois Press, Urbana, IL.

McDonald, R. and Siegel, D. (1986) The value of waiting to invest, *Quart. J. Econ.* **101**, 707–728.

Mooney, H.A. and Bernardi, G. (eds.) (1990) *Introduction of Genetically Modified Organisms into the Environment*, Wiley, Chichester.

Monsanto Company (1999) Bollgard® and Roundup Ready® Cotton Performance – 1996–1997, Online posting, 28 April 1999, http://www.monsanto.com/ag/articles/.bollrr.htm.

Nelson, G.C., Josling, T., Bullock, D., Unnevehr, L., Rosegrant, M. and Hill, L. (1999) The economics and politics of genetically modified organisms in agriculture: Implications for WTO 2000, *Bulletin 809*, College of Agricultural, Consumer and Environmental Sciences. University of Illinois at Urbana-Champaign, IL.

Pindyck, R.S. (1998) Irreversibility and the timing of environmental policy, MIT working paper WP#4047-98-EFA.

Tiedje, J.M., Colwell, R.K., Grossman, Y.L., Hodson, R.E., Lenski, R.E., Mack, R.N., Regal, P.J. (1989) The planned introduction of genetically engineered organisms: Ecological considerations and recommendations, *Ecology* **70**, 298–315.

Sianesi, B. and Ulph, D. (1998) *Species Loss Through the Genetic Modification of Crops – A Policy Framework*, Proceedings of First World Congress of Resource and Environmental Economists, Venice, http://www.feem.it/gnee/libr.html.

Wesseler, J. (1998) Temporal uncertainty and irreversibility – A theoretical framework for the decision to approve the release of transgenic crops, Paper presented at the International Conference on 'Transitions in Agbiotech: Economics of Strategy and Policy', Washington, DC, June 24–25.

Wesseler, J. and Weichert, M. (1998) Der Wert zusätzlicher Informationen bei Investitionsentscheidungen mit einem hohen Grad an Irreversibilität, Proceedings 39. Jahrestagung der Gesellschaft für Wirtschafts- und Sozialwissenschaften des Landbaus, Bonn.

PART C

MANAGING BIOLOGY:
THE ECONOMICS OF MANAGING BIOLOGICAL RESISTANCE

7. Managing Pest Resistance: Timing the Initiation of Refuge Areas

RAMANAN LAXMINARAYAN and R. DAVID SIMPSON

1. Introduction

The application of any pesticide will exert evolutionary pressure in favor of organisms resistant to its toxin. This fact has been recognized for many years, and efforts have been made to understand the optimal use of chemical pesticides so as to efficiently manage the evolution of resistant pests. Renewed interest has arisen in resistance management in recent years, however, with the development of genetically modified (GM) crops. A gene from the bacterium *Bacillus thuringiensis* – frequently abbreviated as 'Bt' – has been inserted in cotton, tobacco, corn and soybean varieties. This gene codes for the production of a protein highly toxic to many insect pests. Unlike chemical pesticides, which, after being sprayed on crops, may be blown or washed off, or carried away by the pests unlucky enough to encounter it first, the pesticide is always present in these 'Bt varieties'. Thus the strategy of managing resistance by optimally timing application of pesticides is no longer possible with the new technology.

The potential for the development of pest resistance to GM crops is a deterrent to their sustainable use, and has been recognized by farmers, seed companies and government regulators. It has been suggested that setting aside refuge areas adjacent to GM cultivars may ensure that the pest-resistant property of GM crops is sustained in the long run (EPA, 1998, Mallet and Porter, 1992; Tabashnik, 1994). Since 1999 the United States Environmental Protection Agency (EPA) has required that farmers planting GM crops incorporating Bt plant refuge areas. This is the first instance in which refuge areas have been required by regulation in the United States (Livingstone et al., 2000).

In response to concerns regarding the potential growth of pest resistance to GM crops, the EPA has required that refuge areas be grown in conjunction with Bollgard cotton, a Bt cotton strain that is effective against the bollworm (EPA, 1998). Farmers have been offered two refuge strategies to choose from.[1] Under the first option, the size of the refuge area (planted with non-transgenic seeds) is roughly 20% of all fields in which transgenic cultivars are grown. Under this option, the refuge area may be sprayed with insecticides, so long as the spray does not contain the same *Bacillus thuringiensis* protein that is expressed by the transgenic plants.

Timothy Swanson (ed.), The Economics of Managing Biotechnologies, 159–169.

Under the second option a refuge area roughly 4% of the entire cultivated area must be planted. This option does not permit the spraying of insecticides in the refuge area.

Some studies have indicated that the refuge areas required by current EPA mandates may not be optimal from a biological standpoint (EPA, 1998; Tabashnik, 1997). Tabashnik reported that a 4% unsprayed refuge area may not be large enough to produce enough susceptible insects to reduce the overall pressure on susceptibility (Tabashnik, 1997). Another study that examined the economic tradeoffs between current yields (that are typically diminished by refuge areas) and future effectiveness of GM crops against pests found that a 20 to 40% unsprayed refuge area was the most optimal from a social standpoint (Hurley et al., 1997).[2] Seed companies, such as Monsanto, have argued that these recommendations are infeasible from the perspective of individual growers and that mandating larger refuge areas than are currently prevalent will likely result in greater non-compliance with EPA's refuge requirements, and thereby increase overall resistance compared to the status quo (Head, 1999).

In this paper we argue that the optimal *initial* refuge area size may well be zero. It will often make little sense to initiate resistance management until resistance becomes an important issue. In other words, all cropland should be devoted to the new GM variety until a threshold condition is reached. Our result might be seen as somewhat analogous to results from the economic literature on fisheries. Just as it is generally the case that a policy of 'maximum sustained yield' is typically not optimal in fisheries, attempting to maintain high levels of pest susceptibility may not be optimal in agriculture. The long-term benefits of doing so by growing large refuge areas may not outweigh the short-term sacrifices.

Refuges are intended to encourage the breeding of susceptible pests. This may seem a strange objective, as the entire point of pest-resistant crops is to reduce pest damage. The argument, however, is that unless a population of 'susceptible' pests is maintained to interbreed with 'resistant' pests, the latter will come to dominate the pest population and the GM crops will lose their relative advantage.[3] The refuge strategy reduces the selective pressure placed on susceptible pests and thereby limits the evolution of pest resistance (Mallet and Porter, 1992; Tabashnik, 1994).

There have been previous economic analyses of resistance and its management. Hueth and Regev (1974) considered the timing of pesticide applications and its effect on resistance. Other papers have considered analogous issues in the management of antibiotic resistance (Goeschl and Swanson, 2000; Laxminarayan, 1999; Laxminarayan and Brown, 2001). Most notably, Hurley et al. (1997) have conducted numerical analyses of optimal refuge strategies.

Our modeling approach differs in at least one important substantive aspect from that of Hurley et al. We might categorize their approach as 'time-invariant dynamic optimization'. By this we mean that they have derived refuge recommendations that would, if implemented in all periods, maximize economic benefits among the set of all time-invariant allocations of land to refuge. Policy makers must, of course, be able to achieve at least as desirable an outcome by choosing a time-variant strategy, and we demonstrate that such a strategy may well differ considerably from a time-invariant one. A time-invariant strategy designed to check the growth of resistance in the long run may be too stringent in the short run. By requiring that the same refuge area be grown in every period, Hurley et al. impose undue restrictions in the initial periods. Effective pesticides are effective precisely because resistance to them is extremely uncommon initially. There is little economic justification, then, for incurring substantial opportunity costs when the threat is as yet distant.

The remainder of the paper is organized as follows. We introduce our model in Section 2. We turn in Section 3 to describing optimal refuge areas. In particular, we consider optimal initial refuge choices. Section 4 concludes the paper.

2. The Model

We develop a simple model of evolution of pest resistance to transgenic plants. The structure is similar to that adopted in earlier literature on bacterial resistance to antibiotics (Laxminarayan and Brown, 2001). The setting is a large area in which monoculture is practiced. The pest population is assumed to be local and both in- and out-migration of pest population is ruled out. Other standard assumptions implicit in deriving the Hardy–Weinberg principle, such as random mating between resistant and susceptible pests, negligible mutation, non-overlapping pest generations and sexual reproduction of pests are all assumed to hold.[4]

The pest population is denoted by D. A number of biological models assume populations to grow logistically (see, e.g., Clark, 1990). We shall suppose that the pest population grows logistically with an intrinsic growth rate of g and a carrying capacity of K per unit of land planted in the crop. Total agricultural land is assumed to be fixed, and is normalized to 1. The total number of new pest organisms hatched (presuming them to be the offspring of egg-bearing insects) in every period is given by $gD(1-D/K)$. From this gross addition we must subtract mortality among pests.

The pest population is divided among 'susceptible' and 'resistant' organisms. The former will be assumed to die on exposure to the toxin embodied in the GM crop. A refuge strategy calls for planting a fraction q of the total land devoted to agriculture in the GM crop. Hence, a fraction $1 - q$ of agricultural land will be devoted to a non-GM variety. We assume that a fraction w of all pests is susceptible to the toxin, and the remaining fraction $1 - w$ is immune.

Resistant pests are immune to the toxin, but bear a fitness cost as a result of their immunity. The fitness cost of resistance is the evolutionary disadvantage placed on resistant pests in the absence of selection pressure (Anderson and May, 1991). There is typically a reason why the entire pest population is not resistant to pesticides in the first place. It is because those organisms that possess specialized mechanisms for resisting a pesticide do so at the expense of attributes that would be more valuable in the absence of the pesticide. Fitness costs account for the relative rarity of resistant organisms in the absence of the pesticide. A pest with a thicker shell, for example, may absorb less toxin, but does so at the evolutionary disadvantage of reduced mobility. A number of empirical studies have demonstrated that pesticide resistance does, in fact, impose high fitness costs in many instances (Alyokhin and Ferro, 1999; Ferrari and Georghiou, 1981; Beeman and Nanis, 1986). If the fitness cost of resistance is high, the selection pressure placed by the GM crop on susceptible pests is strongly counter-acted by the selection pressure placed by nature on resistant pests. In the refuge areas in which GM selection pressure is removed, susceptible pests are placed at an evolutionary advantage, and their numbers will increase relative to resistant ones.

We will denote fitness costs by the constant fraction r. A fraction r of resistant pests will die regardless of whether they live among GM or non-GM crops. It is easily demonstrated that r can be regarded as the difference between mortality rates relative to the mortality of susceptible pests among non-GM crops: 'baseline' mortality can be subsumed in the constants g and K of the population growth equation.

In summary, then, from the intrinsic growth rate we must subtract mortality among susceptible pests, wqD, and mortality among resistant pests, $(1 - w)rD$. Thus we have an expression for the evolution of the pest population,

$$\dot{D} = gD\left(1 - \frac{D}{K}\right) - wqD - (1 - w)rD. \tag{1}$$

It can be shown that the proportion of resistant pests in the population also follows a logistic equation with growth parameter equal to the difference in relative mortality rates between genotypes (see, e.g., Bonhoeffer et al., 1997; Laxminarayan and Brown, 2001). Thus

$$\dot{w} = (q - r)w(w - 1). \tag{2}$$

Since $w < 1$, the larger is the fraction of land planted with the GM crop, q, the greater is the decline in the effectiveness of the toxin in the GM crop.

Subsequent derivations will be more compact if we work in the *population* of resistant pests, R, rather than the *fraction* of the total population that is resistant, $1 - w$. Define

$$R = (1 - w)D. \tag{3}$$

Differentiating totally with respect to time,

$$\dot{R} = (1 - w)\dot{D} - \dot{w}D. \tag{4}$$

Substituting from (1), (2), and (3) in (4),

$$\dot{R} = R\left[g\left(1 - \frac{D}{K}\right) - r\right] \tag{5}$$

and eliminating w from expression (1) as well, we have

$$\dot{D} = D\left[g\left(1-\frac{D}{K}\right)-r\right]-(D-R)(q-r). \qquad (6)$$

3. Optimal Refuge Areas

The proportion of agricultural land set aside as refuge area in each period, $1-q$, determines net crop yield in that period, as well as the effectiveness of the GM crop against pests in succeeding periods. There is, then, an intertemporal tradeoff between, on one hand, crop losses today, and, on the other, more rapidly eroding toxic effectiveness. In the interest of analytical tractability and of generating unambiguous results, we abstract from other costs of production.

Let us suppose that each surviving pest eats an amount α. Normalize gross output per unit area planted to one. Then gross production of GM crops will be q. We will assume that pests disperse uniformly over the cultivated area. A fraction q of the resistant pest population, R, will alight in the GM crop area. Thus the net yield from the GM crop area is given by $q-\alpha qR$. Similarly, the gross yield from the refuge area is $(1-q)$. A fraction $1-q$ of pests will light in the refuge area, where they will consume $(1-q)\alpha$ units of the crop.[5] Net yield in the refuge area is then $(1-q)-(1-q)D\alpha$. Net yield from agriculture is given by the sum of net yields from GM sections and non-GM (refuge) sections of the crop.

$$Y(q, D, w) = 1-\alpha D+\alpha(D-R)q. \qquad (7)$$

In the interest of analytical tractability and clarity we will abstract from the costs of production. We will suppose the objective to be the maximization of the discounted present value of agricultural yield. Let the discount rate be ρ. Then q will be chosen to maximize

$$\int_0^\infty [1-D\alpha+(D-R)\alpha q]\,\mathrm{e}^{-\rho t}\,\mathrm{d}t. \qquad (8)$$

We append the two equations of motion, (5) and (6), with costate variables λ_R and λ_D, respectively, to form the current-value Hamiltonian

$$\begin{aligned} H \;=\; & 1-D\alpha+(D-R)\alpha q \\ & +\lambda_D\left(D\left[g\left(1-\frac{D}{K}\right)-r\right]-(D-R)(q-r)\right) \\ & +\lambda_R R\left[g\left(1-\frac{D}{K}\right)-r\right]. \end{aligned} \qquad (9)$$

The control variable is the fraction of land planted in GM crops, q. As this fraction can never be less than zero or greater than one, we must allow for corner solutions. Let us begin, however, by considering the conditions that will hold if $0 < q < 1$. If q is optimally set to an 'interior' value, then

$$\frac{\partial H}{\partial q} = \alpha(D - R) - \lambda_D(D - R) = 0,$$

or, more compactly,

$$\lambda_D = \alpha. \tag{10}$$

Condition (10) may seem a little strange: the shadow price assigned to the population of pests is positive, implying that one more pest is desirable. Recall, however, that we are working with the *total* pest population, D, and its resistant component, R. A costate variable in an optimal control problem can be interpreted as the *partial* derivative of the objective with respect to the corresponding state variable. Thus λ_D represents the shadow price of an additional pest organism, keeping the population of resistant organisms fixed. This, then, means that λ_D is the value of an additional *susceptible* pest organism.[6] An additional susceptible organism is one that can be killed by the toxin in the GM variety, and hence, one whose consumption of the crop can be eliminated.

An optimal refuge strategy will also satisfy

$$\begin{aligned}\frac{\partial H}{\partial D} &= \rho\lambda_D - \dot{\lambda}_D \\ &= -\alpha(1 - q) + \lambda_D\left(\left[g\left(1 - \frac{2D}{K}\right) - r\right] - (q - r)\right) - \lambda_R g \frac{R}{K}.\end{aligned}$$

As we have just seen that $\lambda_D = \alpha$, a constant, when q lies between zero and one, the above expression can be rearranged as

$$\frac{\alpha}{g}\frac{K}{R}\left(g\left(1 - \frac{2D}{K}\right) - 1 - \rho\right) = \lambda_R. \tag{11}$$

The optimal strategy will also have

$$\frac{\partial H}{\partial R} = \rho\lambda_R - \dot{\lambda}_R = -\alpha q + \lambda_D(q - r) + \lambda_R\left[g\left(1 - \frac{D}{K}\right) - r\right].$$

Using $\lambda_D = \alpha$ again and rearranging the above expression,

$$\rho\lambda_R - \dot{\lambda}_R = \left[g\left(1 - \frac{D}{K}\right) - r\right]\lambda_R - \alpha r. \tag{12}$$

The term in square brackets in (12) is, from (5), $\dot{R}/R$. Thus we can restate (12) as

$$(\rho\lambda_R + \alpha r)R = \dot{\lambda}_R R + \lambda_R \dot{R}. \tag{13}$$

Differentiate (11) totally with respect to time and rearrange it as

$$2\alpha\dot{D} = \lambda_R\dot{R} + \dot{\lambda}_R R. \tag{14}$$

Combining (13) and (14) and rearranging, we have

$$\lambda_R = -\frac{\alpha r}{\rho} + 2\frac{\alpha\dot{D}}{\rho R}. \tag{15}$$

Expressions (11) and (15) must be equal, so equating them and rearranging, we have

$$\rho D + \dot{D} = \frac{\rho K}{2}\left(\frac{g-1-\rho}{g}\right) + \frac{rR}{2}. \tag{16}$$

We could further elaborate expression (16) by using (6) to substitute for $\dot{D}$. The result of that exercise, which renders the left-hand side of expression (16) a quadratic expression in D, is not particularly revealing, however. Happily, however, we can derive clear results for one case of particular interest: what is the optimal refuge strategy at the time a new GM variety is introduced?

An effective pesticide is, by definition, one for which resistance is negligible. Newly developed products, such as transplanted genes that code for the production of Bt toxin, are effective because only a very few of the pests they target are favored with resistant genes. Let us suppose, then, that we are considering a case in which a new pesticide has just been introduced, and R is so small a fraction of D as to be negligible. Divide both sides of (16) by D, the population of pests at the time the new pesticide is introduced, and rearrange it as

$$\frac{\dot{D}}{D} = \frac{\rho K}{2D}\left(\frac{g-1-\rho}{g}\right) - \rho + \frac{rR}{2D}. \tag{17}$$

We can interpret (17) as a condition on the proportional rate of growth to the pest population. The middle term on the right-hand side of (17) is negative, and the first term is negative for growth rates $g < 1 + \rho$. Proceeding under the assumption that R/D is negligible, (17) states that an optimal solution requires that the pest population be declining at a 'fast enough' rate. This rate of decline is increasing in q, the area of land planted in GM crops. Suppose that initial conditions imply that (17) can only be satisfied as an equality if the proportion of land in GM crops were greater than one. This is an impossibility, so we conclude that the optimal strategy calls for setting $q = 1$ until resistance, R, increases to a level at which (17) can be satisfied as an equality for $0 < q < 1$.

Let us give an example. Consider, then, the case of a new GM variety replacing a pesticide with similar characteristics, save that resistance has already developed and been optimally managed for the old pesticide, and resistance to the new GM

Table 1. Optimal initial proportion, q, as a function of intrinsic growth rate, g, and discount rate, ρ.

Block A: $r = 0$										
	g									
ρ	0.05	0.10	0.15	0.20	0.25	0.30	0.35	0.40	0.45	0.50
0.05	0.55	0.29	0.20	0.16	0.13	0.11	0.10	0.09	0.08	0.08
0.10	1.00	0.60	0.42	0.33	0.27	0.23	0.21	0.19	0.17	0.16
0.15	1.00	0.94	0.65	0.51	0.42	0.36	0.32	0.29	0.27	0.25
0.20	1.00	1.00	0.90	0.70	0.58	0.50	0.44	0.40	0.37	0.34
0.25	1.00	1.00	1.00	0.91	0.75	0.65	0.57	0.52	0.47	0.44
0.30	1.00	1.00	1.00	1.00	0.93	0.80	0.71	0.64	0.58	0.54
0.35	1.00	1.00	1.00	1.00	1.00	0.96	0.85	0.77	0.70	0.65
0.40	1.00	1.00	1.00	1.00	1.00	1.00	1.00	0.90	0.82	0.76
0.45	1.00	1.00	1.00	1.00	1.00	1.00	1.00	1.00	0.95	0.88
0.50	1.00	1.00	1.00	1.00	1.00	1.00	1.00	1.00	1.00	1.00
Block B: $r = 0.1$										
	g									
ρ	0.05	0.10	0.15	0.20	0.25	0.30	0.35	0.40	0.45	0.50
0.05			0.60	0.36	0.28	0.24	0.22	0.20	0.19	0.18
0.10			1.00	0.65	0.48	0.40	0.35	0.32	0.29	0.28
0.15			1.00	0.96	0.70	0.57	0.49	0.44	0.40	0.37
0.20			1.00	1.00	0.93	0.75	0.64	0.57	0.51	0.48
0.25			1.00	1.00	1.00	0.94	0.80	0.70	0.64	0.58
0.30			1.00	1.00	1.00	1.00	0.97	0.85	0.76	0.70
0.35			1.00	1.00	1.00	1.00	1.00	1.00	0.90	0.82
0.40			1.00	1.00	1.00	1.00	1.00	1.00	1.00	0.95
0.45			1.00	1.00	1.00	1.00	1.00	1.00	1.00	1.00
0.50			1.00	1.00	1.00	1.00	1.00	1.00	1.00	1.00

variety is negligible. If there is a steady state, it is easily demonstrated that $D = ((g - r)/g)K$.[7] Then (17) can be rearranged as

$$q = \rho \frac{g + \rho + 1 - 2r}{2(g - r)} + r. \tag{18}$$

The results of some computational examples are illustrated in Table 1. In Block A of the table we allow both g and ρ to vary between 0.05 and 0.50 in increments of 0.05, while assuming r is constant at zero – no fitness cost. In Block B we suppose that $r = 0.1$. Note that negative values could result for $g \leq r$ in (18), so we have restricted the range of the table in Block B. If $g \leq r$, the optimal strategy could be wholly to eradicate the pest population, but we presume this to be empirically implausible.

4. Discussion

Since the beginning of 1999, the EPA has required that growers of genetically modified corn, cotton and wheat to follow a refuge strategy intended to slow down the evolution of pest resistance to these crops. This paper uses a simple model of evolution of pest resistance to GM crops to characterize the optimal refuge strategy for GM agriculture. In contrast to earlier work by other researchers, we demonstrate that it may not be optimal to establish refuge areas immediately, even if the evolution of resistance may eventually become a pressing concern.

From a practical perspective, it may be problematic to suppose that farmers will decide, or be called upon, to allocate different amounts of their fields to refuge depending on the current state of resistance. One thing that can be said of the current EPA policy is that it comprises a clear regulatory standard. Our work suggests, however, that a still better policy might be to institute such a clear regulatory standard only after evidence is produced establishing that resistance has, in fact, grown to troubling levels. In this respect, our model points to another analogy to the fisheries literature. With a positive discount rate, the optimal strategy typically will not call for 'maximum sustained susceptibility' analogous to 'maximum sustained yield'. Rather, we should strike a balance between the sacrifice of current output and the stimulation of future resistance.

Acknowledgements

We are grateful to seminar participants at Resources for the Future for useful input, to an anonymous referee for many helpful suggestions, and to Sarah Cline for research assistance.

Notes

[1] Seed companies are responsible for monitoring and enforcement of these refuge strategies. These refuge options vary by crop.

[2] Further, this study does not taken into account the positive environmental amenities provided by buffer zones for providing safe habitat to monarch butterfly larvae etc. and may therefore be a conservative estimate.

[3] The rationale for refuge areas becomes clearer when we note that resistance is often a recessive trait. This means that both parent organisms must carry the resistant gene if their offspring are also be to resistant. If the number of resistant organisms is kept low by maintaining refuge areas, it is very unlikely that two parent pests mating at random will both be resistant.

[4] A *genotype* is a particular genetic configuration. An *allele* is any one of the two or more forms that may compose a gene; for example, alleles for blue or brown eyes are common in many human populations. The Hardy–Weinberg principle of quantitative genetics holds that, for a population satisfying the assumptions we have stated and in which expected mortality is the same across different genotypes, the expected proportion of alleles and of genotypes remains constant from generation to generation.

[5] Implicit in our assumptions are the notions, first, that susceptible pests die so quickly on ingestion of GM crops as to consume only a negligible amount before expiring; and second, that mortality to other causes occurs after consumption. Both assumptions could be relaxed, but neither is important for establishing our general results.

[6] This discussion begs the question of why we did not choose to follow the equivalent approach of defining the state variables as the populations of susceptible and resistant pests, rather than the total and the resistant populations. The answer is that subsequent calculations were rendered somewhat more complicated by the alternative approach.

[7] Note that this expression also applies in the trivial steady state when $r = 0$, and hence susceptibility to the particular pesticide is wholly exhausted. The pest population would then increase to the carrying capacity.

References

Alyokhin, A.V. and Ferro, D.N. (1999) Relative fitness of Colorado Potato Bettle (Coleoptera: Chrysomelidae) resistant and susceptible to the *Bacillus thuringiensis* Cry3A toxin, *J. Econ. Entomol.* **92**(3), 510–515.

Anderson, R.M. and May, R.M. (1991) *Infectious Diseases of Humans: Dynamics and Control*, Oxford University Press, New York.

Beeman, R.W. and Nanis, S.M. (1986) Malathion resistance alleles and their fitness in the red flour beetle, *J. Econ. Entomol.* **79**, 580–587.

Bonhoeffer, S., Lipsitch, M. and Levin, B.R. (1997) Evaluating treatment protocols to prevent antibiotic resistance, *Proc. Natl. Acad. Sci., USA* **94**, 12106–12111.

Clark, C.W. (1973) Profit maximization and the extinction of animal species, *J. Political Econ.* **81**, 950–961.

Clark, C.W. (1990) *Mathematical Bioeconomics: The Optimal Management of Renewable Resources*, John Wiley & Sons, New York.

EPA (1998) The Environment Protection Agency's white paper on Bt plant-pesticide resistance management, EPA, Washington, DC.

Ferrari, J.A. and Georghiou, G.P. (1981) Effects of insecticidal selection and treatment on reproductive potential of resistant, susceptible and heterozygous strains of the southern house mosquito, *J. Econ. Entomol.* **74**, 323–327.

Goeschl, T. and Swanson, T. (2000) Lost horizons: The interaction of IPR systems and resistance management, London.

Head, G. (1999) Reconciling science, economics and practicality in resistance management for transgenic insect-protected crops, Monsanto Corporation, St. Louis, MO.

Hueth, D. and Regev, U. (1974) Optimal agricultural pest management with increasing pest resistance, *Amer. J. Agricult. Econ.* **56**(3), 543–553.

Hurley, T.M., Babcock, B.A. and Hellmich, R.L. (1997) Biotechnology and pest resistance: An economic assessment of refuges, Center for Agricultural and Rural Development, Iowa State University, IA.

James, C. (1998) Global review of commercialized transgenic crops: 1998, ISAAA, Ithaca, NY.

Laxminarayan, R. (1999) Economics of antibiotic resistance, Doctoral Dissertation, University of Washington, Seattle, p. 121.

Laxminarayan, R. and Brown, G.M. (2001) Economics of antibiotic resistance: A theory of optimal use, *J. Environ. Econ. Manag.* **42**(2), 183–206.

Livingstone, M.J., Carlson, G.A. et al. (2000) *Bt Cotton Refuge Policy*, American Agricultural Economics Association Meetings, Tampa Bay.

Mallet, J. and Porter, P. (1992) Preventing insect adaptation to insect-resistant crops: Are seed mixtures or refugia the best strategy, *Proc. Roy. Soc. London B* **250**, 165–169.

Tabashnik, B.E. (1994) Delaying insect adaptation to transgenic plants: Seed mixtures and refugia reconsidered, *Proc. Roy. Soc. London B* **255**, 7–12.

Tabashnik, B.E. (1997) Seeking the root of insect resistance to transgenic plants, *Proc. Natl. Acad. Sci. USA* **94**, 3488–3490.

8. Managing the Risk of European Corn Borer Resistance to Bt Corn *

TERRANCE M. HURLEY, SILVIA SECCHI, BRUCE A. BABCOCK and RICK HELLMICH

1. Introduction

Recent advances in genetic engineering allow genes to be transferred between species. Applications of this technology to agriculture include the insertion of a gene from the soil bacterium *Bacillus thuringiensis* (Bt) into corn and other crops. Corn with the Bt gene produces proteins that are toxic when consumed by the European corn borer (ECB) and other Lepidopteran insects. Since its commercial introduction in 1996, Bt corn has proven to be an effective tool for managing the ECB.

The high efficacy of Bt corn has resulted in rapid and widespread adoption. In 1999, an estimated 25% of United States (US) corn acreage was planted to Bt Corn. Bt corn's high efficacy and rapid adoption foster concerns that the ECB will develop resistance to Bt. The potential for insects to develop resistance to highly effective and widely used pesticides is well documented, including resistance to Bt (Bauer, 1995; Liu and Tabashnik, 1997; McGaughey and Beeman, 1998; Perez and Shelton, 1997; Tabashnik 1994). Concerns of resistance are heightened because Bt occurs naturally and is believed to pose fewer environmental and human health risks than conventional pesticides. If the efficacy of Bt corn declines due to resistance, growers may turn to more hazardous alternatives to achieve sufficient ECB control.

The US Environmental Protection Agency (EPA) has authority over the introduction, use, and registration of plant-pesticides. That is, pesticides produced by a plant due to the introduction of new genetic material. Therefore, the EPA is partially responsible for regulating and registering Bt corn. Through this authority and acting in the public's interest, the EPA has worked with industry and academic scientists to established insect-resistance management (IRM) guidelines for Bt corn. The objective of these guidelines is to preserve the efficacy of Bt and reduce the

* Names are necessary to report factually on available data; however, the USDA, Iowa State University, and the University of Minnesota do not guarantee or warrant the standard of a product, and the use of names implies no approval of the product to the exclusion of others that may be suitable.

Timothy Swanson (ed.), The Economics of Managing Biotechnologies, 171–193.

use of more hazardous pesticides, while not overly burdening the regulated community (US EPA, 1998a). The existing economic literature provides rationale for promoting IRM. Since surviving pests damage crops and propagate, the literature argues pests are an unwanted renewable resource (Hueth and Regev, 1974; Regev et al., 1976, 1983). Pest susceptibility, the converse of resistance, is a valuable nonrenewable resource since susceptible pests can be controlled (Hueth and Regev, 1974; Regev et al., 1983). Capturing this value however results in resistance and a pest that is harder to control. A mobile pest can be viewed as common property (Clark and Carlson, 1990); therefore, farmers have little incentive to account for the effect of pest control decisions on neighbors, which results in too little voluntary IRM.

Currently IRM guidelines for Bt corn are based on a high-dose refuge strategy. The foundations of the high-dose refuge strategy require Bt corn to produce enough toxins to kill all but the most resistant ECB and growers to plant refuge corn. Refuge corn allows susceptible ECB to survive and mate with resistant ECB emerging from Bt corn. If there is a high enough dose and sufficient refuge, most surviving ECB will be susceptible and so will most of their offspring.

The EPA did not originally impose mandatory refuge requirements for most Bt corn. However, mandatory requirements were introduced for the 2000-growing season. In the predominant corn growing regions of the US, these mandatory requirements obligate Bt corn registrants to ensure that farmers planting Bt corn also plant at least 20% refuge corn.[1] Conventional treatments with non-Bt pesticides based on economic thresholds are permitted for controlling ECB on refuge in years of severe infestation.

The new mandatory guidelines for the 2000-growing season represent a departure from previous recommendations. Previous recommendations required more refuge when refuge is treated with conventional pesticides because treated refuge produces fewer susceptible ECB and increases the risk of resistance (Ostlie et al., 1997; Mellon and Rissler, 1998; US EPA, 1998b; ILSI/HESI, 1999). Several arguments provide rationale for not differentiating treated and untreated refuge. At planting, farmers do not know whether conventional pesticide treatments will be economical. In most regions of the US, conventional pesticide treatments for the ECB are uncommon. If refuge is rarely treated, the increased risk of resistance is likely to be small. Therefore, not differentiating treated and untreated refuge benefits farmers by providing greater flexibility without substantially increasing the risk of resistance. However, where conventional pesticide treatments are common, there is a concern that 20% refuge is not enough.

The purpose of this paper is to develop a stochastic dynamic bieconomic simulation model to compare alternative high-dose refuge strategies based on ECB resistance to Bt corn, conventional pesticide use, and the value of agricultural production to farmers and industry. The model that is developed captures seasonal variation in ECB populations using a density-dependent stochastic population model. In addition, field level monitoring data is used to address uncertainty regarding important biological parameters. The model is then used to evaluate the arguments for allowing refuge treatments based on economic thresholds.

The results of the analysis support arguments for allowing refuge to be treated based on economic thresholds in regions with a low historic frequency of conventional pesticide use. Allowing growers to treat refuge in these regions can increase the value of production with a negligible increase or even a decrease in the risk of resistance. However, conventional pesticide use will be higher. Whether the 20% refuge requirement is enough to allow refuge treatments based on economic thresholds in regions with a high historical frequency of conventional pesticide use depends on the primary objectives of the IRM policy. If the primary objective is to reduce conventional pesticide use or increase agricultural production, then higher refuge requirements are not necessary. However, if reducing the risk of resistance is of primary concern, then more refuge is required. Finally, we find that increased agricultural production and reduced conventional pesticide use are usually complementary benefits. Therefore, refuge requirements that result in greater agricultural productivity tend to reduce conventional pesticide use. Alternatively, refuge requirements that result in a decrease in the risk of resistance tend to increase conventional pesticide use.

2. The Model

Two approaches to modeling resistance management are found in the literature. In the economics literature, Hueth and Regev (1974), Taylor and Headley (1975), Regev et al. (1983), and Gorddard et al. (1995) develop dynamic models to evaluate the optimal use conventional pesticides with increasing resistance. These models are deterministic and account for important biological factors with varying degrees of detail. The objective of the analysis is to characterize the time path for pesticide use that maximizes the value of agricultural production. While these models do not explicitly address the high-dose refuge strategy, they do offer insight into how managing resistance can enhance the value of production in the long run. They also demonstrate that the optimal amount of pesticide use will typically vary over time as resistance evolves.

Alstad and Andow (1995), Roush and Osmond (1996), Caprio (1998), Gould (1998), Onstad and Gould (1998a, 1998b), and Peck et al. (1999) provide a sample of the second type of models commonly found in the entomology literature. These biological simulation models vary in spatial and temporal detail and are designed specifically to evaluate the efficacy of the high-dose refuge strategy. Most are deterministic, but some have stochastic components. All focus on describing how fast resistance evolves under alternative refuge requirements. These models have been instrumental in identifying key biological factors that influence the evolution of resistance.

The model we develop extends the one by Hurley et al. (1997), which is founded more in the tradition of the entomology literature and less in the tradition of the economic literature. The model extends both the economic and entomology literature by broadening the set of objectives used to evaluate refuge requirements to include the risk of resistance, value of agricultural production to farmers and industry, and conventional pesticide use. In addition, the model includes a density-dependent random pest population, parametric uncertainty, and conventional pesticide applications based on economic thresholds. While the model makes no attempt to identify optimal IRM requirements, it does help to quantify important tradeoffs that can inform the debate over the merits and deficiencies of alternative IRM requirements.

Consider a simplified production region with a single crop and pest. The region is divided between two varieties of the crop. The first variety, denoted by $i = 0$, is a conventional variety that serves as refuge. The second variety, denoted by $i = 1$, is a Bt variety that is toxic when consumed by susceptible pests. Let $1.0 \geq \phi \geq 0.0$ be the proportion of refuge acreage planted in each season. This value is held constant from one season to the next to facilitate exposition and comparisons with previous refuge recommendations. However, the model can be easily generalized by specify the proportion of refuge as both time and state dependent. The pest reproduces with G generations per season where g denotes the generation in season t. Let $1.0 \geq \iota_{tg}^{i} \geq 0.0$ be the proportion of crop i that receives a conventional pesticide application in season t and generation g. The model allows for conventional pesticide treatments on Bt acreage, as well as refuge acreage because if Bt fails due to the onset of resistance, farmers may turn to conventional pesticides for supplemental control.

The number of pests emerging to damage crops and reproduce is $n_{tg} \geq 0.0$, which depends on the number of pest surviving in the previous generation and environmental factors such as random weather events. Pest populations are notoriously variable over time due to random environmental events such as storms, though not independent from past levels due to reproduction. To capture this random interdependence, let

$$n_{tg} \sim \begin{cases} N_g(n_{tg-1}^S), & \text{for } g > 1, \\ N_g(n_{t-1G}^S), & \text{for } g = 1, \end{cases} \tag{1}$$

where n_{tg}^S is the number of pests that escape control and survive to damage crops and reproduce and $N_g(\cdot)$ is a conditional distribution function. Equation (1) states that the pest population is a random variable that is conditionally distributed based on the number of surviving pests in the previous generation.

The Hardy–Weinberg model characterizes resistance, which is assumed to be conferred by a single allele that is not sex linked.[2] There are two types of alleles:

resistant and susceptible. The proportion of resistant alleles is $1.0 \geq r_{tg} \geq 0$. Each pest posses two alleles, one contributed by its mother and one by its father, and can be one of three genotypes: a resistant homozygote – with two resistant alleles; a heterozygote – with one resistant allele; or a susceptible homozygote – with no resistant alleles. The Hardy–Weinberg model implies the proportion of each genotype is

$$\eta_{tg} = [r_{tg}^2, 2r_{tg}(1 - r_{tg}), (1 - r_{tg})^2], \tag{2}$$

where the vector elements correspond to resistant homozygotes, heterozygotes, and susceptible homozygotes.

The Hardy–Weinberg model assumes no selection pressure – survival rates are the same for all genotypes. Bt crops impose selection pressure on pests with at least one susceptible allele. Let σ_g^i be a 1×3 vector of genotypic survival rates for pests on crop i in generation g with elements corresponding to resistant homozygotes, heterozygotes, and susceptible homozygotes. The survival rate of all genotypes treated with a conventional spray application is σ_g^ι. The vector of genotypic survival rates for each crop, season, and generation is $\rho_{tg}^i = \sigma_g^i + \iota_{tg}^i(\sigma_g^i \sigma_g^\iota - \sigma_g^i)$, which implies the number of pests surviving to damage crop i and reproduce is $n_{tg}^{Si} = \rho_{tg}^i \cdot \eta_{tg}^i n_{tg}$. The vector of genotypic survival rates for the region is $\rho_{tg} = \rho_{tg}^1 + \phi_t(\rho_{tg}^0 - \rho_{tg}^1)$, which implies the number of pests surviving to reproduce is $n_{tg}^S = \rho_{tg} \cdot \eta_{tg} n_{tg}$. Since each surviving pest contributes two alleles, resistant homozygotes contribute two resistant alleles, heterozygotes contribute one resistant allele, and susceptible homozygotes contribute no resistant alleles, the proportion of resistant alleles in the subsequent generation is

$$r_{tg} = \begin{cases} \dfrac{\rho_{tg-1} M \eta_{tg-1}}{\rho_{tg-1} \cdot \eta_{tg-1}}, & \text{for } g > 1, \\[2ex] \dfrac{\rho_{t-1G} M \eta_{t-1G}}{\rho_{t-1G} \cdot \eta_{t-1G}}, & \text{for } g = 1, \end{cases} \tag{3}$$

where M is the 3×3 diagonal matrix $[1.0, 0.5, 0.0]$.

Equations (1–3) and the initial conditions $n_{01} = N_0$ and $r_{01} = R_0$ describe a dynamic stochastic system, which is controlled by the proportion of refuge and conventional pesticide use. To evaluate and compare the performance of this system under alternative control strategies, we focus on measures of the risk of resistance, conventional pesticide use, and the value of production over a fixed time period. The probability that the proportion of resistant alleles exceeds 0.5 within T years measures the risk of resistance:

$$\Theta = \Pr(r_{1T} \geq 0.5), \tag{4}$$

where the probability is defined over the random distribution of pests for $t = 0, \ldots, T$ and $g = 1, \ldots, G$. The expected number of conventional pesticide applications per acre measures pesticide use:

$$\Gamma = E_n \left[\sum_{t=0}^{T-1} \sum_{g=1}^{G} \frac{\phi_t \iota_{tg}^0 + (1 - \phi_t) \iota_{tg}^1}{T} \right], \tag{5}$$

where E_n is the expectation operator defined over the random distribution of pests for $t = 0, \ldots, T$ and $g = 1, \ldots, G$. The expected annualized net present value of production to farmers and industry measures the value of production:

$$\Pi = E_n \left[\frac{\sum_{t=0}^{T-1} \delta^t \pi_t}{\sum_{t=0}^{T-1} \delta^t} \right], \tag{6}$$

where δ is the discount rate and π_t is the annual value of production to farmers and industry in season t. The annual value of production to farmers and industry is

$$\pi_t = \phi \left\{ P_t^0 Y_t^0 \left[1 - D(n_{t1}^{S0}, \ldots, n_{tG}^{S0}) - FC_t^0 - \sum_{g=1}^{G} \iota_{tg}^0 V C_{tg}^0 \right] \right\}$$
$$+ (1 - \phi) \left\{ P_t^1 Y_t^1 \left[1 - D(n_{t1}^{S1}, \ldots, n_{tG}^{S1}) - FC_t^1 - \sum_{g=1}^{G} \iota_{tg}^1 V C_{tg}^1 \right] \right\}, \tag{7}$$

where Y_t^i bushels/acre and P_t^i \$/bushel are the pest-free yields and crop prices; FC_t^i \$/acre is the production cost for items such as seed (excluding industry rents from the sale for Bt seed corn), fertilizer, and labor that are exclusive of the cost of a conventional pesticide application; VC_{tg}^i \$/acre is the cost of a conventional pesticide application; and $D_t^i(n_{t1}^{Si}, \ldots, n_{tG}^{Si})$ is the proportion of pest-free yield lost to pests throughout the season. Equation (7) represents the average net return to farmers plus the proportion of the technology fee collected by industry representing rents paid by farmers to industry for the right to plant the Bt variety.

Equations (4–6) are conditional on the values assigned to the number of generations of pest per season, genotypic survival rates, survival rates for conventional spray applications, number of time periods, prices, pest-free yields, production costs, discount rate, initial pest population, and initial proportion of resistance. While reasonable values are readily available for many of these parameters, others are uncertain. The typical method for addressing this uncertainty is to test the sensitivity of the results to reasonable variations in parameter values. However, if suitable data is available, this uncertainty can be captured explicitly using the estimated distributions for the parameters, such that Equations (4–6) can be rewritten as

$$\Theta = E[\Pr(r_{1T} \geq \Psi)], \tag{4'}$$

$$\Gamma = E\left[E_n\left[\sum_{t=0}^{T-1}\sum_{g=1}^{G}\frac{\phi_t \iota_{tg}^0 + (1-\phi_t)\iota_{tg}^1}{T}\right]\right], \tag{5'}$$

and

$$\Pi = E\left[E_n\left[\frac{\sum_{t=0}^{T-1}\delta^t \pi_t}{\sum_{t=0}^{T-1}\delta^t}\right]\right], \tag{6'}$$

where $E[\cdot]$ is the expectation operator defined over the estimated distribution of uncertain parameters. Combined with Equations (1–3) and (7), Equations (4′–6′) allow for the comparison of alternative IRM requirements based on the tradeoffs between measures of the expected risk of resistance, pesticide use, and value of production.

3. Model Implementation

Implementing the model described in Equations (1–3), (7), and (4′–6′) requires estimates of the conditional distribution of ECB; values or distributions for the exogenous parameters; the proportion of refuge; and $\iota_{tg}^i \forall i \in [1, 2], t \in [1, \ldots, T]$, and $g \in [1, \ldots, G]$. We focus on distributions and parameter values that are characteristic of the Midwestern US when Bt corn is planted to control the ECB.

3.1. Distribution of ECB

ECB populations in the Midwestern US are typically bivoltine (two generations per season). Capturing the variability in these bivoltine populations and intergenerational dependencies requires longitudinal data for both first and second generation ECB under conditions without control. Recent surveys of ECB pressure focus on quantifying ECB tunneling at the end of the season and moth flights. Tunneling data does not allow for distinctions between generations and moth flight data is difficult to calibrate to the field level. An older survey conducted between 1960 and 1969 measured first and second generation larval populations (ECB/plant) at six sites across the Midwest (see Calvin, 1996). Since Mitchell, Hurley, and Hellmich (2000) suggests that state average second-generation ECB populations for Illinois, Minnesota, and Wisconsin were stabile between 1960 and 1990, we use the 1960s survey to estimate the conditional distributions.

Assuming ECB populations are log-normally distributed with parameters μ_g and σ_g, the Midwestern data were pooled and maximum likelihood techniques used for estimation.[3] To capture intergeneration dependencies, we assume $\mu_1 = \beta_{01} + \beta_{11}n_{t-12}^S + \beta_{21}n_{t-12}^{S2}$, $\mu_2 = \beta_{02} + \beta_{12}n_{t1}^S + \beta_{22}n_{t-11}^{S2}$ and $\sigma_g = \beta_{4g}$ for $g = 1, 2$. This specification allows the mean and variance to vary based on pests surviving in

Table 1. Parameter estimates for ECB population models.

Coefficient	Model 1	Model 2
First Generation		
Constant	–3.52[a]	–2.50[a]
	(0.31)	(0.19)
Previous surviving population	1.81[b]	
	(0.72)	
Previous surviving population^2	–0.39	
	(0.27)	
Standard deviation	0.96[a]	1.17[a]
	(0.14)	(0.20)
Second Generation		
Constant	–1.59[a]	–0.66[a]
	(0.30)	(0.21)
Previous surviving population	9.47[b]	
	(4.48)	
Previous surviving population^2	–11.31	
	(10.9)	
Standard deviation	1.11[a]	1.28[a]
	(0.13)	(0.13)
Maximized log-likelihood	5.99	–10.01
Comparison of Model 1 *versus* 2: $\chi^2(4)$	32.00[a]	
Observations	92	92

Notes: Standard errors are in parentheses, [a]denotes a 1% level of significance, and [b]denotes a 5% level of significance.

the previous generation, while the coefficient of variation remains constant. When the coefficient estimates are positive for the linear terms and negative for quadratic terms, the form of μ_g implies that factors such as food scarcity naturally limit ECB populations.

Table 1 reports the maximum likelihood coefficient estimates, the maximized value of the log-likelihood function, and the test for the restriction $\beta_{11} = \beta_{21} = \beta_{12} = \beta_{22} = 0$ which is indicative of intergeneration independence. The log-likelihood ratio test is statistically significant, which suggests intergeneration dependencies are important. As anticipated, the linear terms are positive and quadratic terms are negative. We initialize the model with the average number of first generation ECB per plant, 0.12.

3.2. Pest Survival Rates on Bt Corn and Resistance

Genotypic survival rates on Bt corn relative to refuge are uncertain due to the lack of a confirmed case of ECB resistance to Bt corn. However, Venette et al. (2000) demonstrate how field level monitoring data provides useful information on survival rates. Their method uses information on larvae, L, found in a sample of size N from untreated refuge and larvae, S, found in a sample of size M from an adjacent Bt field to calculate the mean and variance of the survival rate for ECB on Bt corn relative to refuge. While the method they propose focuses on sampling ears of sweet corn, it is more generally applicable to any consistent sampling protocol applied to adjacent fields of refuge and Bt corn.

We extend the method proposed by Venette et al. (2000) to integrate sampling from different sites assuming the sites have the same survival rates and frequency of resistance. Let P be the survival rate of ECB on Bt corn relative to refuge in the season and generation sampled. If the samples are taken in season t and generation g, $P = \rho_{tg}^1 \cdot \eta_{tg}/\rho_{tg}^0 \cdot \eta_{tg}$ for $\iota_{tg}^0 = \iota_{tg}^1 = 0$. Suppose K sites are sampled such that L_k, S_k, N_k, and M_k for $k = 1, \ldots, K$ are the number of larvae found in the samples and the number of samples drawn from adjacent refuge and Bt fields at site k. Assuming that the refuge and Bt samples represent independent draws from Poisson distributions with mean $z_k > 0.0$ and $Pz_k > 0.0$, Bayes rule implies

$$\Pr(P, z \mid L, S, N, M) \propto \Pr(P, z) \prod_{k=1}^{K} \frac{\mathrm{e}^{-N_k z_k}(N_k z_k)^{L_k}}{L_k!} \frac{\mathrm{e}^{-M_k P z_k}(M_k P z_k)^{S_k}}{S_k!}, \quad (8)$$

where z, L, S, N, and M are vectors containing the elements of z_k, L_k, S_k, N_k, and M_k for $k = 1, \ldots, K$ and $\Pr(P, z)$ represents prior beliefs about the distribution of P and z. Let

$$\Pr(P, z) = \Pr(P) \prod_{k=1}^{K} \Pr(z_k)$$

and $\Pr(z_k)$ be an improper uniform prior for $k = 1, \ldots, K$. After integrating over z_k for $k = 1, \ldots, K$, Equation (8) can be rewritten as

$$\Pr(P \mid L, S, N, M) \propto \Pr(P) \prod_{k=1}^{K} \frac{P^{S_k}}{(N_k + M_k P)^{L_k + S_k + 1}}. \quad (9)$$

Equation (9) is an improper distribution for the relative survival of ECB on Bt corn and provides information on the distribution of σ_g^0, σ_g^1, and r_{tg}. To use this distribution, priors for σ_g^0, σ_g^1, and r_{tg} and sampling data for L, S, N, and M are needed. Following the analysis reported in ILSI/HESI (1999), assume $\Pr(\sigma_1^0 = \sigma_2^0 = [1.0, 1.0, 1.0], \sigma_1^1 = \sigma_2^1 = [1.0, \sigma_{RS}, 0.0], r_{tg} = R_0) = \Pr(R_0, \sigma_{RS})$. These assumptions imply that all ECB have normal survival rates on refuge, resistant homozygotes have normal survival rates on Bt corn, and susceptible homozygotes do not survive on Bt corn. It also implies that the heterozygote survival rate on Bt corn is uncertain, as is the frequency of resistant alleles. As with Hurley et al. (1999), we choose $\Pr(R_0, \sigma_{RS})$ to be uniformly distributed such that $\Pr(R_0 < 4.38 \times 10^{-3}) = 0.95$ and $0.1 \geq \sigma_{RS} \geq 0.0$.

Sampling data is taken from two sources. Monsanto Company provided data collected by university and industry collaborators from eight Midwestern states and 104 different sampling sites. Drs. Robert Venette and William Hutchison from the University of Minnesota provided data collected from four sites in Minnesota. The data that are used were collected in 1997. While 1998 and 1999 data are available, aggregation across sites is hard to justify because increased resistance may have already developed in regions with higher Bt corn adoption rates. Since 1997 represents the first year that Bt corn was extensively planted, most samples of second generation larvae would have been exposed to the selection pressure of Bt for only one generation prior to sampling.

A total of 8,814 larvae were found in 6,670 samples taken from refuge fields. A total of 36 larvae were found in 8,640 samples taken from Bt fields. Assuming that all larvae that were found in Bt fields possessed a resistant allele, the estimated average frequency of resistant alleles from Equation (9) is 4.4×10^{-3}. This estimate is fourfold higher than estimates for the tobacco budworm reported in Gould et al. (1997) and exceeds the 95% confidence interval recently estimated for Midwestern ECB using the F2 screen developed by Andow and Alstad (1998).[4] Alternatively, if we assume none of the 36 larvae possessed a resistant allele because resistance was not confirmed and there are other reasonable explanations for their presence on Bt corn, the estimated frequency of resistant alleles from Equation (9) is 1.1×10^{-3}. The 95th percentile for this estimate is 3.6×10^{-3}. Since the later estimates are more consistent with Gould et al. (1997) and other recent estimates, we assume none of the 36 larvae found on Bt corn were resistant. Table 2 summarizes the distribution of the initial frequency of resistance and heterozygote survival rate used in the analysis.

3.3. Costs, Revenues, and Pest Damage

Point estimates were used for information on costs, revenues, and pest damages because reasonable estimates are readily available. A summary of the benchmark assumptions is reported in Table 2. US Department of Agriculture National Agricultural Statistics Service (NASS) and Economic Research Service (ERS) data provide reasonable estimates for the real price, pest-free yield, and production cost of refuge corn, which are held constant over time. The real price of corn, \$2.35, is the monthly average from 1991 to 1996 deflated to 1992.[5] The average Iowa yield from 1991 to 1996 was about 123 bushels per acre. Assuming an average annual ECB yield loss of 6.4% (Calvin, 1996) implies a pest-free yield of 130 bushels per

Table 2. Summary of (a) parameter values and (b) distributions.

(a)

Parameter	Benchmark value/ other values
Biological Parameters	
Generations of pests per cropping season	2
Survival rate of ECB on refuge corn for all genotypes	1.0
Survival rate of ECB on Bt corn for resistant homozygote	1.0
Survival rate of ECB on Bt corn for susceptible homozygote	0.0
Survival rate for conventional pesticide applications 1st generation	0.20
Survival rate for conventional pesticide applications 2nd generation	0.33
Initial pest population (pests/plant)	0.12
Economic parameters	
Planning horizon (years)	15
Interest rate	0.04
Price of corn per bushel	$2.35
Pest-free yield for Bt corn and refuge (bushels/acre)	130
Production cost for Bt and refuge corn ($/acre)	$185.00
Constant marginal yield loss for first generation (pests/plant)	0.055
Constant marginal yield loss for second generation (pests/plant)	0.028

(b)

Parameter	Mean	Standard deviation	95th percentile	Correlation
Initial frequency of resistant alleles	1.1×10^{-3}	1.1×10^{-3}	3.6×10^{-3}	–0.49
Heterozygote survival on Bt corn	0.027	0.026	0.083	

acre. Excluding returns to management, the average production cost, $185, comes from 1995 ERS corn budgets deflated to 1992 prices and is assumed to include scouting costs. The cost of a conventional pesticide treatment, $14 an acre, is taken from Mason et al. (1996).

The pest-free yield is assumed to be the same as refuge for the benchmark simulation because we have no evidence to suggest that Bt yields are lower in the absence of ECB. Most of the increased cost of Bt seed is sunken research and development. While farmers pay a $10 an acre technology fee, only part of this fee reflects an increase in marginal cost of growing Bt corn. The remainder represents rents paid to industry by farmers in order to use Bt corn. The percentage of the technology fee that represents an increase in the marginal production cost is proprietary information, but likely to be small because no special handling is required once the Bt gene has been inserted into the plant. Therefore, we presume the entire technology fee represents industry rents and that any difference in the marginal cost is negligible.

Damage estimates for the ECB vary depending on a variety of environmental and management factors. For instance, damages will be higher when corn is stressed and in early or late-planted corn. Depending on a plant's stage of development, estimates indicate a marginal yield loss ranging from 2 to 6% pests/plant (Mason et al., 1996). Since our interest is in evaluating the average seasonal damage of the ECB over a production region, we assume $D_t^i(n_{t1}^{Si}, n_{t2}^{Si}) = \mathrm{Min}\{d_1 n_{t1}^{Si} + d_2 n_{t2}^{Si}, 1.0\}$ where $d_1 = 0.055$ and $d_2 = 0.028$ based on Mason et al. (1996).

A time frame of reference and discount rate are also needed in order to compare alternative IRM requirements. The time frame used for the analysis is 15 years, which is based on recommendations made by the scientific advisory panel convened by the EPA in 1998 (US EPA, 1998b) and the value used in ILSI-HESI. A real interest rate of 4.0% is used for discounting the value of production.

3.4. Spray Applications

Originally, separate refuge recommendations were made based on whether refuge was treated with conventional pesticides. Recently, the EPA mandated a single recommendation that allows growers to treat refuge based on economic thresholds. The method for calculating the economic thresholds has not been specified. To incorporate economic thresholds into the model, we use the methodology offered by Mason et al. (1996), which implies

$$\iota_{tg}^i = \begin{cases} 1.0, & \text{for } n_{tg} > \dfrac{VC_{tg}^i}{P_t^i Y_t^i d_g \sigma_g^i \eta_{tg}(1 - \sigma_g^t)}, \\ 0.0, & \text{for } n_{tg} \le \dfrac{VC_{tg}^i}{P_t^i Y_t^i d_g \sigma_g^i \eta_{tg}(1 - \sigma_g^t)}. \end{cases}$$

The method is based on a cost benefit analysis that compares the value of improved ECB control to the cost of the pesticide treatment. The calculation does not factor in risk or population dynamics. However, the calculation is practical and a common starting point for many farmers. Typical ECB survival rates for conventional pesticide applications are 0.20 for the first generation and 0.33 for the second generation (Mason et al., 1996).

4. Results

The economic and environmental tradeoffs of increasing refuge to reduce the risk of resistance are evaluated by comparing the risk of resistance, pesticide use, and the value of production as refuge increases from 0 to 100%. To calculate these tradeoffs, Monte Carlo integration is used to evaluate Equations (4′–6′) for the benchmark parameter assumptions. First, we characterize the economic and environmental tradeoffs when refuge treatments are made based on economic thresholds. We then compare treated *versus* untreated refuge. Finally, we explore the sensitivity of the results for a treated refuge to factors that increase the frequency of conventional pesticide applications.

4.1. Tradeoffs for Treated Refuge

Figure 1 presents the benchmark simulation results. The value of production increases from \$112.57 to \$119.43 as refuge increases from 0 to 19% and then falls to \$110.76 as refuge increases to 100%. Conventional pesticide use falls from 0.082 to 0.0047 applications per acre as refuge increases from 0 to 22%, but then increases to 0.10 applications per acre as refuge increases to 100%. As refuge increases from 0 to almost 40%, the risk of resistance falls from 1.0 to 0.0.

These results illustrate how managing resistance can increase production and reduce conventional pesticide use in the long run. With the full adoption of Bt corn and no refuge the value of production increases by 1.6%, while conventional pesticide use declines by 20%. However, with 19% refuge, the value of production increases by 7.8%, while conventional pesticide use falls by 90%. Therefore, over 15 years, there is only a modest increase in the value of production and decrease in conventional pesticide use when no refuge is planted. When no refuge is planted, both farmers and the environment lose substantial benefits from Bt corn in the long run due the rapid evolution of resistance. By planting refuge, the evolution of resistance is slower, which extends the efficacy of Bt corn and provides better ECB control in the long run with fewer conventional pesticides.

The results also suggest that the long-run costs of obtaining reductions in the risk of resistance are relatively small. With 19% refuge, there is more than a 1 in 5 chance of resistance developing within 15 years. When there is 27% refuge, there is only a 1 in 20 chance. By increasing refuge from 19 to 27%, conventional pesticide use is virtually unaffected, while the value of production falls by less than 0.2%. The value of production decreases negligibly because of the density dependence of ECB populations. This density dependence allows Bt corn to suppress ECB populations over time provided resistance does not develop too fast. With lower average populations, the cost of planting refuge is reduced along with the need for conventional pesticides. However, if not enough Bt corn is planted, ECB suppression is weak, which raises the cost of planting refuge and the need for conventional

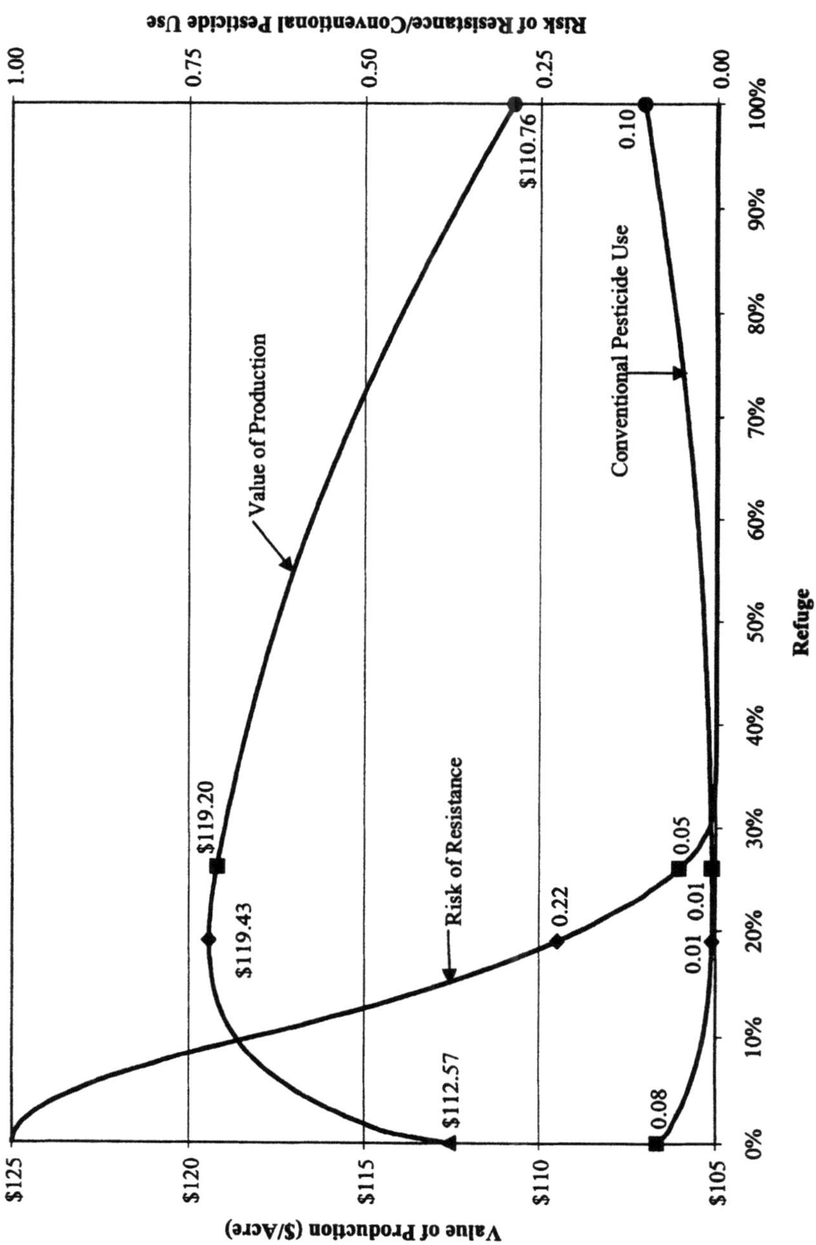

Figure 1. Benchmark tradeoffs between agricultural production, conventional pesticide use, and risk of resistance.

pesticides. Therefore, the value of production decreases at an increasing rate as refuge increases above 19%.

It is also important to note the complementary relationships between the value of production, conventional pesticide use, and the risk of resistance. Increasing refuge between 0 and 19% produces the complementary results of increasing the value of production and decreasing pesticide use and the risk of resistance. Increasing refuge between 19 and 22% decreases pesticide use and the risk of resistance, but also decreases the value of production. Increasing refuge above 22% decreases the risk of resistance, but also decreases the value of production and increases pesticide use. Therefore, increases in the value of production and decreases in pesticide use tend to be complementary benefits that are usually obtained through an increase in the risk of resistance.

4.2. Treated Versus Untreated Refuge

Figure 2 illustrates the effect of allowing conventional pesticide treatments on refuge based on economic thresholds. In Figure 2, the value of production and risk of resistance are compared for treated and untreated refuge as refuge increases from 19 to 35%. At 19% refuge, the value of production is maximized for both treated and untreated refuge. As refuge approaches 35%, the risk of resistance approaches 0 for treated and untreated refuge. Conventional pesticide use is also shown for when refuge is treated using economic thresholds.

Allowing conventional pesticide treatments based on economic thresholds has a number of notable impacts. As suspected, allowing refuge treatments increases the risk of resistance and conventional pesticide use, though only modestly. It also increases the value of production, but again rather modestly. The value of production is less sensitive to increasing refuge when refuge treatments are allowed because farmers have more flexibility and can use conventional pesticide applications to enhance control on refuge in years of severe ECB infestations.

The implications of these results are that allowing refuge treatments decreases the cost of resistance management to farmers and industry, while increasing the cost to the environment in terms of increased conventional pesticide use. Without refuge treatments, 25.7% refuge is required to reduce the risk of resistance to 1 in 20. With 25.7% refuge, the value of production is $119.16 without treatments. This same value of production could be achieved by increasing refuge to 26.9% and allowing growers to treat using economic thresholds. With 26.9% treated refuge, the risk of resistance becomes 1 in 25. Therefore, allowing refuge treatments provides the opportunity to either increase the value of production, reduce the risk of resistance, or both. Again, these benefits are not free because allowing refuge treatments increases conventional pesticide use.

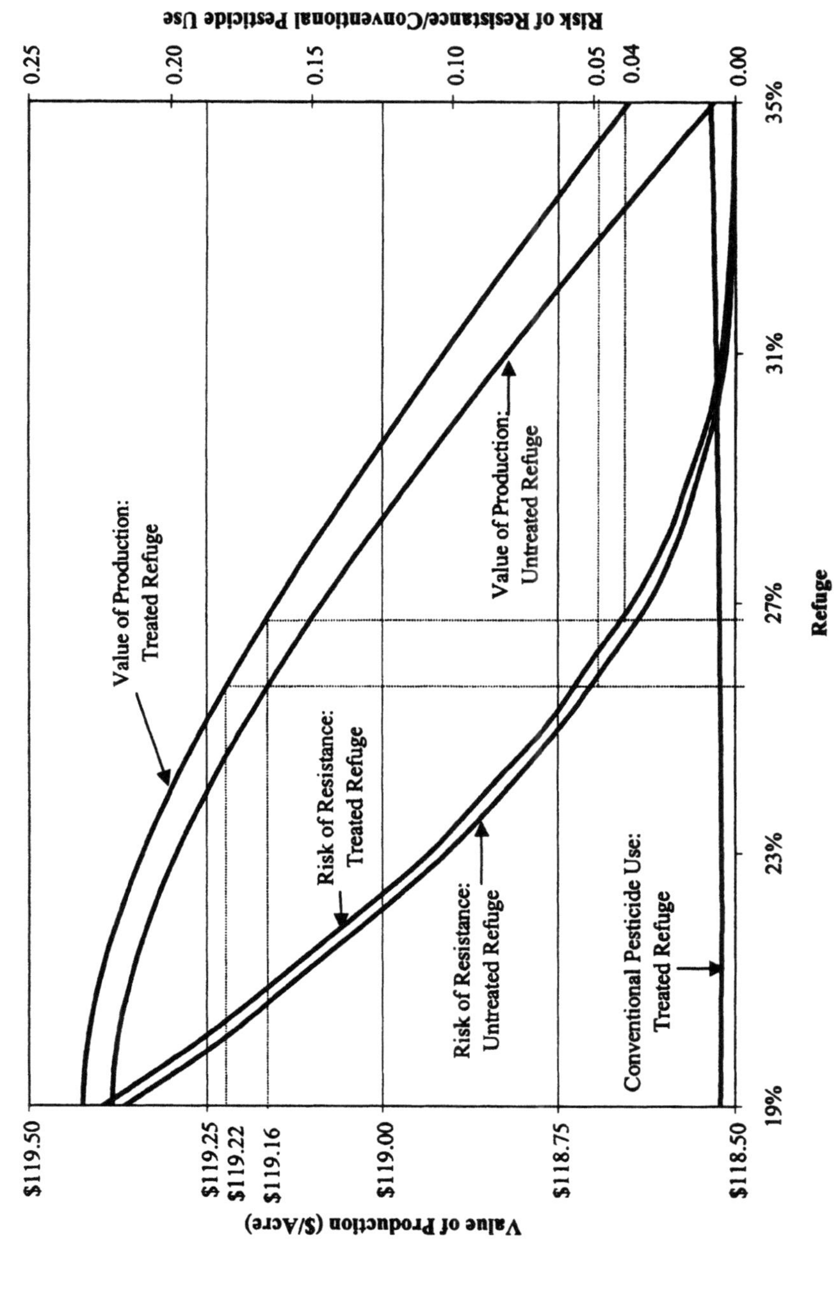

Figure 2. Value of production, conventional pesticide use, and risk of resistance with and without refuge treatments based on economic thresholds.

4.3. Model Sensitivity

It is important to keep in mind that the results reported in Figures 1 and 2 depend on a variety of assumptions. While we find that the qualitative results of the model are robust to changes in the value of different parameters, more specific results, such as the proportion of refuge that maximizes the value of production, are more sensitive. Therefore, it is the qualitative results that are important to remember and not specific numbers.

The benchmark simulations imply that an acre of land would have received a conventional spray application in 1 out of every 10 years prior to the introduction of Bt corn. A result that is consistent with grower surveys on pesticide use in the Midwestern US. In these regions, allowing refuge treatments has little impact on the risk of resistance and value of production because conventional pesticide applications are likely to be even less frequent with the widespread adoption of Bt corn. Treatments will be less frequent because ECB populations are likely to be lower on average due the high efficacy of Bt corn.

Allowing refuge treatments will have a greater impact in regions where the frequency of conventional pesticide treatments has been historically high. Two categories of factors can result in more frequent pesticide treatments. On the revenue side, if pest-free yield is higher, damages more severe, ECB infestations more frequent and severe, or the survival rates for conventional pesticides are lower, economic thresholds will be lower and treatments more frequent. On the cost side, lower pesticide and application costs reduce economic thresholds resulting in more frequent conventional treatments. We now explore the sensitivity of our results for treated refuge to both revenue and cost factors. For revenue factors, we increase the pest-free yield from 0 to 100%. For cost factors, we reduced the cost of a conventional pesticide treatment from 0 to 90%.

Tables 3 and 4 report the sensitivity of the simulation results to increases in crop revenues and decreases in conventional pesticide treatment costs. The first column reports the average frequency of conventional pesticide treatments when no Bt corn is planted. This frequency of treatments is a measure of the historic frequency of treatments prior to the introduction of Bt corn and increases with increasing crop revenues and decreasing conventional treatment costs.

The second column reports the percentage of refuge that maximizes the value of production without regard for resistance or conventional pesticide use. The third column reports the percentage increase in the proportion of refuge that maximizes the value of production, while constraining the risk of resistance to less than 1 in 20.

The fourth column reports the elasticity of the value of production with respect to the risk of resistance for increasing refuge to reduce the risk of resistance to less than 1 in 20. This elasticity is the percentage decrease in the value of production

Table 3. Sensitivity of simulation results to an increase in crop revenues.

Average annual frequency of conventional pesticide treatments without Bt corn	Percentage of refuge that maximizes the value of production	Percentage increase in refuge required to reduce the risk of resistance to 1 in 20	Elasticity of the value of production with respect to the risk of resistance	Elasticity of conventional pesticide use with respect to the risk of resistance
0.10	18.9	38.0	2.4×10^{-3}	0.017
0.12	18.8	39.2	2.1×10^{-3}	0.025
0.14	18.9	39.2	1.9×10^{-3}	0.053
0.15	18.9	39.8	1.7×10^{-3}	0.073
0.17	18.8	41.0	1.6×10^{-3}	0.076
0.20	18.8	42.7	1.5×10^{-3}	0.105
0.22	18.8	43.5	1.4×10^{-3}	0.135

Table 4. Sensitivity of simulation results to a decrease in the cost of conventional pesticide treatments.

Average annual frequency of conventional pesticide treatments without Bt corn	Percentage of refuge that maximizes the value of production	Percentage increase in refuge required to reduce the risk of resistance to 1 in 20	Elasticity of the value of production with respect to the risk of resistance	Elasticity of conventional pesticide use with respect to the risk of resistance
0.10	18.9	38.0	2.4×10^{-3}	0.017
0.12	18.9	39.0	2.4×10^{-3}	0.030
0.15	18.9	39.8	2.4×10^{-3}	0.060
0.18	18.8	41.8	2.5×10^{-3}	0.091
0.22	18.9	43.1	2.5×10^{-3}	0.139
0.29	18.9	47.0	2.6×10^{-3}	0.171
0.40	19.2	54.3	2.8×10^{-3}	0.232
0.53	21.5	53.2	2.7×10^{-3}	0.295
0.69	22.3	61.4	2.8×10^{-3}	0.317
0.97	24.5	69.3	2.9×10^{-3}	0.347

divided by the percentage decrease in the risk or resistance and is a measure of the cost of reducing the risk of resistance in terms of the value of production.

The fifth column reports the elasticity of conventional pesticide use as refuge increases to reduce the risk of resistance to less than 1 in 20. This elasticity is the percentage increase in the conventional pesticide use divided by the percentage decrease in the risk of resistance and measures the cost of reducing the risk of resistance in terms of increased pesticide use.

The percentage of refuge that maximizes the value of production is relatively insensitive to higher crop revenues or lower application costs that result in more frequent conventional pesticide treatments. This is not the case for the percentage of refuge that reduces the risk of resistance to less than 1 in 20. As the frequency of conventional pesticide treatments increases, the size of refuge needed to reduce the risk of resistance to less than 1 in 20 also increases substantially.

The elasticity of the value of production is small and relatively insensitive to increases in treatment frequency due to either revenue or cost factors. This result suggests that the cost of increasing refuge to reduce the risk of resistance in terms of the value of production is relatively small. It is also interesting to note that if treatments are more frequent due to higher revenues, then the relative cost of increasing refuge to lower the risk of resistance is lower. Alternatively, if treatments are more frequent due to lower treatment cost, then the relative cost of increasing refuge to lower the risk of resistance is higher.

The elasticity of conventional pesticide use is more substantial and increases as revenue rises or treatment costs fall. Therefore, the cost of reducing the risk of resistance in terms of conventional pesticide use is relatively higher than the cost of reduced agricultural production. The cost will also be higher in regions where the historic frequency of treatments is higher.

5. Conclusion

Bt corn offers growers a powerful new tool for controlling the European corn borer (ECB), a significant agricultural pest in the Midwestern United States (US). Unfortunately, the high efficacy and widespread adoption of Bt corn could result in the rapid development of ECB resistance to Bt. If ECB resistance to Bt develops, growers will lose a valuable new technology for controlling the ECB and may turn to other more hazardous pesticides to obtain sufficient control.

The US Environmental Protection Agency (EPA) is concerned about resistance and would like to preserve Bt corn as a reduced risk pesticide. Industry and academic scientists have developed a high-dose refuge strategy to combat ECB resistance to Bt corn. The foundations of this strategy are for Bt corn to express enough toxins to kill all but the most resistant ECB and for growers to plant a proportion of their acreage to refuge corn where Bt is not used for control. Refuge corn slows the evolution of resistance and serves to preserve the efficacy of Bt corn.

Recently, the EPA mandated that farmers in the Midwestern US plant at least 20% refuge corn with their Bt corn. This 20% refuge requirement allows growers to treat refuge with conventional pesticides in years of severe ECB infestation using economic thresholds. This mandate represents a departure from previous recommendations which required farmers to plant more refuge when the refuge was going to be treated with conventional pesticides. Therefore, concerns have emerged regarding whether 20% refuge is enough when conventional pesticide treatments are allowed based on economic thresholds.

We develop a stochastic dynamic bioeconomic simulation model to evaluate the effect of refuge treatments based on economic thresholds on agricultural productivity, conventional pesticide use, and the risk of resistance. We find that allowing refuge treatments should not substantially increase the risk of ECB resistance to Bt corn throughout most of the Midwestern US, but could increase the value of production to farmers and industry and the use of conventional pesticides. The reason for this result is that conventional pesticide treatments for the ECB have been historically low due to high application costs and poor efficacy. With the widespread adoption of Bt corn, average ECB populations are likely to fall such that refuge treatments will be even more unlikely. Infrequent refuge treatments have little impact on the risk of resistance.

Whether conventional pesticide treatments for refuge based on economic thresholds should be allowed in regions with historically high frequencies of pesticide use depends on the primary objectives of the policy. Refuge treatments should not be allowed without higher refuge requirements if the primary goal is to limit the risk of resistance. However, if the primary goal is to reduce conventional pesticide use or improve agricultural production, then allowing treatments with current refuge requirements could be sufficient.

The model we develop provides a useful framework for comparing alternative resistance management strategies based on a range of different policy objectives. An important weakness the model still shares with others is a failure to account for factors that influence farmer adoption of Bt corn and compliance with insect-resistance management requirements. Currently, more research is needed to understand adoption and compliance behavior. Once a better understanding is obtained, the model can be augmented by specifying the proportion of refuge planted as a function of farmer adoption and compliance incentives. Models that more explicitly consider farmer adoption and compliance behavior will provide more reliable estimates of the economic and environmental tradeoffs of using refuge to manage ECB resistance to Bt corn.

Notes

[1] There is a separate 50% refuge requirement for Bt corn in areas of the US where cotton is predominantly grown due to the potential interactions between Bt corn and Bt cotton.

[2] The Hardy–Weinberg model lies at the foundation of population genetics due to its remarkable ability to predict gene frequencies and heritability. The principle is an extension of Mendelian inheritance and is used extensively by population biologist to describe the inheritance of genetic traits such as resistance. Examples exploring ECB resistance to Bt corn are found in Gould (1998), Onstad and Gould (1998a, 1998b) and Roush and Osmond (1996). The fundamental assumptions of the model are (i) a diploid pest, (ii) sexual reproduction, (iii) non-overlapping generations, (iv) random mating, (v) large populations, (vi) negligible migration, (vii) negligible mutation, and (viii) no selection pressure (Hartl, 1988).

[3] A gamma distribution was also explored, but the predicted population did not fit as well.

[4] Personal communication with Dr. D.A. Andow of the University of Minnesota.

[5] Depending on the rate of adoption of Bt corn, there could be supply-side price effects that are not treated and depend on refuge size.

References

Alstad, D.N. and Andow, D.A. (1995) Managing the evolution of resistance to transgenic plants, *Science* **268**, 1894–1896.

Andow, D.A. and Alstad, D.N. (1998) The F2 screen for rare resistance alleles, *J. Econ. Entomol.* **91**, 572–578.

Bauer, L.S. (1995) Resistance: A threat to insecticidal crystal protiens of *Bacillus thuringiensis, Florida Entomol.* **78**(3), 414–443.

Calvin, D.D. (1996) Economic benefits of transgenic corn hybrids for European corn borer management in the United States, Public Interest Document Supporting the Registration and Exemption from the Requirement of a Tolerance for the Plant Pesticide *Bacillus thuringiensis* subsp kurstaki Insect Control Protein as Expressed in Corn (*Zea mays L.*) (Kent A. Croon).

Caprio, M.A. (1998) Evaluating resistance management strategies for multiple-toxins in the presence of external refuges, *J. Econ. Entomol.* **91**(5), 1021–1031.

Clark, J.S. and Carlson, G.A. (1990) Testing for common versus private property: The case of pesticide resistance, *J. Environ. Econ. Manag.* **19**, 45–60.

Gorddard, R.J., Pannell, D.J. and Hertzler, G. (1995) An optimal control model for integrated weed management under herbicide resistance, *Austr. J. Agricult. Econ.* **39**(1),71–87.

Gould, F. (1998) Sustainability of transgenic insecticidal cultivars: Integrating pest genetics and ecology, *Ann. Rev. Entomol.* **43**, 701–726.

Gould, F., Anderson, A., Jones, A., Sumerford, D., Heckel, D.G., Lopez, J., Micinski, S. and Leonard, R. (1997) Initial frequency of alleles for *Bacillus thuringiensis* toxins in field populations of *Heliothis virescens, Proc. Natl. Acad. Sci. USA* **94**, 3519–3523.

Hartl, D.L. (1988) *A Primer of Population Genetics*, second edition, Sinauer and Associates, Sunderland, MA.

Hueth D. and Regev, U. (1974) Optimal agricultural pest management with increasing pest resistance, *Amer. J. Agricult. Econ.* **56**(), 543–552.

Hurley, T.M., Babcock, B.A. and Hellmich, R. (1997) Biotechnology and pest resistance: An economic assessment of refuges. Working Paper 97-WP 183, Center for Agricultural and Rural Development, Ames, IA.

Hurley, T.M., Secchi, S., Babcock, B.A. and Hellmich, R. (1999) Managing the risk of European corn borer resistance to transgenic corn: An assessment of refuge recommendations, CARD Staff Report 99-SR 88.

ILSI/HESI (1999) *An Evaluation of Insect Resistance Management in Bt Field Corn: A Science Based Framework for Risk Assessment and Risk Management*, ILSI Press, Washington, DC.

Liu, Y.B. and Tabashnik, B.E. (1997) Experimental evidence that refuges delay insect adaptation to *Bacillus thuringiensis*, *Proc. Roy. Soc. London B* **264**, 605–610.

Mason, C.E., Rice, M.E., Calvin, D.D., Van Duyn, J.W., Showers, W.B., Hutchison, W.D., Witkowski, J.F., Higgins, R.A., Onstad, D.W. and Dively, G.P. (1996) *European Corn Borer Ecology and Management. North Central Regional Extension*, Publication No. 327, Iowa State University, Ames, IA.

McGaughey, W.H. and Beeman, R.W. (1988) Resistance to *Bacillus thuringiensis* in colonies of Indianmeal moth and almond moth (Lepidoptera: Pyralidae), *J. Econ. Entomol.* **81**, 28–33.

Mellon, M. and Rissler, J. (1998) *Now or Never: Serious New Plans to Save a Natural Pest Control*, Union of Concerned Scientists, Cambridge, MA.

Mitchell, P.D., Hurley, T.M. and Hellmich, R.L. (2000) Economic evaluation of Bt corn refuge insurance, Working Paper 00-WP 243, Center for Agricultural and Rural Development, Ames, IA.

Onstad, D.W. and Gould, F. (1998a) Do dynamics of crop maturation and herbivorous insect life cycle influence the risk of adaptation to toxins in transgenic host plants?, *Environ. Entomol.* **27**, 515–522.

Onstad, D.W. and Gould, F. (1998b) Modeling the dynamics of adaptation to transgenic maize by European corn borer (Lepidoptera: Pyralidae), *J. Econ. Entomol.* **91**, 585–593.

Ostlie, K.R., Hutchison, W.D. and Hellmich, R.L. (1997) Bt corn and the European corn borer, NCR publication 602, University of Minnesota, St. Paul, MN.

Peck, S.L., Gould, F. and Ellner, S.P. (1999) Spread of resistance in spatially extended regions of transgenic cotton: Implications for management of *Heliothis virescens* (Lepidoptera: Noctuidae), *J. Econ. Entomol.* **92**(1), 1–16.

Perez, C.P. and Shelton, A.M. (1997) Resistance of *Plutella xylostella* (Lepidoptera: Plutellidae) to *Bacillus thuringiensis* in Central America, *J. Econ. Entomol.* **90**, 87–93.

Regev, U., Gutierrez, A.P. and Feder, G. (1976) Pests as a common property resource: A case study of alfalfa weevil control, *Amer. J. Agricult. Econ.* **58**, 186–197.

Regev, U., Shalit, H. and Gutierrez, A.P. (1983) On the optimal allocation of pesticides with increasing resistance: The case of the alfalfa weevil, *J. Environ. Econ. Manag.* **10**, 86–100.

Roush, R. and Osmond, G. (1996) Managing resistance to transgenic crops, in N. Carozzi and M. Koziel, eds., *Advances in Insect Control: The Role of Transgenic Plants*, Taylor and Francis, London, pp. 271–294.

Tabashnik, B.E. (1994) Evolution of resistance to *Bacillus thuringiensis*, *Ann. Rev. Entomol.* **39**, 47–79.

Taylor, C.R. and Headley, J.C. (1975) Insecticide resistance and the evaluation of control strategies for an insect population, *Canad. Entomol.* **107**, 237–242.

US Environmental Protection Agency (1998a) The environmental protection agency's white paper on Bt plant-pesticide resistance management, Biopesticides and Pollution Prevention Division, Office of Pesticide Programs, Office of the Assistant Administrator for Prevention, Pesticides and Toxic Substances.

US Environmental Protection Agency (1998b) FIFRA Scientific Advisory Panel, Subpanel on *Bacillus thuringiensis* (Bt) Plant-Pesticide and Resistance Management, February 9–10, Docket No. OPP 00231.

Venette, R.C., Hutchison, W.D. and Andow, D.A. (2000) Early detection of insect resistance to *Bacillus thuringiensis* in transgenic crops: Practical and statistical considerations for a field-based method, *J. Econ. Entomol.* **93**(4), 1055–1064.

PART D

MANAGING INDUSTRY:
THE ECONOMICS OF MANAGING INDUSTRIAL IMPACTS

9. Monopolisation and the Regulation of Genetically Modified Crops: An Economic Model

ALISTAIR MUNRO

1. Introduction

The introduction of genetically modified organisms (GMOs) into the market place has brought with it controversy, notably over the potential scientific risks of the new biotechnologies, but also over perceived threats, such as the monopolisation of food supply, which have a strong economic dimension. Nevertheless compared to discussion of the scientific issues, economic analysis of regulation has been far less common, despite the fact that some features of the new biotechnology industries (e.g. vertical restraint and monopolisation) are familiar features of many other industries and thus have a long history, both of economic analysis and economic regulation by national governments.

Many of the fears about monopolisation have been propelled by the rapid adoption of GM varieties in countries, such as the USA, where governments have placed relatively few regulatory hurdles in the path of the new biotechnologies. Beyond this lies a worldwide rise in concentration in the seed industry, coupled with greater integration between agro-chemical, seed and life-science firms. Hayenga (1998), for instance, reports C4 (the percentage of the market supplied by the largest four producers) figures of 70% for the US corn market, 47% for the purchased soybean seed market and in excess of 90% for cotton seed.

The forces behind this rise in concentration are unclear. One possibility is that it arises from the inefficiencies of incomplete integration in the presence of asset specificity – along Williamsonian lines. A second hypothesis is that the worldwide tightening of intellectual property rights legislation has raised the appropriability of biotechnology innovations. In the past weakness of property rights meant that any potential monopoly power gained through concentration could not be exploited because of the ease of entry through imitation. Enforceable property rights might then lead to the monopoly gains from concentration being reaped. A third possible source of concentration is regulation itself, where costs for regulatory approval tend to be fixed. This establishes an 'entry price' for competing in the regulated market; in turn limiting the feasible number of suppliers.

Whatever the sources of increased concentration, it is clear that, as with other forms of monopolisation, there is at least the possibility of a reduction in con-

Timothy Swanson (ed.), The Economics of Managing Biotechnologies, 197–217.

sumer welfare. Consequently, this paper puts forward two variants of a simple model of crop production, each one tailored to understanding a particular aspect of transgenic food technology and its regulation. The first variant of the model considers the issue of the introduction of a new technology owned by a monopolist which lowers the costs of production. The argument here is quite standard: if the old technology remains available and there is free entry, then the introduction of the new technology can harm consumer welfare. However, in many agricultural industries effective competition between transgenic and non-transgenic technologies relies on the availability to farmers of traditional plant varieties in sufficient numbers. So in the second variant of the model the issue of predatory pricing by the monopolist is considered and it is shown that, if storage costs are sufficiently high, the introduction of the new technology can raise prices and make consumers worse off. Finally, I consider some of the public policy implications of these models for the prudential regulation of transgenic technologies.

2. Background

Transgenic foods are foods where genes from other species have been introduced into a plant or animal, usually to create or enhance specific properties which would not be feasible through traditional breeding methods. The most well-known examples include the Flavr-Savr™ tomato, designed to have a longer shelf-life compared to traditional varieties and the various Roundup Ready™ plants such as soyabean, maize and sugar beet, created by Monsanto. In the latter case, genes from bacteria which confer resistance to Roundup™ (Monsanto's glyphosate herbicide) were introduced into farmed plant species. As a result, the plants have a higher tolerance for glyphosate which can therefore be used at higher doses to combat weed growth. The herbicide can also be used at times in the growth cycle (such as after crop emergence) which would previously have been disastrous for yields. In addition, the transgenic varieties also change the feasible set of crop rotations: traditional maize varieties are sensitive to glyphosate, residues of which remain in the soil. Maize engineered to be glyphosate-tolerant can be planted in rotation with soybean crops for example, when previously this was not feasible.

By 1995, 60 plant species had been genetically modified and over 3,000 field tests of their use conducted worldwide in over 32 countries, including the USA, Japan and most of the nations of the EU. Between 1992 and 1995, 95 field trials of GMOs were conducted in France, 59 in Belgium, 58 in the UK, 51 in the Netherlands and 22 in Germany. The major crops modified in Europe include oilseed rape, maize, potatoes, tomatoes and sugar beet. Meanwhile in the USA and Canada, soya, maize and cotton are also among the crops most subject to modification, while in Japan, rice has been the subject of transgenic experiments.

Table 1. Major transgenic crops in the USA, farmer-reported planting, 2000.

Crop	Percentage of total acreage		
	Herbicide-resistant only	Insect-resistant (Bt) only	Total
Corn	18	5	25
Soybean	52	n/a	52
Cotton	22	18	56

Source: National Agricultural Statistics Service (2000).
Note: Row totals include plantings of crops containing 'stacks' of inserted genes.

Table 2. The main economic risks from transgenic food technology.

Characteristic	Associated risk, cost or externality
Monopolisation	1. Reduced choice and higher prices
	2. Higher risks of crop failure
	3. Reduction in biodiversity
'Genetic pollution'	4. Accelerated pest evolution
	5. Transfer of antibiotic resistance to environment
	6. Transfer of herbicide resistance to weeds
	7. Low-level Bt toxicity creating increased resistance

Introduction of GMOs into the home has been gradual to date, with few crops grown commercially outside the USA and the rest of the Americas. The most widespread transgenic food is vegetarian cheese, production of which involves a genetically modified yeast. Meanwhile 2% of US soybean production was transgenic in 1996, rising to 15% in 1997 and as Table 1 shows, by 2000, the majority of soybean and upland cotton plantings for the USA were transgenic. In Europe, Bt corn has been grown in France and Spain since 1998. Other GMOs used commercially include cotton, tomatoes, maize and bacteria modified to produce bovine somatotropin.

Table 2 lists some of the economic concerns associated with transgenic crops. The economic consequences of some of these risks have been considered extensively elsewhere. Sianesi and Ulph (1998), for instance, examine the impact of reductions in crop variety on habitat diversity and hence on the number of wild bird species. They produce a model in which a rising tax (which eventually chokes off reductions in variety) is the optimal means of supporting diversity. In a series of papers, Goeschl and Swanson (1996, 1997, 1998), examine carefully how best to value and maintain biodiversity, including the advantages of *in situ* rather than *ex situ* conservation of gene stocks. Meanwhile, Munro (1997) considers the general impact of economic behaviour on evolutionary pressures, but does not consider the particulars of the new biotechnology. This paper is complementary to the previous work, more concerned with some of the conventional sources of market failure, which nevertheless pose a risk with the new technologies.

3. Monopolisation and Vertical Restraint

The models which follow are built around the example of engineered resistance to a herbicide, the form of transgenic crop which is most widespread. In a typical example of this, a gene (or genes) are introduced into a plant which confers resistance to a herbicide manufactured by the same agro-chemical company. Typically the firm selling the transgenic seed does not have monopoly rights to the generic herbicide. However its ownership of the seed variety may give it the power to impose vertical restraints, forcing buyers of the seed to also use its own brand herbicide. In the case of Monsanto's 1996 Roundup Ready™ gene agreement, signed by farmers if they wished to buy Roundup Ready™ soybean seed, farmers agreed to use only Roundup™ glyphosate herbicide on the crop and not competing brands. To enforce this and other aspects of the agreement, they had to also to agree to the right of Monsanto to inspect and test the field crops for a period of up to three years and pledge not to save, re-use, re-plant or sell the seed (RAFI, 1997).

Although the biotechnology industry was increasingly concentrated prior to the introduction of transgenic foods, and although most of these large firms are producing genetically modified food, the model which follows assumes perfect competition in both seed and pesticide industries prior to an innovation. Thereafter, a single firm dominates the market. These assumptions place an upper bound on the potential costs from the risk of monopolisation. Actual costs are likely to be lower.

I shall suppose that prior to the innovation, farms buy herbicide and seeds from competitive industries at a price per unit of p^h and p^s respectively. The farms themselves are in a competitive industry and face a constant marginal cost of production, q, of $c_n = c(p^h, p^s)$ where $c(\ldots)$ is the unit cost function for the farms and the subscript n refers to the non-GM status of the crop. Demand is given by $p = a - q$ where p is price, q is total farm production and $a > c_n > 0$. For each crop, after all costs have been incurred there is a strictly positive probability γ of crop failure, possibly due to disease, in which case output to market is zero. I suppose that for each seed type the probability of crop failure is an independent variable[1] and that, in the competitive state, the number of varieties of seed is sufficiently large that $(1 - \gamma)q$ reaches the market with certainty. Thus, in the competitive equilibrium,

$$p = c_n/(1 - \gamma),$$

$$q = (a(1-\gamma) - c_n)/(1-\gamma)^2. \quad (1)$$

We suppose now one firm innovates, to produce a complementary seed-herbicide combination such that, for all input prices,

$$c_g(p^h, p^s) < c_n(p^h, p^s), \quad (2)$$

where c_g is the typical farm's unit cost function with the new GM technology. Following Arrow (1962) there are two cases to consider, depending on whether the cost reductions from the innovation are sufficient to destroy all competition in the seed and herbicide industries.

Case 1. Monopoly production of the seed and pesticide

Any farm which uses the genetically modified product still faces a probability γ of crop failure, hence in the competitive equilibrium,

$$p^1 = c^1/(1-\gamma). \quad (3)$$

Clearly if there is only supplier in the industry, the following condition must be satisfied:

$$c_g < c_n, \quad (4)$$

where the right-hand side of this equation is the pre-innovation cost function evaluated at the pre-innovation input prices.

If γ is sufficiently large, there is the possibility that some farms may find it profitable to be fringe producers, making profits when the genetically modified crop fails. For this not to be profitable, the following condition must be satisfied:

$$\gamma(1-\gamma)a + (1-\gamma)^2 p^1 - c_n < 0. \quad (5)$$

The a in this equation is the price received when only one unit is available in the market. Simplifying yields

$$(c_g + \gamma a)/(1-\gamma) < c_n. \quad (6)$$

Case 2.

If $(c_g + \gamma a)/(1-\gamma) \geq c_n$ then fringe production is potentially profitable. Let p^H be the price when the genetically modified crop fails. In the competitive equilibrium for fringe producers, expected profits are zero:

$$p^H \gamma(1-\gamma) + (1-\gamma)^2 p^1 - c_g = 0. \quad (7)$$

Meanwhile for producers of the GM crop, expected profits are also zero, so that

$$p^1 = c_g/(1-\gamma). \quad (8)$$

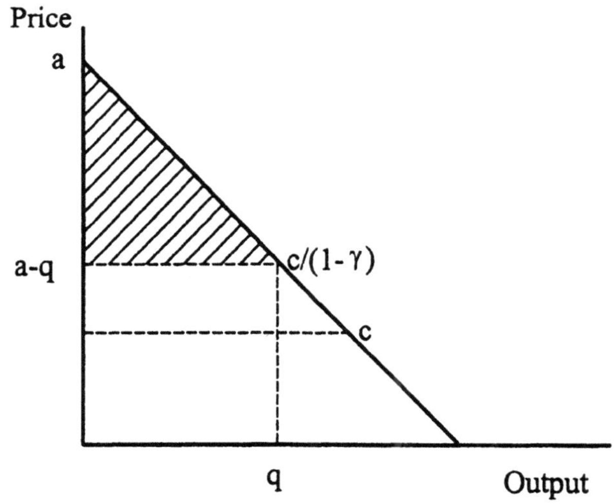

Figure 1. Welfare gains from innovation.

Note that, in general the prices charged by the profit maximising supplier GM seed and its complementary herbicide will differ between these two cases, hence c^1 and p^1 will also differ.

3.1. PROFIT MAXIMISATION

The GM firm's problem is to choose prices p^s and p^h to maximise profits,

$$(p^s - c^s)s + (p^h - c^h)h, \tag{9}$$

where s and h are the production of seed and herbicide respectively and c_i $(i = 1, 2)$ is the constant marginal cost of production of product i. In turn, the demand for s and h are determined by the prices charged, so that $s = s(p^s, p^h)$, $h = h(p^s, p^h)$.

First-order conditions are,

$$(p^s - c^s)s_1 + s + (p^h - c^h)h_1 = 0, \tag{10a}$$

$$(p^s - c^s)s_2 + h + (p^h - c^h)h_2 = 0, \tag{10b}$$

where subscripts indicate partial derivatives.

3.2. CONSUMER WELFARE

Let the consumer surplus be V, which under these conditions is the shaded area illustrated in Figure 1. At a price, p, $V = [a - p]^2/2$, hence V is strictly convex in prices. As a result, a sufficient condition for expected consumer surplus to rise in the wake of the introduction of the GMO is that the mean price does not rise. That is,

$$\gamma p^H + (1-\gamma)p^L \neq p^0. \tag{11}$$

In Case 1, $p^H = a$ and $p^L = c_g/(1-\gamma)$ which gives us

$$\gamma a + (1-\gamma)c_g = \frac{(1-\gamma)\gamma a + (1-\gamma)^2 c_g}{1-\gamma} < \frac{c_n}{1-\gamma} = p^0, \tag{12}$$

where the penultimate step in this argument follows from Equation (5) – the condition which defines Case 1. In Case 2, by Equation (7) the expected value of post-innovation prices is equal to the pre-innovation price. Hence (11) is also satisfied, giving the following proposition:

PROPOSITION. *Expected consumer surplus rises in the wake of the GM innovation.*

This proposition is quite general, turning on the quasi-convexity of the indirect utility function and the constraint on the monopolist posed by the competitive fringe. Since producer surplus is zero before the innovation and positive afterwards, it is also true that the sum of expected surpluses also rises. The result is unsurprising (at least to economists) since the argument is closely based on Arrow's (1962) analysis of the benefits of cost-cutting technology. Since the price chargeable by the monopolist is bound above by the cost of the old technology, post-innovation prices cannot rise in this model and hence consumers cannot lose from the introduction of the new technology.

In order for consumers to lose, some other factor must therefore be important: either greater risk or the ability of the monopolist to eliminate competition from the existing technology or an externality. I consider each of these factors, in turn.

3.3. Risk

Crop yields are inevitably stochastic. Some risks are specific to a particular farm or region or not specific to particular varieties, but other risks, such as diseases are often selective in the damage they do, harming yields from varieties which have a specific vulnerability in the germplasm, while being resisted by other varieties. If the introduction of new technologies means that a narrower range of crop varieties is grown then in theory, the variance of aggregate yields should rise as a result. The evidence to date on this is mixed. Wright (1997) concluded cautiously that

> The hypothesis that greater worldwide uniformity of germplasm due to the increased dominance of high-yield varieties is not associated with greater relative yield fluctuations cannot be rejected at present.

For instance, Hazell (1989) reports an increase in the coefficient of variation from 2.8 to 3.4% for cereal yields (outside of mainland China) for a period running from the 1960s to 1983. Against this, Singh and Byerlee (1990) point to a declining variability in wheat yields for the 1951–1986 period.

In the specific case of Bt varieties, there are reasons for anticipating reductions in risk: damage to corn by the European corn borer is stochastic, varying from year to year, but positively correlated across farms within a region for a given year. Bt corn varieties reduce yield losses to the borer, leading, as Gianessi and Carpenter (1999) point out, to it being recommended for planting as a form of insurance.

However, let us suppose for the moment that the new technologies do lead to greater variance in aggregate crop production levels. As noted in Section 3.2, if the competitive fringe of producers using the old technology survives, then the welfare gains from the new lower-cost technology will still be positive. In addition, even if fringe producers do not survive, as long as the mean price is lower, then any higher variance of prices raises welfare still further, because indirect utility functions are quasi-convex. Any costs attached to higher variability in prices must therefore be costs to producers or due to the absence of the competitive fringe. On the first of these, note that in the models presented so far is no producer surplus to consider. However, in general producer surplus is, like consumer surplus, convex in prices and hence greater randomisation (for a given mean) raises expected producers surplus. This last result depends on the assumption that producers are neutral with regard to risk in income or have access to competitive insurance markets. However, when farmers are largely dependent on farm incomes and markets are incomplete as is often the case in developing countries, such assumptions are probably incorrect. It is then more reasonable to suppose aversion to farm income risk, especially for societies where significant drops in household income imply malnutrition or worse (see Newbery and Stiglitz, 1981). Where this is the case consumer welfare may rise with the new technology (since the average price falls), but producer welfare may rise or fall.

To summarise, the risks from monopolisation are likely to be largest in poor countries. In richer countries, where the consequences of crop failure are more easily moderated, there is less of a clear argument for control of the new biotechnology on monopoly grounds (though see next section). However, with regard to the hypothesis of increased variability in food supply and to its economics consequences there is a case for much further research.

4. Predatory Pricing

The results of the previous section rest on the existence of free entry using the old technology, which caps the price the monopolist can set. However, in the case of agriculture, an essential factor, namely seed, is subject to degradation if stored

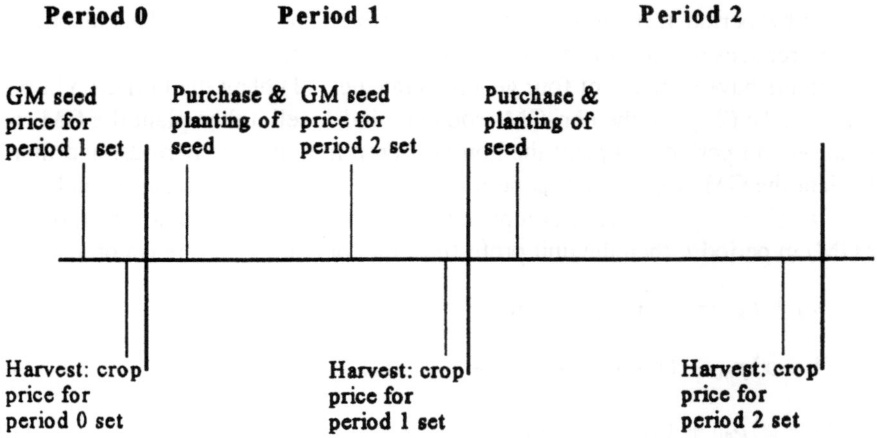

Figure 2. The timeline for a predatory pricing model.

for sustained periods. Meanwhile planting and harvesting merely to maintain the health of the seed base may be financially unsustainable for small farmers. In this section I adapt the model to consider this possibility.

Suppose that there are three time periods, 0, 1 and 2. Figure 2 summarises the timeline for decisions. The GMO technology is introduced without announcement at time 0, after the previous year's crop production has been set, but not yet sold.[2] Farms can either sell the non-GMO crop for seed, keep it for seed,[3] or sell it to consumers. In period 1 farms must choose whether to stay in the industry or exit (or enter), then which technology to use. Crop production is then set, the product harvested and farmers make the same decisions about what to do with the harvest as at the end of period 0. Period 2 is a re-run of period 1, except all product is consumed at the end. While this adds an artificial end to the model, it provides a convenient means of analysing the problem faced by agents involved.

Prior to the introduction of the GMO, a proportion of the crop is held back each year in order to provide seed for the following season. Suppose that to produce 1 tonne of the crop, α tonnes of the previous year's crop are required.[4] In addition to the opportunity cost represented by this input, there is an additional cost of c_n per tonne of product. With free entry and perfect competition, prior to introduction of the GMO, the zero profit condition is $p_0 = c_n + \alpha p_0$, or $p_0 = c_n/(1-\alpha)$. Consumer demand is then $(a - p_0)$, so that the total harvest is $(a - p_0)/(1-\alpha)$. Note that p_0 will not, in general, be the equilibrium price once the availability of the new technology is announced.

A key issue in models of predation is the ability of firms to commit to particular time paths for output or prices. I shall assume that the firm supplying the GM seed will set its price for period 2 so as to maximize profits in that period. In other words, it is unable to commit to a second period price in period 0, when it chooses its price for period 1. It chooses a price for period 1 knowing this. This inability to commit reduces its power to price in a predatory manner.

Farmers have a choice of four generic strategies: (1) plant the GM crop in both time periods; (2) plant the non-GM crop in period 1, sell it then plant the GM crop in the second period; (3) plant the non-GM crop in both time periods, and finally (4) plant the GM crop in the first period and the non-GM crop in the second.

If p_{it} ($i = n, g, t = 0, 1, 2$) represents the price of the ith seed type (non-GM or GM) in period t, then the unit profit from the four strategies are (in order),

$$p_{n0} + p_{g1} - c_{g1} + p_{g2} - c_{g2}, \tag{13}$$

$$(1 - \alpha)p_{n0} + (1 - \alpha)p_{n1} - c_n + p_{n2} - c_n, \tag{14}$$

$$(1 - \alpha)p_{n0} + p_{n1} - c_n + p_{g2} - c_{g2}, \tag{15}$$

$$p_{n0} + p_{g1} - c_{g1} + p_{n2} - c_{n2} - \alpha p_{n1}. \tag{16}$$

Note that the on-farm cost of growing the GM variety is potentially time period-dependent, because the monopolist can set different prices for its seed in the two periods. On the other hand, for simplicity, it is assumed that the non-seed costs of growing the non-traditional variety are constant.

From these equations, it follows, for example, that in the second period the farmer will switch to the GM crop from growing the traditional variety, provided $p_{g2} - c_{g2} > p_{n2} - c_n - \alpha p_{n1}$. If this is the case then there will be no seed market at the end of the first period. For the remainder of the section, I shall make the simplifying assumption that consumers have no preference between GM and non-GM varieties. Thus $p_{gi} = p_{ni} = p_i$ ($i = 1, 2$). Then, for example, no non-GM seed will be planted if $c_n + \alpha p_{n1} > c_{g2}$.

Suppose that with the GM crop, the unit costs in period t are $c_g + \alpha_g r_t$ where r_t is the price of the GM seed and α_g is the amount of GM seed which must be planted to produce 1 unit of the crop. If the marginal cost of producing the seed is k, then in the final period the monopolist will choose r_2 to maximise profits $\pi_2 = (r_2 - k)\alpha_g[a - c_g - q_{n2} - \alpha_g r_2]$. It seems reasonable to suppose that, post-development costs, the GM seed is produced using a technology similar to the main crop. If this is the case then $k \cdot c_g/(1 - \alpha_g)$. In the interests of reducing parameters, I shall take this relationship to be exact, in which case the optimal price is

$$r_2 = \frac{1}{2\alpha_g}(c - q_{n2}). \tag{17}$$

If $q_{n2} = 0$ (i.e. no non-GM is grown in the second period) then (17) produces a consumer price and a unit cost of $p_2 = c_{g2} = (a + c_g/(1 - \alpha_g))/2$. This is the monopoly price – the price that would be set if the firm owned the farms directly and bought its seed at a marginal cost of $c_g/(1 - \alpha_g)$.[5]

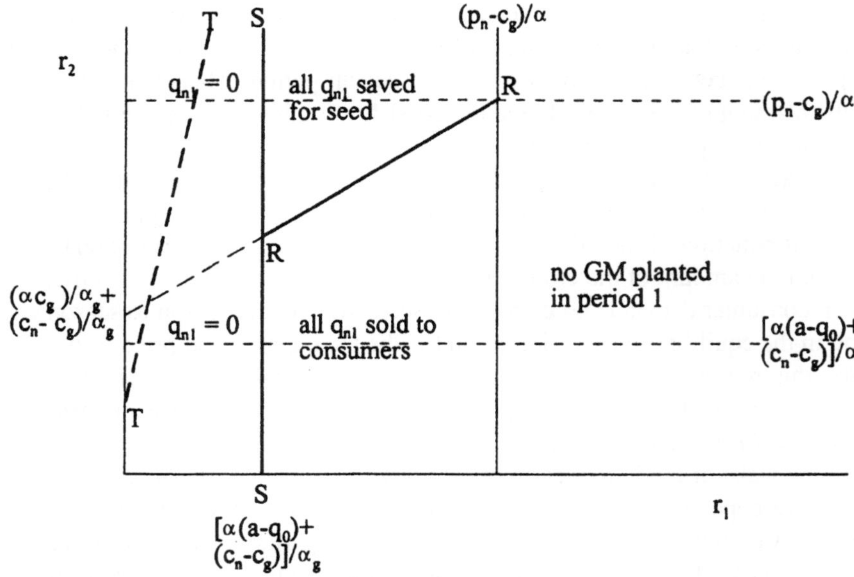

Figure 3. The feasibility and desirability of predation.

Figure 3 is a device for understanding the outcomes in this model. The axes show the prices of the GM seed in the first and second periods. Note first that $(p_0 - c_g)/\alpha_g$ is the value of r_1 at which the unit cost to the farmer of the GM technology is equal to the unit cost of the non-GM technology. Hence no GM seed will be planted in period 1 if r_1 exceeds $(p_0 - c_g)/\alpha_g$.

If r_1 is below this critical price then some GM crops will be sown in period 1. Some non-GM crops may also be sown. Whether any non-GM crops sown in period 1 are sold directly to consumers or used for seed depends on the relative profit of each strategy. If GM crops are sown in each period then given perfect competition amongst farmers the price of output is determined by the price of seed. That is,

$$p_1 = (r_1\alpha_g + c_g),$$

$$p_2 = (r_2\alpha_g + c_g). \tag{18}$$

Now whether the non-GM crop harvested in period 1 is sold directly to consumers or used for seed depends on the profitability of these alternative strategies. Thus, if $p_2 > c_n + \alpha p_1$ all the crop will be used for seed. Using (18) this yields the boundary condition

$$r_2 = \frac{(c_n - c_g) + \alpha(\alpha_g r_1 + c_g)}{\alpha_g}. \tag{19}$$

If r_2 exceeds this value, then all of any non-GM crop will be kept for seed in period 1. If r_2 is less than this condition, then all the crop will be sold directly to consumers. In Figure 3 the line segment RR depicts the relevant section of the equation. Note that unlike the standard model of predatory pricing, the use of the output in one period as an input in the subsequent period means that, conditional on the planting of the non-GM seed in period 1, it is a high price in period 1 which eliminates final period competition.

However, the amount of non-GM seed sown in period 1 will depend on the price of the crop in period 1, relative to the price in period 0. Let us suppose that prior to the introduction of the GM strain, the market was in a long-run stationary state in which the amount of the crop grown each season was constant and sufficient to match consumer demand and demand for seed without any season to season alteration in the equilibrium price. Let q_n and p_n be the quantity and price respectively under this regime. With free entry, the unit cost of production equals the price: $\alpha p_n + c_n = p_n$, hence $p_n = c_n/(1-\alpha)$. Thus, the total harvest in period 0 is $q_0 = q_n = (a(1-\alpha) - c_n)/(1-\alpha)^2$.

In the wake of the introduction of GM seed, $p_0 = a - q_0 + \alpha q_{n1}$. Whether q_{n1} is positive depends on the returns from using the non-GM variety still. GM and non-GM varieties will only be planted in the same period if the price of non-GM seed adjusts so that there is no cost advantage in planting the GM seed. In other words, $r_1\alpha_g + c_g = p_0\alpha + c_n$. It follows that, given the value of r_1 set by the GM producer, $q_{n1} = [(q_0 - a)/\alpha] + (r_1\alpha_g + c_g - c_n)/\alpha^2$. Or, no non-GM seed will be planted in period 1 if

$$r_1 < \frac{1}{\alpha_g}\left[\frac{-\alpha^2 a}{1-\alpha} + c_n\left(1 + \frac{\alpha}{(1-\alpha)^2}\right) - c_g\right] \tag{20}$$

which can be written more simply as $r_1 < [\alpha(a - q_0) + (c_n - c_g)]/\alpha_g$. This is the condition depicted by the vertical line segment SS in Figure 3. Note that the right-hand side of (20) is decreasing in the costs of producing the GM crop, increasing in the quantity of the crop required to produce 1 tonne of the crop in the following year and increasing in the cost of producing the non-GM crop. The equation is a necessary condition for the predatory price to be optimal, but to be sufficient we need to know that a strategy of predation is more profitable than a non-predatory approach and that storage of the non-GM seed through period 1is not an economically feasible means for producers of the non-GM crop to remain in production.

I tackle the issue of profitability first. It can be seen from Figure 3 that the critical issue is whether a marginal increase in r_1 is profitable along the boundary SS, i.e. at points where planting of non-GM seed in period 1 is just zero. Let the GM producer's profits be $\pi = \pi_1 + \pi_2$ where π_i $(i = 1, 2)$ is the profit in period i. Then,

$$\frac{d\pi}{dr_1} = \frac{\partial \pi_1}{\partial r_1} + \frac{\partial \pi_2}{\partial r_2}\frac{dr_2}{dq_{n1}}\frac{dq_{n1}}{dr_1} + \frac{\partial \pi_2}{\partial q_{n1}}\frac{dq_{n1}}{dr_1}. \tag{21}$$

Note first that in period 2, r_2 will always be chosen optimally (conditional on previous decisions), hence by the envelope theorem, the middle term on the right-hand side of (21) is zero. The sign of the final term on the right-hand side of (21) depends on whether the expression is evaluated above or below the intersection of SS with RR. If below, then no non-GM seed is saved in period 1, so $\partial\pi_2/\partial q_{n1} = 0$. Hence, the optimality of pricing to eliminate all non-GM seed depends only on the sign of $\partial\pi_1/\partial r_1$ – in other words, on whether the monopoly price of the GM seed lies below $\alpha(q_0 - a) + (c_n - c_g)/\alpha_g$. If it does, then in the language of industrial organisation, the non-GM variety is *blockaded* – in other words the cost advantage of the GM variety is sufficiently great that, even if the GM seed is priced to maximise monopoly profits, competition from the alternative technology is not sustainable.

If we now consider the segment of SS above its intersection with RR, any planting of the GM crop in the first period will have negative implications for profits for the second period, because the crop will be used for seed. Hence, the final term in (21) is negative. As before therefore, if the monopoly price is less than $\alpha(q_0 - a) + (c_n - c_g)/\alpha_g$, the non-GM crop is blockaded and no predation is necessary to eliminate its non-GM rival.

It follows that the interesting case, as far as predation is concerned, arises when $\partial\pi_1/\partial r_1$ is positive along SS at the monopoly value of r_2. When this is the case the balance of losing profits from decreasing r_1 *versus* the gain in profits from decreased competition in period 2 must be determined. Formally, elimination of the non-GM variety will be optimal provided,

$$\alpha_g\left[a + \frac{2c_n(\alpha\alpha_g - 1 + \alpha)}{(1-\alpha)} + \frac{c_g}{1-\alpha_g}\right] - \frac{\alpha_g^2}{2\alpha^3}\left(a - \frac{c_g}{1-\alpha_g}\right) \tag{22}$$

is negative. The first term in this expression represents the gain of first period profits from increasing r_1; the second term is the loss of profits from the second period if some non-GM seed is planted in period 1 and then used for seed for period 2. The important part of this equation is the ratio α_g^2/α^3. Every non-GM seed produced in period 1 becomes $1/\alpha$ of the crop in period 2. Hence there is a large cost to the GM producer of allowing non-GM crops to be planted. It is this factor that encourages predation.

4.1. Storage Costs

The alternative to planting seed is to store it. Suppose that storage costs are b, so that 1 tonne of the seed stored becomes $1 - b$ at the end of the storage period. Pursuing this strategy, a solitary farm that stores seed at the end of period 0, then plants it for harvest in period 2 makes profits per unit of

$$-\alpha p_0 + (1-b)(p_2 - c_n) = -[r_1\alpha_g + c_g - c_n] + (1-b)[r_2\alpha_g + c_g - c_n].$$

This strategy yields zero profits when

$$r_2 = \frac{r_1\alpha_g + (c_g - c_n)b}{(1-b)\alpha_g}. \tag{23}$$

Thus storage effectively puts a cap (which may not be binding) on the GM producer's second period price. In Figure 3, Equation (23) is depicted by the broken line segment TT.[6] A drop in storage costs makes TT shallower in slope while raising the intercept (on the assumption that $c_g < c_n$). If storage is costless, so that $b = 0$, then (23) reduces to $r_2 = r_1$, in which case predation can never be optimal.

In short, therefore, if there is no advantage to the GM crop then it will not be planted. If the cost or yield advantage is sufficiently large then it will eliminate the old varieties at a price to farmers which is always below the unit cost of the technologies it replaces. However at intermediate cost savings there is the possibility of effective predation by the GM producer in which initially the unit cost of the GM crop is sufficiently below the unit cost of the older varieties for the older varieties to be eliminated. However, once elimination has occurred the price charged to farmers for GM seed will lead to unit costs of production exceeding their original values. This last outcome is constrained by a low cost of storage, but in theory if storage costs are high enough then successful predation can still occur.

4.2. Simulation

Most of the relevant parameters in the critical equations (such as (22)) are not known with any certainty. Nevertheless a small amount of simulation provides some flesh on the bare bones of the analytical results. I begin by normalising the demand system by setting $a = 1$. A reasonable figure for a monopoly price is 10–50% above its competitive value, which would imply a range for $c_n/(1-\alpha)$ from 0.833 to 0.5. As noted earlier, for many crops values of α are typically small, usually under 0.1, so I take c_n to lie in the range 0.5 to 0.9, while a lies between 0.02 and 0.1.

Some preliminary estimates of the cost advantages from GM crops are available from US field trials and farm data. Fernandez-Cornejo and McBride (2000) provide an accessible survey. Most of the evidence suggests that yields from herbicide-resistant varieties of soybean, cotton and corn are at or below traditional varieties. The gains to farmers come from simpler and cheaper methods of weed control, combined with the ability to plant more closely. Hence for herbicide resistance, $\alpha \approx \alpha_g$, but $c_g < c_n$. For Bt varieties of cotton and corn, the gains come from lower pest damage, leading to both lower pesticide costs and higher yields, in other words $\alpha < \alpha_g$, and $c_g \leq c_n$. Typical cost savings are of the order of 5–15% though as Fernandez-Cornejo and McBride (2000) point out, in some regions of the USA in particular years, the returns to GM technology appears negative, though not

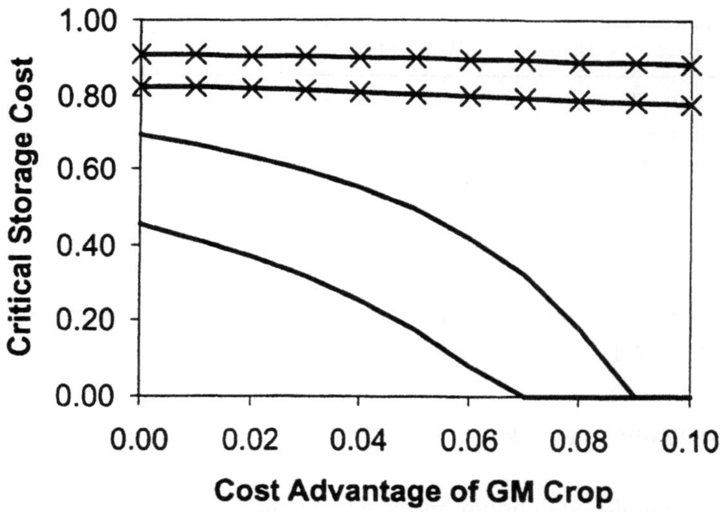

Figure 4. Constraints on predation: critical storage cost values.

significantly so. According to field trials (see, e.g., Gianessi and Carpenter, 1999), the returns are higher for Bt traits than for herbicide resistance. This may be of significance in the tropics where pest damage can reach very high levels. In the simulations I take reductions in a and in costs to be between 1 and 15%.

Figure 4 depicts critical values of b above which predatory pricing is not constrained by the possibility of storage of non-GM during period one for planting in period 2. The top two lines are for values of $c_n = 0.7$; the bottom two curves are for $c_n = 0.9$. For each pair of curves, the top one represents $\alpha = 0.2$, with $\alpha = 0.4$ in the bottom curve. The bottom axis shows the reduction in non-seed costs associated with the GM variety. So, for instance, with a cost advantage for the GM crop of 0.04, $c_n = 0.9$ (implying a cost saving of 4.4%) and $\alpha = 0.4$, storage costs of greater than 25% per time period imply no constraint on the ability of the GM producer to price in a predatory fashion. Note that, for these particular parameter values, predation is indeed the optimal strategy for the GM producer. For higher values of cost savings, where the curves in Figure 4 join the x-axis, production of the non-GM varieties is blockaded.

The curves depicted in Figure 4 are fairly typical. Higher values of a reduce the critical values of b (because each seed is less productive) and greater cost savings lower the critical value. It is also the case (not shown) that reductions in the value of α_g, relative to α reduces the critical value of b, although the effect is small. To put these figures in context, note that Cromwell et al. (1992) report typical storage costs of 40% per annum in some African countries. Such figures would make predation feasible for multiplication rates of 25 at cost advantages for GM crops of only 4–

Table 3. Prices in the post-GM market.

Price	Non-GM	With GM crop
Period 0	$c_n/(1-\alpha)$	$(-a\alpha(1-\alpha)+c_n)/(1-\alpha)^2$
Period 1	$c_n/(1-\alpha)$	$\min(p_{gm}, -[a\alpha/(1-\alpha)]+c_n[1+\alpha/(1-\alpha)^2])$
Period 2	$c_n/(1-\alpha)$	$p_{gm}=(a+c_g/(1-\alpha_g))/2$

5%. For temperate zone crops such as wheat grown in richer countries, however, much lower storage costs would limit monopoly power.

4.3. Consumer Welfare

Table 3 sets out the equilibrium prices without and without the introduction of the GM crop on the assumption that predation or a blockade is optimal.

Prices in the first two periods are below prices in the non-GM scenario. If entry is blockaded then the monopoly price in period 2 is also below the non-GM price and so consumers gain from the innovation as they did in Section 3. If entry is prevented only by predatory pricing in period 1, then the price in period 2 is above that in the non-GM scenario. Consumer welfare is decreasing in the price level, so obviously if the weight attached to period 2 in consumer welfare is sufficiently large then consumer welfare will have fallen overall.[7] However, if the reduction in prices achieved in periods 0 and 1 is sufficiently large, then overall consumer welfare improves, even with the charging of the monopoly rate in the final period. Again this is a familiar lesson: consumers gain from predatory price wars, while the wars are in progress. It is only when competition has been eliminated that consumers suffer.

Figure 5 provides some evidence from simulation in which welfare for each period is equally weighted. Shown are figures for net welfare gain for the case of $c_n = 0.9$ and 0.7. The top two curves represent $c_n = 0.9$, $\alpha = 0.4$, and $c_n = 0.9$ with $\alpha = 0.2$ in descending order. The bottom curve shows the case of $c_n = 0.7$ and $\alpha = 0.2$. Note that $\alpha = 0.4$ is not depicted in this case because it is only marginally different from the $\alpha = 0.2$ case. As might be expected, when $c_n = 0.9$, a higher cost saving is associated with a larger welfare gain, which becomes positive for the largest values of cost saving where the monopoly price is below the pre-GM competitive value. However, for smaller cost savings, the net welfare effect is negative. When $c_n = 0.7$, the monopoly price is much larger than the competitive price and in this case, welfare gains are never positive. Comparing Figures 4 and 5 it can be seen that it is perfectly possible for storage costs to be such that predation is feasible and welfare gains are negative.[8]

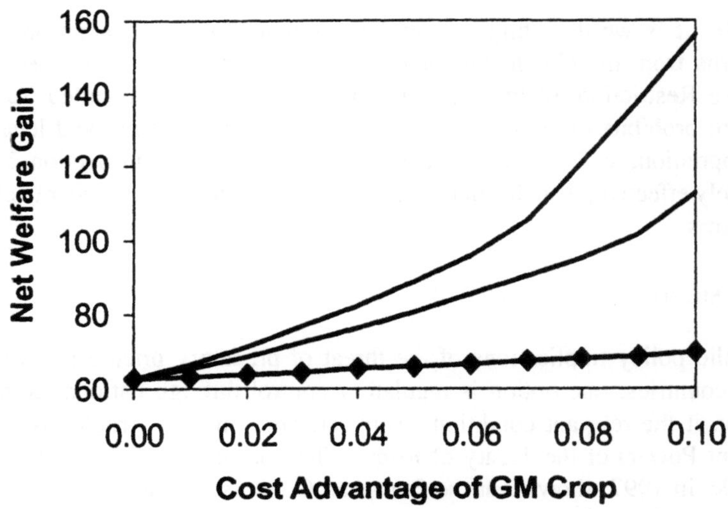

Figure 5. Net welfare gains from GM.

5. Discussion

It is obvious from the preceding that there is nothing automatic about the process of monopolisation – a number of assumptions must be satisfied. In particular it is worth emphasising that as long as competing seeds remains readily available, for the owner of GMO patent rights, the power to set prices is heavily circumscribed. In addition, with only a limited number of crops and few in direct competition with one another,[9] the current situation is unlikely to last – more competition between transgenic varieties is likely to be a feature of the next ten years, especially given the fact that the new technologies lower the costs of developing new varieties, compared to traditional breeding techniques.

Worldwide, there is also an established series of *ex situ* seed-banks and storage facilities and this again limits monopoly power, particularly in countries with well-organised systems and where storage costs are low. In other countries, where storage costs are high[10] the power of the GMO producer may be stronger. However, Goeschl and Swanson (1998) raise an important point about the limited value of *ex situ* storage when there is rapid evolutionary change in weeds and pests and hence rapid change in the optimal genotype of a crop. It is not clear how this affects the monopoly power of the GMO producer since, presumably rapid and localised change in the optimal genotype would also limit the extent of the market penetration of a standardised variety in the first place.[11] If though pricing was sufficient low to capture the market, the opportunity costs would be consequently that much higher.

Finally it is worth noting the implicit assumption of full appropriability of the returns from the GM technology. If this is not possible, either because Genetic Use Restriction Technologies (GURTs), such as the so-called Terminator Gene, are prohibited (e.g. by law or social norm), or economic and legal means of appropriation, such as the contractual devices mentioned in Section 2, are not completely effective, then this may limit the ability of the GM producer to eliminate competitors.

5.1. Remedies

The public policy implications of the threat of predatory pricing are unclear. In the EU countries, one option is regulatory control through national competition policy or, if the relevant conditions were satisfied, through Article 86 (Abuse of Dominant Power) of the Treaty of Rome. The fine of Euro 8.8m levied on Irish Sugar Plc. in 1997, shows that the European Commission has been willing to act against agricultural firms engaging in predatory pricing (McDonald and Dearden, 1999). A less reactive and more structural approach to promoting competition would be to relax the regulations governing development and deployment of new transgenic varieties.[12] Such an approach would likely prove contentious and serves to illustrate the difficulties of marrying competition and environmental policy. A third option, less likely to run into the same controversy is greater support from the public finances for *ex situ* and *in situ* conservation.

A final option is reducing the cost of storage. As already noted, this can involve investment in storage facilities, in transport infrastructure or in direct subsidies. If any policy reduces storage costs b, this shifts the line TT in Figure 3 down and to the right. However, if the constraint placed by storage on the GM producer's pricing policy is non-binding in the absence of positive intervention in the storage market, then small changes in b may have no effect on prices of GM crops.

6. Conclusion

A previous draft of this paper was entitled "Should we ban the terminator gene?" The answer from this paper is 'not necessarily' but there are clear indications that under some circumstances GM crops may lower welfare. Monopolisation in itself though is not a cause of welfare loss, since the usual route to monopoly control for a new technology is via greater efficiency. The risks of welfare loss arise through the possibility of greater variability in output and through the possibility of predation. Predation is most likely to be a threat to welfare when: (1) there is only one GM producer; (2) the advantage of its product over traditional varieties is not large; (3) there is pre-existing uniformity in the varieties being grown; (4) the monopoly price of the crop is high relative to the competitive value; (5) there is an absence of publicly supported *in situ* and *ex situ* conservation, and (6) total storage costs are high.

Acknowledgements

I am grateful for comments from the participants at the Biotechnology, Environmental Policy and Agriculture Workshop on the Management of the Impacts of Biotechnologies, Rome, May 2000, particularly Tim Swanson and my discussant, Gardner Brown.

Notes

[1] Sources of risk which are highly correlated across all cultivars, such as drought, flood or storm damage are therefore ignored. This maximizes the increase in risk, consequent upon the monopolisation of production.

[2] Arguably, the announced price for the following season's seed should be after the harvest. With no random element to the harvest in the model, the only significant difference produced by making this alternative assumption would be to the price in period 0, when the sudden introduction of the new technology would lead (possibly) to a second market opening for the produce from period 0. The qualitative results of the story would not be changed.

[3] Obviously only the latter is feasible for some crops, such as hybrid maize, in which case the power of the would-be monopolist is strengthened.

[4] For many crops, the multiplication factor ($= 1/\alpha$) is high, making seed costs a small fraction of overall costs. Cromwell et al. (1992), for instance, list multiplication factors of 25 for wheat and 50 for rice.

[5] This result, familiar in the vertical restraint literature, arises because of the lack of substitution possibilities in the farm's production function.

[6] Note that the relationship between the intercept of this equation and the other lines on the figure is drawn arbitrarily.

[7] While clearly not in the model as it stands, period 2 could be viewed as the 'long-run' with periods 0 and 1 representing the initial phases of adjustment to the new technology. If this is the case, placing a higher weight on period 3 welfare, compared to welfare in periods 0 and 1, might well be justified.

[8] Note that placing a greater weight on period 1 rather than period 2 raises welfare, but may still not yield a net welfare gain.

[9] For instance, the National Corn Growers Association (USA) website lists only four GM varieties (from three companies) of corn with approval for import into the European Union as of mid-1999. Two of the varieties express insect toxins, one has a gene for glufosinate tolerance and one variety has both features.

[10] Cromwell et al. (1992) note that "the cost of installing and operating controlled environment stores in tropical stores is very high and this is seldom an economic approach if real costs are to be passed on to farmers" (p. 41).

[11] In fact, the localised nature of the optimal variety is a general factor limiting monopoly power.

[12] In addition to the four varieties of transgenic maize approved for import into the European Union, the NCGA website lists another seven varieties awaiting approval.

References

American Soybean Association (1999) http://199.217.150.25/asa/documents/biotechsum.htm.

Arrow, K.J. (1962) Economic welfare and the allocation of resources for inventions, in R.R. Nelson (ed.), *The Rate and Direction of Inventive Activity: Economic and Social Factors*, Princeton University Press, Princeton, NJ, pp. 609–625.

Coase, R. (1972) Durability and monopoly, *J. Law Econ.* **15**, 143–149.

Cromwell, E., Friis-Hansen, E. and Turner, M. (1992) The seed sector in developing countries: A framework for performance analysis, Overseas Development Institute, Working Paper 65.

Department of the Environment Transport and the Regions (1999) *Environmental Risks of Herbicide-Tolerant Oilseed Rape: A Review of the PGS Hybrid Oilseed Rape.*

Ellstrand, N.C. (1988) Pollen as a vehicle for the escape of engineered genes? *TREE* **3**(4), S30–32.

Fernandez-Cornejo, J. and McBride, W.D. (2000) Genetically engineered crops for pest management in U.S. agriculture: Farm level effects, Agricultural Economic Report No. 786, Resource Economics Division, United States Department of Agriculture.

Gianessi, L.P. and Carpenter, J.E. (1999) *Agricultural Biotechnology: Insect Control Benefits*, Biotechnology Knowledge Centre, Washington, D.C.

Goeschl, T. and Swanson, T. (1996) Market integration and crop genetic resource conservation, mimeo, Cambridge University, Faculty of Economics.

Goeschl, T. and Swanson, T. (1997) A methodological framework for valuing in situ conservation of crop genetic resources, mimeo, Cambridge University, Faculty of Economics.

Goeschl, T. and Swanson, T. (1998) Protecting the Red Queen: *In* and *ex situ* conservation of genetic resources and long term food supply, mimeo, University College London.

Gul, F., Sonnenschein, H. and Wilson, R. (1986) Foundations of dynamic monopoly and the coase conjecture, *J. Econ. Theor.* **19**, 423–455.

Hayenga, M. (1998) Structural change in the biotech seed and chemical industrial complex, *AgBioForum* **1**(2), 1–10.

Hazell, P.B. (1989) Changing patterns of variability in world cereal production, in J.R. Anderson and P.B. Hazell (eds.), *Variability in Grain Yields: Implications for Agricultural Research and Policy in Developing Countries*, John Hopkins University Press, Baltimore, MD, pp. 13–34.

Krimsky, S. and Wrubel, R.P. (1996) *Agricultural Biotechnology and the Environment*, University of Illinois Press, Urbana, IL.

McDonald, F. and Dearden, S. (eds.) (1999) *European Economic Integration*, Addison Wesley Longman, Harlow.

Mikkelsen, T.R., Andersen, B and Jorgensen, R.B. (1996) The risk of transgene spread, *Nature* **380**, 31.

Munro, A. (1997) Economics and biological evolution, *Environ. Res. Econ.* **9**, 429–449.

National Agricultural Statistics Service (2000) *Prospective Plantings*, United States Department of Agriculture.

National Corn Growers Association (1999) http://www.ncga.com/02profits/know_where/know_grow_approved.htm.

Newbery, D. and Stiglitz, J. (1981) *The Theory of Commodity Price Stabilization*, Clarendon Press, Oxford.

Norris, C.E., Simpson, E.C., Sweet, J.B. and Thompson, J.E. (1999) Monitoring weediness and persistence of genetically modified oilseed rape (Brassica napus) in the UK, in P.J.W. Lutman (ed.), *Gene Flow and Agriculture: Relevance for Transgenic Crops*, British Crop Protection Society Symposium Proceedings No. 72, pp. 75–81.

Nottingham, S. (1998) *Eat Your Genes*, Zed Books, London.

Portney, P.R. (1992) Trouble in Happyville, *J. Policy Anal. Manag.* **11**(1), 131–132.

RAFI (1997) http://www.rafi.cca/communique/19972.html.

Rey, P. and Tirole, J. (1986) The logic of vertical restraints, *Amer. Econ. Rev.* **76**, 921–939.
Roller, S. and Harlander, S. (eds.) (1998) *Genetic Modification in the Food Industry*, Blackie Academic and Professional, London.
SCIMAC (1999) http://www.ukasta.org.uk/scimac/gui8.html.
Sianesi, B. and Ulph, D. (1998) Species loss through the genetic modification of crops – A policy framework, mimeo, University College London.
Singh, A.J. and Byerlee, E. (1990) Relative variability in wheat yields across countries and over time, *J. Agricult. Econ.* **41**(1), 21–32.
Timmons, A.M. et al. (1996), Risks from transgenic crops, *Nature* **380**, 487.
Tirole, J. (1989) *The Theory of Industrial Organization*, MIT Press, Cambridge, MA.
Treu, R. and Emberlin, J. (2000) Pollen dispersal in the crops maize, oil seed rape, potatoes, sugar beet and wheat: Evidence from publications, Report for the Soil Association.
Williams, J.C. and Wright, B.D. (1991) *Storage and Commodity Markets*, Cambridge University Press, Cambridge.
Williamson, M. (1996) Can the risks from transgenic crops be estimated?, *Tibtech* **14**, 449–450.
Wright, B.D. (1997) Crop genetic resource policy: The role of *ex situ* genebanks, *Austr. J. Agricult. Res. Econ.* **41**(1), 81–115.

10. The Diffusion of Benefits from Biotechnological Developments: The Impact of Use Restrictions on the Distribution of Benefits

TIMO GOESCHL and TIMOTHY SWANSON

1. Introduction

In the management of the research and development (R&D) process, society is attempting to solve a particular form of a public goods supply problem. The information generated by the R&D process has the character of a public good, i.e. it is non-rival and non-excludable. In the absence of regimes that ensure a flow of rents to the creator of this information, its public good character poses a strong disincentive for private investment in R&D. The rationale for the creation of property rights in innovations is that such regimes will have the effect of encouraging investments in R&D, and hence the supply of information resulting from it.

This paper analyses the impact of such property rights regimes, in the specific context of the plant breeding sector. Plant breeding forms an essential part of R&D in the agricultural sector. It has been shown to be a major source of agricultural productivity growth (Traxler et al., 1995; Huffman and Evenson, 1993; Schmidt, 1984; Thirtle, 1985; Evenson and Kislev, 1973). At the same time, plant breeding poses an even more formidable problem to society than usual R&D processes because the R&D output, namely seeds, has a self-reproducing property. This makes it very difficult for the innovator to control the dissemination of innovative traits. Additionally, cross-breeding offers competitors the potential for accumulating others' innovations within their own R&D output. In essence, the ease of transfer of traits between crops makes it very hard to protect the proprietary information contained in improved varieties (Swanson, 1996). The result is that in the absence of intervention, very little private R&D would be carried out in plant breeding in comparison to the social benefits that are generated through such crop improvements.

There have been at least two responses to this problem. One has been a sequence of *legal forms* of rent protection through intellectual property rights (IPRs) in cultivars. These rights have gone through a series of changes, commencing with several domestic forms of IPRs in various countries since the 1930s, and culminating in the development of international regimes for plant variety protection (the UPOV convention) in 1970. This has led to a pronounced increase in private plant breeding

Timothy Swanson (ed.), The Economics of Managing Biotechnologies, 219–248.

in those countries where effective protection has been afforded to these legal claims (Butler, 1996).

Another response has been the use of *technological forms* of IPR protection. This form of protection makes use of a particular feature of hybrid vigor to enforce effectively claimed property rights in innovative plant varieties. Hybrid varieties contain a very diverse set of genetic material, and their reproduction will generate highly distinct offspring. In essence, the precise reproduction of an F1-hybrid requires possession of the knowledge of the inbred lines that were used in its production. Since innovators need not disclose these lines, unauthorized reproduction of the seed can be effectively prevented. The application of hybridization in the commercial seed sector suggests that the availability of use restriction (via hybrid varieties) is an effective and preferred means for protecting innovative plant varieties (Butler and Marion, 1985; Butler, 1996). However, for biological reasons, this response has only been commercially viable for a small set of the most important crops, namely maize and sorghum.

The primary distinction between the legal and technological forms of property right protection is the extent to which the domestic regime concerned (where the innovative product is being utilized) is crucial to the determination of the effectiveness of the regime. *Legal forms* of protection are entirely dependent upon the effort and resources of the domestic regime for effect. If the state chooses not to respond to the property right holders complaints, or its courts refuse to enforce the claimed rights, then the regime is entirely without effect. Users may reproduce, resell and effect R&D with impunity, if the domestic regime does not expend the resources necessary to prevent these acts. On the other hand, the commitment of the domestic regime is irrelevant to the effectiveness of the *technological forms* of protection. Even if the domestic regime is openly hostile to the claims of the property rights holder, technological forms of protection enforce the claim of right within the boundaries of that state as much so as in any other. Hence, technological forms of protection are state-neutral while legal forms of protection are state-dependent for effectiveness.

The economic impact of these two distinct property right regimes is the subject of this paper. Clearly, one important aspect of this impact is its capacity for rent creation and appropriation, i.e. the incentives for undertaking R&D. The other aspect concerns how the resulting information diffuses once these innovations are made (Ordover, 1991). Evenson and Kislev (1973) pointed out in their early study on wheat and maize that the international diffusion of innovations in hybrids may be inferior to that of open-pollinating varieties. On the one hand, this might seem to be an unavoidable consequence of rent creation and appropriation, as the creation of impediments to the costless diffusion of information is precisely the form that effective property rights in innovation must take. On the other hand, the creation of a barrier to costless diffusion need not, and in fact should not, necessarily equate with the observation of dramatically reduced flows of information. It should instead result in increased payments to the innovator to generate increased flows of information. The observed impact of reduced diffusion is an unwanted by-product of the second-best policy of creating these property rights.

Where the impacts fall from the implementation of this second-best policy is important. Agriculture is a sector of fundamental importance for all developing countries. More than 4 billion people in the developing world directly depend on some type of farming activity for their sustenance. The welfare effects of inhibited diffusion to these people are of significant consequence. Especially in the case of global agriculture, the impacts of restricted diffusion must be carefully balanced against the beneficial impacts of increased rates of innovation.

The purpose of this paper is to review some evidence from the diffusion of innovations in hybrids and non-hybrids in order to establish the impact of various forms of property rights regimes (legal and technological) on the diffusion of innovation. The particular evidence that we will be looking at is the diffusion of yield gains to developing countries. In this, the paper is closely related to a number of recent efforts to quantify the extent of international spill-overs from R&D and of international convergence of productivity growth in both agriculture (McCunn and Huffman, 2000; Schimmelpfennig and Thirtle, 1999; Huffman and Evenson, 1993) and economies as a whole (Coe and Helpman, 1995; Coe et al., 1997; Bayoumi et al., 1999; Frantzen, 1998, 2000; Barro and Sala-i-Martin, 1995). These papers have examined, theoretically and empirically, the direction, volume and growth impacts of the flow of innovations between countries.

Our paper is closest in spirit to the empirical work by Coe et al. (1997). There they examine how countries, which are primarily receivers of R&D outputs, benefit from R&D undertaken abroad. Developing countries provide minimal R&D investment in the world economy generally (Coe et al., 1997).[1] Within agriculture, their share in R&D expenditure is higher, but it is significant only in the public sector (Pardey et al., 1991) and – in contrast to R&D in other sectors – may contribute little to technological progress generally (Evenson and Kislev, 1973).[2] This suggests that the agricultural sector may not differ from other sectors of the world economy to an extent that would render the analytical framework of Coe et al. inapplicable.

The novel aspect contained in this paper is that it presents evidence on international diffusion under two different regimes of IPR protection, one 'legal' and the other 'technological'. This is a crucial distinction because the legal form of protection has received little support throughout the developing world in the past, and hence it has had little effect on restricting the flow of information between developed and developing countries. It is argued here that the legal form of protection in effect generated rents from restrictions enforced in the developed world, while allowing costless diffusion of innovations between developed and developing countries. The *technological form* of protection in effect generated increased rates of return to the innovators but at the cost of the potentially restricted diffusion of innovation. The purpose of the paper is to assess the incidence of this potentially restricted diffusion, which countries bear this cost and to what extent?

The organization of the paper is as follows. In Section 2 we present evidence on the unconditional convergence in international crop yields in hybrids and non-hybrids based on the eight most widely cultivated crops.[3] The absence of convergence in hybrid crop yields motivates a panel study across these eight crops in Sections 3 and 4 – in order to measure the diffusion rates of innovation under the contrasting property right regimes. In Section 5 we discuss the results of the panel study and provide a number of possible explanations for the results before concluding.

2. Convergence in International Cultivar Yields

We commence this study with an examination of the 'rate of convergence' in international agriculture. The present study examines the development of crop yields in the eight most widely cultivated crops, namely barley, cotton, maize, millet, rice, sorghum, soybeans and wheat. The time period examined includes most of the time period for which F1 hybrid varieties have been widely available (from 1961 to 1999) and comprises all developing countries for which complete and apparently genuine yield data could be obtained. The eight crops have been grouped into two categories, one containing the two crops for which hybrid seeds have been available on a widespread basis, namely maize and sorghum; the other containing the six remaining crops.

2.1. Data and Variables

The yield data for the eight crops examined are annual yield data from the FAO Statistical Database (FAOSTAT). The data record the harvested production per unit of harvested area for crop products based on the annual harvest data and the area harvested. Data are recorded in hectogramme (100 grammes) per hectare (hg/ha). The data is not always fully reliable. Specifically, all countries were omitted for which no complete time series of yield data was available for the 39 year period or whose yield data showed an obvious lack of reliability. Even with a stringent application of these rules, the panel size is never below 39 countries, with soybeans particularly affected by the rule of complete time series.[4]

There are obvious objections to the exclusive use of yield data in order to examine the productivity development and the scope for spill-overs between developed and developing countries. The variable that is customarily used in such contexts is the total factor productivity (TFP) of an economy (Grossman and Helpman, 1997; Frantzen, 2000) in macro-economic settings or of the agricultural sector (Schimmelpfennig and Thirtle, 1999) for sector-wide analyses rather than the observed yield. Since yield development comprises both factor mobilization and productivity increases, inferring TFP dynamics from yield development is problematic. Empirically, however, in the absence of crop-specific TFP data, yield data may proxy well for TFP: results for the US indicate that TFP growth accounts for almost all of the real output growth in agriculture (Ball et al., 1997). This suggests that at least in the case of developed countries, changes in factor inputs may be insubstantial to the empirical exercise. But even for less developed countries, the practice of using yield as a proxy for TFP can be defended on the grounds that if the agricultural production function is very elastic regarding factor input substitution (in other words, if there is a high degree of complementarity between inputs), then the factor input mix is largely determined by the state of technology (Evenson and Kislev, 1973).

A second point is that the empirical assessment of R&D spill-overs normally relies on data of R&D capital in different countries in order to estimate the volume of externalities, thus imputing R&D output by reference to R&D inputs. In this paper, we use the transformations of yield data as a proxy for R&D output, in the same vein as they are used as a proxy for TFP.[5] There are two reasons for this choice. One is the obvious data limitation regarding crop-specific R&D capital:[6] little sufficiently disaggregated data is available. Secondly, R&D capital that may be employed for imitation and adaptation of genetic improvement in cultivars is often only partially based in the receiver country: International agricultural R&D plays a vital role in the diffusion of agricultural innovations and is not included in country specific estimates, thus biasing the results. This is to emphasize again that yield data may provide a good proxy for the actual amount of innovation in crops in developing countries.

2.2. Yield Development in Major Crops

Table 1 presents some summary data about the eight crops examined: the crop with the highest global acreage in 1999 has been wheat with 214.2. million hectares and the least significant crop included in this study is cotton with 34.3 million ha in 1999. As shown, the growth rates of yields in these crops have been lower in developing as opposed to developed countries[7] for five out of eight crops, with the exception of rice, soybeans and wheat. Correspondingly, the relative yield gap between developed and developing countries, i.e. the percentage by which developing countries lag behind the yield in developed countries in the specific crop, has decreased only for these three crops while it has widened for the five others. Yet, even for those crops for which the gap has narrowed, wide differences in global agricultural productivity persist up to this very moment in time. Across all eight crops, average yields in developing countries are about 57% lower than the crop yields in developed countries.[8] Developing countries are therefore still far off the productivity frontier in this primary sector. There are, however, significant

Table 1. Acreage, global distribution, growth and relative yield gap in 8 major crops.

Crop	Global acreage in million ha in 1999	Average growth rate in developed countries, 1961–1999	Average growth rate in developing countries, 1961–1999	Relative yield gap in 1961	Relative yield gap in 1999
Barley	58.6	1.53%	1.03% (40)	–57%	–59.9%
Cotton	34.3	2.45%	1.54% (60)	–24%	–47.4%
Maize	139.2	2.27%	1.42% (95)	–65%	–72.4%
Millet	37.2	0.93%	0.41% (46)	–49%	–57.4%
Rice	153.1	0.85%	1.24% (60)	–64%	–57.9%
Sorghum	44.8	2.08%	0.54% (64)	–48%	–67.2%
Soybeans	72.1	1.24%	1.58% (32)	–46%	–40.0%
Wheat	214.2	1.75%	1.89% (54)	–60%	–54.5%

differences in the relative yield gaps between the eight crops. In 1999, the smallest yield gap existed in soybeans at around 40% while the greatest gap could be found in maize at around 72%.

The growth rates in developed countries reflect the dramatic technological improvements in agricultural production. Yields in developed countries have been expanding at different rates, ranging from 0.85% per annum in the case of rice and 2.45% per annum in the case of cotton. This translates into a total increase of yield of around 157% in the case of cotton, a more than doubling of the yield, and of around 40% in the case of rice over the 39 year period being assessed. This growth at the productivity frontier suggests an expanding set of production possibilities over time, albeit expanding at different rates for different crops.

2.3. Testing for Convergence

The presence of convergence of growth rates across countries is a central concept in the theory of growth and is of great theoretical and empirical interest (Baumol et al., 1989; Grossman and Helpman, 1997). There are two different forms of convergence that are generally examined: One is unconditional convergence of growth rates, or also known as σ-convergence, the other convergence conditional on some factor such as human capital endowment or the saving rate, also known as β-convergence.

Various examples of σ-convergence and β-convergence have been examined in the empirical growth literature (Barro and Sala-i-Martin, 1995; Grossman and Helpman, 1990). In agriculture, McCunn and Huffman (2000) examine TFP data for US agriculture and find evidence that supports the presence of conditional convergence. This result implies different steady-state TFP in the long run, conditional on R&D spill-ins from other states, private R&D and farmer education.

As mentioned above, there are at least two quite different convergence concepts present in the literature, namely β- and σ-convergence. The first applies to a situation where a poor economy exhibits a faster growth rate than a more developed one, resulting in a process of catching-up over time. This is the concept present in Baumol (1986), DeLong (1988), and Barro (1991). The second applies to the development of cross-sectional dispersion in a characteristic variable over time, such as per-capita income. This is the concept employed by Easterlin (1960), Streissler (1979), and Lichtenberg (1994). Naturally, the convergence tests employed by each concept differ markedly such that presented with a set of data, these concepts can come to quite different conclusions about the presence or absence of convergence (see Lichtenberg, 1994; Barro and Sala-i-Martin, 1995).

In order to test for β-convergence, it is customary to regress the average growth rate of a suitable productivity indicator, AVG, on the country's initial productivity value, Y, at the initial point t_0.[9]

$$\mathrm{AVG} = c + \beta \cdot \log(Y_{t_0}) + \varepsilon. \tag{1}$$

The test of convergence involves a significance test (t-test) on β where a value significantly below zero indicates a negative relationship between initial productivity and subsequent growth rates and thus convergence in the sense of poorer countries catching up (Barro and Sala-i-Martin, 1995). We will refer to this test as the test for 'mean-reversion' as there is – strictly speaking – no convergence taking place (Lichtenberg, 1994).

For σ-convergence to occur, it is sufficient that the cross-sectional variance within the sample decreases over time. The usual method is therefore to regress the dispersion in yields on a time trend using the regression equation

$$\mathrm{var}(\ln y_t) = \phi_1 + \phi_2 t + \varepsilon_t, \tag{2}$$

where ε_t is a normally distributed random variable with mean zero and variance s. Again, for the sample to pass the test of convergence, the parameter ϕ_2 estimated from the sample must be significantly below zero, thus implying that dispersion within the sample has been decreasing over time (Lichtenberg, 1994). We will refer to this test as the test for convergence.

We use the standardized[10] yield data as measurements of partial productivity of agriculture across all countries. We then separately estimate the parameters β and ϕ_2 for the yields of hybrid and non-hybrid crops. For the mean-reversion test, we compute the average growth rate over the observation period from 1961 to 1999 and use the logarithm of the standardized yield level in 1961, Y_{1961}, as the explanatory variable. For the convergence test, we compute the dispersion through the sample variance of the logarithms of the standardized partial productivity data.

Table 2. Regressions for mean-reversion in crop yields, 1961–1999.

Crop	Hybrid crops	Non-hybrid crops
β^a	–0.0053	–0.0170
	(0.003818)	(0.00246)***
R^2	0.016	0.157
(number of observ.)	(119)	(259)

[a]The figure in parentheses is the standard error. * indicates statistical significance (parameter $\neq 0$) at a 10% level, ** at a 5% level, and *** at a 1% level.

2.4. Results on General Convergence

This section examines the presence of various forms of convergence by technology (hybrid *versus* non-hybrid), region and crop. Table 2 reports the regression results from Equation (1) where the sample excluded developing countries with negative growth rates in yields.[11]

For non-hybrid crops, the hypothesis that countries with lower initial productivity have a high growth rate over time is not rejected by the values of the coefficients: there is a negative relationship between initial yield levels and a country's average growth rate over the following 38 years. This relationship is particularly strong in cotton, rice and wheat, somewhat less so in millet, soybeans and barley. What is striking is that there are two crops for which absolute convergence is rejected by the data, namely sorghum and maize.

Figures 1 and 2 present the graphical result for the regression reported in Table 2.

Examining the evidence on a crop specific basis provides further support for the aggregate result. Each of the non-hybrid crops individually exhibits the characteristics of the aggregate, namely that there has been a statistically negative correlation between the productivity of land (yield) in agriculture in 1961 and the subsequent growth rate until 1999. In both maize and sorghum, there is no such correlation.

Although these findings do not imply that countries are converging to the same steady state growth rate, it is evidence of catching-up in yields taking place in non-hybrids, but not in hybrids. This in turn implies a narrowing of the gap in agricultural productivity that has been occurring in non-hybrid crops only.

The test for σ-convergence is a more stringent criterion. It involves testing for decreasing dispersion in countries' yields over the time period from 1961 to 1999. Table 3 report the results for the aggregate sample of hybrids and non-hybrids for the 39 year period.

For both hybrid and non-hybrid crops, the convergence parameter ϕ_2 is statistically significant and positively sloped.[12] This indicates – within the entire group

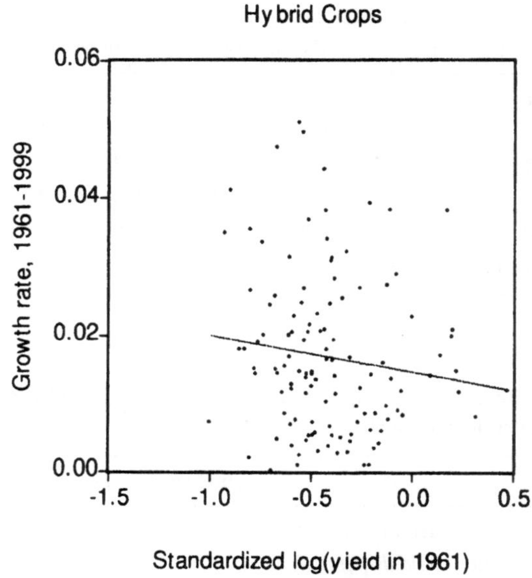

Figure 1. Initial yield *versus* growth rate in hybrids, 1961–1999.

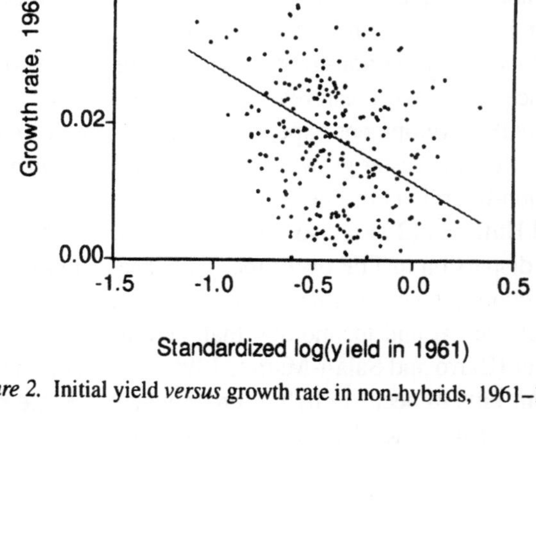

Figure 2. Initial yield *versus* growth rate in non-hybrids, 1961–1999.

Table 3. Regressions for convergence of crop yields, 1961–1999.

Crop	Hybrid crops	Non-hybrid crops
Φ_1	0.002603	0.003793
	(0.000138)***	(0.000110)***
Φ_2	7.41E-05	2.11E-05
	(6.23E-06)***	(4.99E-06)***
R_2	0.793	0.326
(number of observ.)	(39)	(39)

Table 4. Results of convergence tests for hybrid and non-hybrids crop yields in developing countries, 1961–1999.

Crop	Hybrid crops	Non-hybrid crops
β-convergence (*mean-reversion*)	No	Yes
σ-convergence (*yield dispersion*)	No	No

of developing countries – a growing dispersion in crop yields over time in both hybrids and non-hybrids and the absence of convergence to the same steady state rate of growth, i.e. failure to observe so-called σ-convergence. This implies that the yields are not converging to the same steady state growth rate across the developing world.

Combining the tests for β- and σ-convergence, we arrive at a differentiated picture for hybrid and non-hybrid crops. The results are summarized in Table 4.

In hybrids, there has been both no mean reversion of crop yields and no decrease in yield dispersion over time. In non-hybrids, yield dispersion also failed to occur, but there is evidence for global mean reversion, suggesting that countries trailing behind in productivity development experience some degree of catching-up with the frontier by experiencing faster growth than the average country.

The result on non-hybrids is in line with similar productivity studies based on TFP. McCunn and Huffman (2000) also find evidence for catching-up, but no evidence for reduced dispersion in TFP over time. This idea is intuitively appealing as one would expect countries with below-average productivity to experience higher marginal returns on capital and technology than countries operating close to the productivity frontier (Barro and Sala-i-Martin, 1995, ch. 1). It is curious, therefore, that this mechanism fails to occur in hybrid crops where the rate of innovation at the frontier is higher than in non-hybrid crops (see Table 1). This would suggest a greater scope for yield increases in developing countries for hybrids than in non-hybrids.[13]

3. A Model of Diffusion

Mean reversion is obviously a weak criterion of economic development, implying little more than the scope for catching-up with more developed countries. Conversely, where mean-reversion fails to occur, it indicates a significant problem in growth performance. The absence of mean-reversion for hybrid crop yields (as opposed to non-hybrids) merits further analysis.

The usual approach to analyzing the growth performance of a sector is to make the convergence process conditional on other variables. This normally results in multivariate models of convergence (Lichtenberg, 1994). When convergence differentials appear at the crop-specific level, as is the case here, there are a number of empirical difficulties in replicating the multivariate approach. The most relevant limitation is the absence of appropriately disaggregated data on total factor productivity and research capital and investment at the crop-specific level (Pingali and Heisey, 1999). The scope for exploring the role of other factors in an explicit fashion is therefore limited. However, the available data enables us to examine the development of crop yields in developing countries from another relevant perspective, namely as a process of diffusion of innovations from the technological frontier to countries lagging behind.

The examination is conducted as a panel study of the yield developments in the eight crops over the period from 1961 to 1999. To estimate the rate of diffusion from the frontier, we adopt a specification used in a related study by Barro and Sala-i-Martin (1995, 1996) that examines the impact of imitation on growth for developing countries that do not innovate themselves.

3.1. A Leader-Follower Model of Technological Diffusion

In this section, we present a simple model of technological diffusion based on a theoretical framework commonly used in the context of growth and innovation. For the details of the model, we refer the reader to Barro and Sala-i-Martin (1995, ch. 10). In this model, the source of technological progress are the constant returns to scale to innovation in intermediate goods in the spirit of Romer (1990). The particular multi-economy setting that we explore here is a leader-follower model in which there is a technological frontier at which innovations occur which are imitated by countries off the frontier. Countries have different endowments of inputs that allows them to produce final output and to generate new products through innovation (leader) or imitation (follower). As Barro and Sala-i-Martin (1995) show, this setting allows a straightforward estimation of a convergence model.

Assume that a follower country's cost of imitation, v, is an increasing function of the ratio of the number of intermediate goods in the follower (N_F) and the leader country (N_L) such that

$$v = \psi\left(\frac{N_F}{N_L}\right), \tag{3}$$

with $\psi' > 0$ and $\psi'' < 0$. The usual conditions apply.[14] Barro and Sala-i-Martin then show that it is possible to derive the optimal ratio of goods in equilibrium, $R = N_F/N_L$, as a function of the factor endowments and the cost of imitation only. In other words, there is an R^* that is unique and optimal. If $N_F/N_{L'} = R^*$, then the leader and the follower country are in a steady-state characterized by a constant growth rate of N_F and $N_{L'}$. If $N_F/N_{L'} < R^*$, then $v < v^*$, i.e. imitation becomes cheaper as there is an abundance of useful products available to be copied. The result is a model with common convergence characteristics: an economy grows proportionately faster the further below it is from its steady state. Barro and Sala-i-Martin then formulate the result as a log-linear approximation such that

$$\gamma_F = \gamma_L - \mu \cdot \log\left[\frac{R}{R^*}\right], \tag{4}$$

with γ denoting the growth rate of the follower and leader country, respectively, and μ denoting the speed of convergence. This can be transformed directly into

$$\gamma_F = \gamma_L - \mu \cdot \log\left[\frac{y_F/y_L}{(y_F/y_L)^*}\right]. \tag{5}$$

This gives the growth rates for a country of the frontier (a follower) as the growth at the technological frontier (a leader country) minus the 'friction' induced by the fact that imitations do not diffuse without cost as μ is positive transformation of the cost of imitation (Equation (1)).

In the context of diffusion of innovations in crops, there are structural factors that inhibit the diffusion of innovations and that are likely to remain constant over time (Evenson and Kislev, 1973). The most important structural factor is agro-ecological barriers to diffusion that will limit the amount of innovations useful in a follower country. One way to interpret these barrier is to see them as equivalent to intrinsic productivity differences between leader and follower country in a particular crop. This can be accommodated within the given model as a statement that R^* is constant over time and specific to each follower country, implying a specific transmission ratio of innovations from the frontier to follower countries. With R^* constant, this means then that for each country, we can transform (3) to estimate the following model

$$\gamma_F = \gamma_L - \mu \cdot \log\left(\frac{y_F}{y_L}\right) - a_F. \tag{6}$$

In practice, the model has to be transformed for further estimation to remove spurious correlation in the dependent and independent variables. We therefore transform Equation (6) into an error-correction model that has a correction mechanism component, captured in β, and a structural component, captured in a_i. This results in

$$\Delta G_{it} = a_i + \beta_i \cdot G_{i,t-1} + \varepsilon, \tag{7}$$

with $G_{it} = \log(y_{i,t}/y_t^*)$ such that β is an estimator of the catch-up rate of the country i to the development at the frontier, denoted by an asterisk (*) and $a_i < 0$ represents a measure of the country-specific structural barrier to an innovation from outside. Equation (7) states that differences in the growth path of crop yields can originate from two sources: (1) Inherent and persistent problems in the follower country to keep up with the yield dynamics in the leader country. These are captured in a country-specific estimation of a_i. (2) Problems in the diffusion of innovations from the leader to the follower country. This is captured in the catch-up parameter β.

4. Estimating the Diffusion of Innovations

In this section, we estimate the rates of diffusion of innovations under both property right regimes (hybrid and non-hybrid crops). A particular feature of this analysis is that it enables us to decompose the development of the yield gaps in each of the crops into its three basic components: (1) innovation at the frontier, (2) the diffusion process of these innovations to developing countries, and (3) the country-specific factors that impact on the capacity for yield growth such as specific agro-ecological conditions. The first will tend to increase yields as the set of technological possibilities expands. The second will decelerate the speed at which these gains reach developing countries, and the third part will determine the long-run capacity of a country to experience yield growth in a particular crop at a rate above or below the growth rate at the frontier. We turn to all of these implications in Section 5, while in Section 4 we focus on the impacts on diffusion.

4.1. Estimating the Econometric Model

The method we use to estimate the parameters b and a_i is a fixed-effect panel estimation model that allows for heterogeneity among the countries through variable intercepts (Hsiao, 1986). In order to empirically estimate the Barro–Sala-i-Martin diffusion model consistently. To recapitulate, the model has the form

$$\Delta G_{it} = a_i + \beta \cdot G_{i,t-1} + \varepsilon, \tag{8}$$

where G is the gap in logarithms of the yield between the specific country and the lead country, Δ signifies the change in the gap and ε is a normally distributed random variable with $E(\varepsilon) = 0$ and a known variance. The intercept term a_i denotes the long-term difference in productivity growth in equilibrium between the frontier and the developing country i. As laid out in Section 1, the interpretation of a_i is to regard it as a country-specific intercept that captures the agro-ecological and institutional factors that influence the overall productivity development of a crop in the individual country. In this it captures the content of the hypotheses that claim country-specific factors are responsible for the disproportionate yield gap that exists in the case of maize and sorghum. The coefficient β that is to be estimated then reports the diffusion coefficient of the particular crop.

Empirically, we enter the yields of the frontier and the country i in the form of $\log(y_{it})$ with y denoting the yield of country i. Then we estimate the diffusion coefficient β according to the model above as a GLS-regression correcting for the residuals being cross-section heteroskedastic by down-weighting each pool equation by an estimate of the cross-section residual standard deviation.[15]

4.2. Econometric Results

Each of the estimations delivers a coefficient β that is statistically highly significant and we also report a parameter $\tilde{a}$ that denotes the average intercept for all countries in the estimation. The Durbin–Watson coefficients and inspection of the Ljung-Box Q-diagram indicate that serial correlation is not present in this estimation, thus strengthening our claim that the results provide an analysis that is independent from the trends at the frontier.

Before interpreting the results, it is convenient to perform some algebra in order to bring the model into a simpler form. Re-arranging (4), we arrive at the following equation for the growth rate of yield, $\Delta\hat{y}t$ in the average developing country:

$$\Delta\hat{y}_t = \Delta y_t^* - (1+\beta)\cdot G_{t,t-1} + \hat{a} + \varepsilon. \tag{9}$$

This formulation reveals the separate components that drive the growth rate of yields in the average developing country: the first component is the yield gain at the frontier Δy^*. This reflects the expansion of the set of technological possibilities. The second component captures the extent to which an innovation can diffuse in the country. We define the gap G to take on positive values. Therefore, we would expect that the coefficient β is negative (indicating that innovations do not have a negative effect on growth) and that the closer the coefficient is to -1, the more rapid the gains dissipate from the frontier to the average developing country. The third parameter, $\hat{a}$, summarizes the country-specific growth lags as an average. A positive value would indicate that on average, developing countries have a higher 'intrinsic' rate of yield growth in this crop.

Table 5. Regressions for diffusion of innovations in different crops.

Crop	Hybrid crops	Non-hybrid crops
β^a	–0.242	–0.312
	(0.008158)***	(0.007711)***
α	–0.33611	–0.28171
R^2	0.136	0.167
(number of observ.)	(6004)	(8664)
DW-statistic	2.438933	2.374825

The figure in parentheses is the standard error. *** at the 1% level.

4.3. Interpreting the Results: Diffusion

The results reported in Table 5 show that hybrids have had a lower rate of diffusion of innovations from the frontier to developing countries than non-hybrids. While non-hybrids carried over roughly 69% of the gap opened by an innovation into the next year, hybrids carried over about 76%. This means that developing countries retained 7% more of the yield gap in hybrids relative to non-hybrids. This explains an important part of the cumulative yield gap that has developed in hybrids.

The results also indicate that structural effects, such as agro-ecological conditions, have contributed to inhibiting yield growth of hybrids in developing countries. The parameter a is the mean of the individually estimated parameters a_i. The means computed for hybrids and non-hybrids indicate that in hybrids, the average developing country has had a greater negative long-term deviation from the growth rate of the frontier than in non-hybrids. The combination of structural and diffusion effects has therefore resulted in considerable differences for the diffusion of innovations under the two systems.

For individuals cultivating different crops, an important criterion for evaluating crops is the loss of yield suffered as a result of slow diffusion. This gives a different angle on the problem of diffusion as it looks at the present value of the cumulative process of an innovation arriving at developing countries in a delayed fashion.[16] Figure 3 reports the multiplier to the initial shock that results from lagging behind the technological frontier for these plant varieties (present value being determined at a 10% discount rate).[17] The curve depicts the present value of the total accumulated losses as a multiple of the initial 'loss'. Looking at the estimates for hybrid and non-hybrid crops, the graph shows that the differences between the rates of diffusion result in the economic loss being about a third higher in hybrids as opposed to non-hybrids.

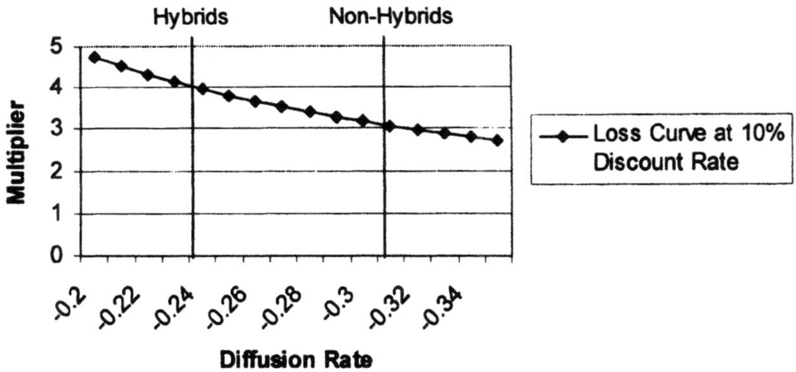

Figure 3. Loss multiplier as a function of the diffusion rate.

4.4. Interpreting the Results: Country-Specific Lags

A second set of important differences arises from the country-specific data on 'individual growth capacity'. This reveals firstly that, on average, developing countries would experience slower growth in all crop yields as the coefficient $\hat{a}$ is below zero for all crops. However, these impediments to growth are quite different between crops, ranging from rice, a crop with good intrinsic growth potential in developing countries at $\hat{a} = -0.230$, to wheat, with high average barriers to growth at $\hat{a} = -0.384$. This captures whether the history of diffusion of each crop has brought it to countries where the local conditions are beneficial or adverse to the successful cultivation of the plant.[18] Interestingly, there is no correlation between parameter estimates of $\hat{a}$ and β, which indicates that the processes of diffusion are disjoint from the effects of local conditions.[19]

Another informative statistic is how diverse countries are in their experience. It shows that in sorghum, cotton and soy, there is a wide dispersion of local coefficients, indicated by the variance-to-mean ratio $r = \sigma_i^2/\mu_i > 5$, while the experiences are fairly similar between countries for the other crops, where $r < 4$.

5. Discussion

This paper has put forward three sets of observations on crop yield developments across developing countries. The first observation set out the average yield gaps between developed and developing countries. The yield gaps are large across the entire range of crops (developing country yield gaps between 40 and 60%) but the two outliers in the group clearly are maize (72% gap) and sorghum (67% gap). The second observation consisted of a test for 'absolute convergence' across both hybrid and non-hybrid varieties – a test for whether countries with lower yields at the beginning of the period of observation (in 1961) have experienced higher average growth rates over the ensuing period (between 1961 and 1999). This exercise showed that the hybrid crops do not exhibit this absolute convergence property, indicating the presence of unusual convergence-limiting factors for these crops. We then examined the rates of diffusion from the technological frontier to developing countries. This part of the study found significant differences in the rates of diffusion of innovation between hybrid and non-hybrid crops, and that diffusion had been particularly slow for the hybrid varieties. Based on a standard innovation-diffusion model (Barro and Sala-i-Martin, 1995), this can be interpreted as evidence for higher imitation costs in these two crops.

In each part of the study, the hybrid crops (maize and sorghum) have been highlighted as the distinctive crops in the relationship between developed and developing country crop yields. In this section we set out our hypothesis concerning the reasons underlying these observations on these crops' relative performances.

In the context of the diffusion model by Barro and Sala-i-Martin (1995), differences in the diffusion coefficients are indicative of differences in the cost of imitation, i.e. the costliness of transferring an innovation from its developed country origin to its developing country context. An important determinant of the cost of imitation in the context of crops is how readily the value-adding traits in a novel variety from the leader country can be identified and extracted by the follower. This in turn is significantly influenced by the form of property right protection afforded to the value-adding traits within the innovative variety.

There is no question that property rights may be claimed in innovative plant varieties,[20] but the capacity to protect these claims varies. At present there are two principal forms of protection for claims of property rights to innovations in plant varieties: (1) 'legal protection', which is dependent for effect on the resources expended on monitoring and enforcement by the follower country; and (2) technological protection', which is independent of the resources expended by the follower for effectiveness. It is probably fair to assert that – over the past 40 years – there has been little legal protection afforded to intellectual property rights claims to innovations in plant varieties throughout most of the developing world. Since most developing countries have had little to gain from expending resources on the enforcement of property rights for the benefit of innovators situated primarily overseas, there have been minimal incentives for such expenditures. Therefore, it is likely that the primary route available for the effective protection of intellectual property right claims in developing countries has been technological.

Currently, technological protection is available only in the form of modern hybrid varieties, and thus limited in practice to the 'outbreeding crops': maize and sorghum. Hybridization affords protection to improved plant varieties of these species, because the seed from them that is sold to farmers represents a relatively diverse gene pool and subsequent re-plantings generate widely divergent varieties. The other crop species reproduce asexually, and hence the parents and offspring are identical in genetic structure. Sales of improved varieties from these species may be copied perfectly (and almost costlessly) from purchased seed, unless national laws effectively prevent such practices.

It is for this reason that we are able to claim that the technologically protected species act in effect as a case study on the impact of effective or 'strong' property rights in innovation. They are to be contrasted with the impacts of innovations in the non-technologically protected species that act, in developing countries, as 'weakly' protected innovations. Maize and sorghum are distinguished by the unique capacity for the technological protection of innovations in their sectors; they represent the lines down which strong property rights protection has been in effect over the past 50 years.[21]

We believe that the observed differences in yield growth and diffusion across crop varieties noted in the previous sections are attributable to the distinctive property right regimes that were available for claiming rights to innovation in these varieties. The observations are consistent with the idea that strong property right regimes have resulted in varying costs of imitation across countries, which increase with the distance of the country from the technological frontier. This increasing cost of imitation translates into the observed consequence that innovations are impacted, slow to diffuse, especially for the two crops afforded effective protection. The ultimate outcome is that the two crops in which strong property rights exist are the only two which do not exhibit absolute convergence. The poorer countries fail to 'catch up', only for those crops where strong intellectual property rights regimes prevail. Finally, this failure to catch up is captured in aggregate terms in the relative lags between the yields in developing and developed countries. All of the observations on crop yields and changes across the past 40 years are consistent with the hypothesis that strong property rights protection over innovations inhibits their diffusion across the developing world.

If this is the case then it provides significant evidence in the general debate about the global impact of enhanced property right regimes. These observations imply that the receipt of benefits from strong property rights protection is inversely related to the distance of the particular country from the technological frontier. This would imply that, even if innovation occurs more rapidly under strong property rights protection, countries far from the frontier might prefer the combined rate of innovation/diffusion inherent within a weaker form of property rights regime. All intellectual property rights regimes would entail an inherent trade-off between innovation and diffusion, and the preferred regime would depend upon the perspective (i.e. technological level) of the country concerned (Krugman, 1979; Lai, 1999).

6. Conclusion

This paper has examined the development of yield in developing and developed countries in the eight most important agricultural crops over a period of almost 40 years. Our results indicate that although yield growth has been impressive, problems in global distribution of agricultural productivity persist and give cause for concern. Our results also indicate that there are significant differences in both the dynamics of yield growth in the developed countries and the diffusion of these gains to developing countries between crops that require explanation. Evidence on the convergence of yields in developing countries shows that convergence occurs in all non-hybrid crops but does not occur with respect to the hybrids. We explore the reasons for this difference further by estimating the diffusion coefficients of innovations from the yield frontier to developing countries. We conclude that the failure of convergence in the hybrid crops can be explained by the exceptionally low rate of diffusion in these crops.

Hybrid seed varieties have been available for two crops alone over the past 50 years. This has led to higher than average growth of yields in these crops through the mobilization of private R&D efforts. At the same time, our results indicate that the technological protection of property rights claims afforded by hybridization has a negative effect on the rate of diffusion of these innovations. The existence of this innovation-diffusion trade-off highlights the problematic international welfare implications inherent in choosing a particular regime of intellectual property protection. In other papers we have demonstrated how the perception of the benefits from property rights regimes varies with the country's initial position relative to the technological frontier (Göschl and Swanson, 2000).

The case of crop varieties is a possibly unique setting within which the debate over the impacts of enhanced property right regimes might be tested. This initial evidence indicates that there is an inherent trade-off between enhanced rates of innovation (and thus growth) and enhanced rates of diffusion (and hence distribution). This means that there are frictions within the system of technological dissemination that inhibit the flows of beneficial information, and that enhanced property rights regimes will work most prominently against the interests of those states furthest from the frontier. Although the results are preliminary, they give cause for concern about the promotion of property rights regimes with such profound distributional implications.

Acknowledgements

This research has its origins in a research grant by the Department for International Development to investigate the impacts of genetic use restriction technologies on developing countries. We are grateful to Robert Carlisle for encouraging us to explore this area. We are particularly grateful to James Symons for helpful discussions on the econometrics, and grateful to Mark Rogers for helpful discussions, without implicating them in any way in all of the remaining errors.

Appendix A: Convergence Results – Disaggregation by Region and Crop

The aggregation present in the regressions in Section 2.4 erases the fine structure of the data with regard to the geographical and crop-specific occurrence of convergence. We shortly examine the question whether there is significant convergence that the aggregation overlooks. Examining the evidence on convergence on a regional basis, we observe a very small number of regions with a decrease in regional yield dispersion. Table 6 reports the regression results by world and by region. In hybrids, there has been a reduction in yield dispersion on a regional basis in the Caribbean at a 10% significance level and in South Asia at the 5% significance level.[22] In non-hybrids, yields in Central America have been converging at a 1% level of significance and in South East Asia at a 5% level.[23] In all other cases, yield dispersion has rather increased. This implies that although the aggregate perspective overlooks isolated cases of convergence, the overall picture delivered at the global level is fairly accurate.

Breaking the data down to the crop specific level by region, there are again some instances of yield dispersion decreasing at a regional level in some crops. Table 7 lists the instances of significant convergence. Out of 64 possible cases, only eight exhibit convergence, of which a further two disappear when serial correlation is corrected. Cotton is the crop for which convergence has been most prevalent and accounts for half of the robust cases of convergence. By region, there is less of a discernible pattern, although South Asia and Sub-Saharan Africa both feature more than one case of yield convergence in a specific crop. The absence of convergence at a less aggregated level is – again – evidence that the aggregation does not grossly misrepresent the picture available from the yield data. It also means that the explanation for the failure to observe a decrease in yield dispersion in both types of crops and mean-reversion in hybrids does not lie at the level of regional disaggregation or crop specificity.

Appendix B: Robustness of Diffusion Results

In this Appendix, we briefly review the robustness of our diffusion results to changes in the econometric methodology adopted. There are three main points to be addressed: (1) the sensitivity of the results to the definition of the frontier, (2) the sensitivity of the results to the time period chosen, and (3) the sensitivity of the results to the inclusion of other explanatory variables, and the argument that diffusion problems are a function of agro-ecological factors only.

Table 6. Regression of cross-sectional variance of yields on trend by region.

Reference area/coefficient	Hybrid crops	Non-hybrid crops
Central America		AR(1)
ϕ_1	0.002393	0.007539
	13.12968***	23.47693***
ϕ_2	–9.09E-06	–0.000118
	–1.116526	–8.617558***
R_2	0.033470	0.868437
Caribbean	AR(1)	AR(1)
ϕ_1	0.003946	–0.001135
	14.56941***	–1.102844
ϕ_2	–2.26E-05	0.000183
	–1.933255*	4.239138***
R^2	0.308384	0.744380
South America		
ϕ_1	0.002841	0.002226
	16.81862***	21.90425***
ϕ_2	3.40E-05	6.01E-06
	4.496879***	1.322669
R_2	0.359680	0.046344
Sub-Saharan Africa		AR(1)
ϕ_1	0.001662	0.004390
	10.59792***	5.750232***
ϕ_2	4.24E-05	-1.95E-07
	6.053543***	–0.006375
R_2	0.504442	0.536273
North Africa and Middle East		
ϕ_1	0.003924	0.004443
	8.258243***	14.87574***
ϕ_2	0.000104	3.11E-05
	4.912760***	2.326613**
R_2	0.401349	0.130710

Table 6. Continued

Reference area/coefficient	Hybrid crops	Non-hybrid crops
South Asia	AR(1)	AR(1)
ϕ_1	0.001071	0.000695
	6.066385***	1.688054
ϕ_2	−1.91E-05	3.55E-05
	−2.623612**	2.192337**
R_2	0.739817	0.871075
South-East Asia	AR(1)	AR(1)
ϕ_1	0.002483	0.002612
	3.747586***	4.446791***
ϕ_2	1.01E-05	−1.30E-05
	0.362224	−0.541726
R_2	0.379966	0.504564
East Asia		AR(1)
ϕ_1	0.002509	0.001654
	8.530568***	1.884630*
ϕ_2	−1.49E-05	5.00E-05
	−1.131658	1.375955
R_2	0.034352	0.460016

B.1. The Frontier

The model assumes that innovations occur only at the frontier and then diffuse to the follower countries. We justify this approach by reference to the differential in biotechnological capacity between developing and developed countries. In order to test how sensitive the results are to the choice of the frontier, we ran the estimations against individual known biotechnological leader countries for specific crop (in the case of maize, the USA, in the case of wheat, the United Kingdom, in the case of rice, Japan). The results generally show a deterioration in the statistical significance of the estimators, but no significant difference with respect to the absolute levels of the estimators. The decrease in the significance can be attributed to the fact that an individual country's yield curve contains more variance than a pool of countries. The only exception to this is the case of rice. We estimated the diffusion process with respect to two yield frontier countries, Japan and South Korea. This led to a significant increase in the diffusion rate.[24] Our hypothesis is that diffusion in rice probably took place by innovations from other paddy-cultivating

Table 7. Evidence on decrease in yield dispersion, by region and by crop, 1961–1999.

Crop/Region	Maize	Sorghum	Barley	Cotton	Millet	Rice	Soybeans	Wheat
Caribbean	+++ (++)	N/A	N/A	N/A	N/A	+++ (+++)	N/A	N/A
Central Am.	0/0	0	N/A	++ (0)	N/A	- - - (0)	N/A	N/A
South America	+++ (++)	0	+++ (++)	- - - (- - -)	N/A	0 (0)	+++	+++
Sub-Saharan. Africa	+++	+++ (++)	- - - (- - -)	- - - (- - -)	++	0 (0)	N/A	+++
North Africa and Middle E.	+++ (++)	+++	+++	+++ (+++)	+	+++ (+++)	N/A	0
South Asia	- - - (- - -)	- - - (0)	+++ (+++)	N/A	- - - (- - -)	0	N/A	- - - (0)
South-East Asia	++ (0)	N/A	N/A	- - - (- - -)	N/A	+++ (+++)	0/0	N/A
East Asia	0	N/A	N/A	N/A	N/A	N/A	N/A	N/A

Note: A 'plus' sign indicates a positive slope of the time trend at 10% (+), 5% (++) and 1% (+++) level of statistical significance, and vice versa for the 'minus' sign. The result in parentheses indicates the result when correcting for serial correlation where present. The convergence parameter has only been computed for regions where the sample contained more than three countries per crop. N/A therefore indicates where the sample size was smaller than three.

countries at the frontier being disseminated to a greater extent to countries with comparatively low yields. Innovations in dry-land cultivation were hence restricted in their diffusion. Japan and South Korea both use paddy cultivation for rice, which makes the explanation of separate technological systems plausible and bring rice closer in line with the average of non-hybrid crops. This suggests a solution to the rice paradox of non-diffusion from the frontier and simultaneous presence of absolute convergence because the results on absolute convergence and on diffusion seem to contradict each other. Closer inspection of the diffusion results reveals that although the diffusion coefficient is at the lower end, rice has the best result for the country-specific growth rate. Our interpretation of this result is to postulate that rice may be particular in that crop improvements in the frontier countries may have little to contribute to paddy cultivation systems that are in use in many of the developing countries.

B.2. Time Period

We performed a test on the robustness of the coefficients by dividing the panel into four sub-periods, from 1961 to 1970, 1971 to 1980, 1981 to 1990, and from 1991 to 1999. We then repeated the panel estimation on the four sub-periods. The coefficients are listed in Table 8. It shows that the coefficients changed considerably

Table 8. Coefficients of diffusion and structural effects in hybrid and non-hybrid crops, 1961–1970, 1971–1980, 1981–1990, and 1990–1999.

	1961–1970	1971–1980	1981–1990	1991–1999
Hybrids				
β	–0.383772	–0.655827	–0.811600	–0.730338
α	–0.46343	–0.86339	-1.18209	-1.07869
R_2	0.263923	0.367706	0.440946	0.429292
Non-hybrids				
β	–0.529839	–0.732226	–0.692318	–0.850695
α	–0.44816	–0.64504	–0.64333	–0.79662
R_2	0.322702	0.399394	0.387185	0.458522

over the time period, indicating in general an increase in the rate of diffusion to developing countries in both hybrids and non-hybrid crops. It also shows that with the exception of the period between 1981 and 1990, non-hybrids have consistently outperformed hybrid crops in terms of diffusion.

The general increase in the rate of diffusion is intuitively appealing as it reflects the general improvements in technology and infrastructure. Technological progress resulted in improved techniques for replicating innovations, thus accelerating transfer of value-adding traits. The development of global infrastructure on the other hand speeded up the direct transfer of innovations from industrialized to developing countries.

However, the performance differential between hybrids and non-hybrids persisted. Period 1980–1990: Multinational companies entering adaptive breeding, but apparently not for long.

How about the fixed effects? Again, true to hybrids disadvantaged there. It also shows that the frontier developed technologically away from DCs across the board. Non-existence of international public breeding? What is important from the perspective of this paper is that the ranking of the crops in terms of diffusion performance is left unaltered by this procedure. In both cases, maize, sorghum and rice are the crops with the lowest diffusion within each sub-period. This confirms the overall picture of relative performance that the first estimation aimed to sketch.

B.3. The Effects of Land Expansion

One alternative explanation for yields in maize remaining low is that the model we estimated omits a crucial variable included in other models that measure the impact of R&D on productivity (see, e.g., Evenson and Kislev, 1973), namely the changes in the acreage of the crop.

The argument is that the above regression results for yield development overlook the fact that the quality of land inputs may degrade as cultivation expands. If expansion into marginal lands was an important factor determining yields, then maize yields – as one of the most successful crops in terms of expansion in developing countries – may have been depressed as a result of that. In order to test this alternative explanation, we estimated an augmented model of the following form

$$\Delta G_{it} = a_i + \beta \cdot G_{i,t-1} + \gamma \cdot \Delta L_{it} + \varepsilon, \qquad (10)$$

where L is the acreage of the crop. Empirically, we enter ΔL_{it} as the log of the area harvested in country i at time t. On the basis of the hypothesis above, we would expect the coefficient γ to be negative in order to capture that the expansion of land under cultivation should have a negative effect on the yield development of the crop in country i.

The econometric results show that the impact of land changes (measured by the coefficient of the log of harvested area) is not significant for any of the crops with the exception of maize (at a 1% level) and sorghum (at a 10% level). This means that – with the exception of maize and sorghum – the difference in the speed of yield growth between developing countries and the lead countries is explained very well by the diffusion of technological innovations, but not by changes in the crop area harvested. In the case of maize and sorghum, the impact of land changes is significant, but it has their impact is minuscule: the change in the productivity gap has an effect several times the magnitude of a change in the acreage. The absolute value of the coefficient of acreage is -0.010 for maize and -0.006 for sorghum.[25] It is also noteworthy that inclusion of land effects has barely any impact on the coefficient of the productivity gap. This points to the fact that the processes of diffusion and intensification are disjoint. By including acreage, we have identified a significant, but small additional factor that worked exclusively in the case of maize and sorghum, but does not have any bearing on the argument regarding the diffusion process.

B.4. Agro-Ecology

A common concern about the diffusion of agricultural innovations is the transferability of knowledge between different agro-ecological zones. A first indicator would be that the correlation coefficient between the yield gaps in 1961 and 1999 is 0.56, suggesting a persistence of gaps over time that could be caused by intrinsic problems in transferring biological technology. These concerns have been with the diffusion literature since its inception (see, for example, Evenson and Kislev, 1973; Myren, 1969). There is significant debate, however, over the extent of transferability, with some authors claiming that it is overestimated (for example, in the case of wheat, see Maredia and Eicher, 1995) or underestimated (Morris, pers. comm.).

Our results do not refute the idea that agro-ecological specificity may be a factor and the country-specific intercepts do pick up structural differences between countries that reflect intrinsic suitability for cultivation. Although we are not able to distinguish between agro-ecological and other structural factors that may influence the long-run rate of growth in our specification, agro-ecology is taken into account. Despite this, important differences between the rates of diffusion of different crops persist.

Notes

[1] According to UNESCO estimates cited by Coe et al. (1997), the industrialized countries accounted for 96% of total world R&D expenditures (UNESCO, 1993).

[2] There are several factors that suggest that the contribution of developing countries' agricultural R&D to the development at the technological frontier is significantly less than their share of global R&D expenditures in agriculture. The first reason is the location-specificity of biological capital which requires adaptive breeding to capture the value-adding traits if they are developed for other agro-ecological conditions. This means that much of developing countries' R&D expenditures will go towards obtaining the private good components of research developed elsewhere rather than providing innovations themselves. The second factor is that much of developing countries' R&D is directed towards maintenance rather than output-enhancement as pest and pathogen adaptation in the more tropical conditions experienced in many developing countries occurs more frequently than in developed countries (Pardey, Roseboom and Anderson: Public-Sector Ag Research; in Pardey et al., 1991). The third factor is that developing countries may unnecessarily duplicate research efforts undertaken in similar agro-ecological settings that could be transferred at little cost ('Go it alone' approach) (Maredia and Eicher, 1995; see Eicher, 1990, for the example in Sub Saharan Africa).

[3] The criterion applied here is the global acreage of a crop.

[4] Soybeans were only introduced into many countries over the last 40 years and hence have starting dates after 1961.

[5] We are currently undertaking further research based on selected R&D indicators.

[6] Evenson and Kislev (1973) use the publication of crop-specific scientific papers in the *Plant Breeding Abstracts* as a proxy for R&D activity in a specific country. It is difficult to extend this approach to all eight crops examined.

[7] The classification of countries into 'developing' and 'developed' follows Pardey et al. (1991) rather than the FAO classification.

[8] This estimate gives equal weight to each country and is based on the country classification adopted in Pardey et al. (1991) and taken up by the wider literature on agricultural R&D. A comparison of yields on an area-weighted basis directly based on FAO data and its classification of developing and developed countries produces an even more dramatic picture while leaving the ranking of crops basically unaffected (see Göschl and Swanson, 2000).

[9] For a more thorough treatment of β-convergence tests, see Barro and Sala-i-Martin (1995, ch. 11).

[10] Standardization here refers to an indexation of yields in different crops in order to make yields (measured in hg/ha) comparable between different crops.

[11] The reason for excluding countries with negative growth rates is that they are unlikely to contribute meaningful data to the exercise. Their inclusion does not fundamentally affect the results, but renders the interpretation of the results somewhat less cogent.

[12] McCunn and Huffman do not report evidence on serial correlation. This type of correlation is widespread in our data and is intuitively obvious (data generated certainly not i.i.d.). We compute the

Ljung-Box Q-statistic for both regressions and observe that serial correlation is a serious problem in the case of non-hybrids. Correcting for serial correlation in the non-hybrid regression, the parameter estimate for ϕ_2 is unchanged, but – as expected – the t-statistic decreases. However, ϕ_2 is still significant at the 5%-level of significance.

[13] For further analysis on convergence, and disaggregation across crops and regions, see Appendix A.

[14] The $\psi(1)$ is sufficiently large in order to prevent the possibility of total imitation in the limit and the $\psi(0)$ is sufficiently small such that there is a minimum of imitation going on at any point in time.

[15] The presence of heteroskedasticity tends to lead to higher diffusion coefficients. This weighting procedure corrects for that. The White test for cross-section heteroskedasticity is performed for all estimations and reports consistent parameters for all crops.

[16] This curve is constructed on the assumption that the demand curve for agricultural output has constant and equal demand elasticity in the developed and the developing countries.

[17] The curve is fairly robust against changes in the discount rate. A higher rate pushes the curve down slightly, and vice versa.

[18] There are for each crop countries in which the intrinsic growth rate of the yield is basically equal or above that prevalent in the frontier countries. In the case of barley, this holds for Zimbabwe; in the case of cotton, for Israel and Syria; in the case of maize, for Chile; in the case of millet, for China; in the case of rice, for Egypt and Korea; in the case of sorghum, for Egypt and Israel; in the case of Soybeans, for Ethiopia; and in the case of wheat, for Egypt and Zimbabwe.

[19] The correlation coefficient between $\hat{a}$ and β is 0.02.

[20] The so-called UPOV convention provides that each member state should provide property right protection in innovative plant varieties. It is also possible to take traditional forms of patent rights in innovative seeds.

[21] An interesting issue in future is the impact of technological change that affords technological protection to other crop species, so-called genetic use restriction technologies, and the welfare implications for various countries (Göschl and Swanson, 2000).

[22] The negative slope of the time trend in hybrid yields in Central America and East Asia is not significant.

[23] The decrease in Sub-Saharan Africa is not significant.

[24] The diffusion coefficient is in this regression is −0.31 (0.013854)** and the average country-specific lag −0.29578.

[25] An interesting side effect is the change in the country-specific intercepts when including the effect of changes in the acreage. It shows that in Argentina (M+S), Egypt (M+S), Israel (S), Mexico (S), Peru (S), Chile (M) and China (M), intrinsic growth rates would have been higher than at the frontier if expansion had not put marginal lands into use for maize (M) and sorghum (S) cultivation.

References

Alston, J.M. and Venner, R.J. (1998) The effects of U.S. plant variety protection act on wheat genetic improvement, Paper presented at the Symposium on 'Intellectual Property Rights and Agricultural Research Impact' sponsored by NC 208 and CIMMYT Economics Program, El Batan, Mexico, March 5–7.

Alston, J.M., Pardey, P.G. and Smith, V.H. (1998) Financing agricultural R&D in rich countries: What's happening and why?, *Austr. J. Agricult. Res. Econ.* **42**(1) 51–82.

Ball, V.E., Bureau, J.-C., Nehring, R. and Somwaru, A. (1997) Agricultural productivity revisited, *Amer. J. Agricult. Econ.* **79**, 1045–1063.

Barro, R.J. (1991) Economic growth in a cross-section of countries, *Quart. J. Econ.* **106**, 407–443.

Barro, R.J. and Sala-i-Martin, X. (1995) *Economic Growth*, McGraw-Hill, New York.

Baumol, W.J. (1996) Productivity growth, convergence, and welfare: What the long-run data show, *Amer. Econ. Rev.* **76**, 645–661.

Baumol, W.J., Blackman, S.A.B. and Wolff, E.N. (1989) *Productivity and American Leadership: The Long View*, The MIT Press, Cambridge, MA.

Bayoumi, T., Coe, D.T. and Helpman, E. (1999) R&D spillovers and global growth, *J. Internat. Econ.* **47**(2), 399–428.

Butler, L.J. (1996) Plant breeders' rights in the U.S.: Update of a 1983 study, in J. Van Wijk and W. Jaffe (eds.), *Proceedings of a Seminar on The Impact of Plant Breeders' Rights in Developing Countries*, Santa Fe Bogota, Colombia, March 7–8, University of Amsterdam, Amsterdam, pp. 17–33.

Butler, L.J. and Marion, B.W. (1985) *The Impact of Patent Protection on the U.S. Seed Industry and Public Plant Breeding*, Food Systems Research Group Monograph 16, University of Wisconsin-Madison, Madison, WI.

Byerlee, D. and Moya, P. (1993) *Impacts of International Wheat Breeding Research in the Developing World, 1966–1990*, CIMMYT, Mexico, D.F.

Capalbo, S.M. and Antle, J.M. (1989) Incorporeality social costs in the returns to agricultural research, *Amer. J. Agricult. Econ.* **71**, 458–463.

CIMMYT (1999) *A Sampling of Impacts 1999. New Global and Regional Studies*, CIMMYT, Mexico, D.F.

Coe, D.T. and Helpman, E. (1995) International R&D spillovers, *Eur. Econ. Rev.* **39**(5), 859–887.

Coe, D.T., Helpman, E. and Hoffmaister, A. (1997) North-South R&D spillovers, *Econ. J.* **107**, 134–149.

DeLong, B. (1988) Productivity growth, convergence, and welfare: Comment, *Amer. Econ. Rev.* **78**, 1138–1154.

Easterlin, R.A. (1960) Regional growth of income: Long-run tendencies, in S. Kuznets, A.R. Miller and R.A. Easterlin (eds.), *Population Redistribution and Economic Growth, United States, 1970–1950. II: Analysis of Economic Change*, The American Philosophical Society, Philadelphia, PA.

Eicher, C.K. (1990) Africa's food battles, in C.K. Eicher and J.M. Staatz (eds.), *Agricultural Development in the Third World*, 2nd edn., Johns Hopkins Studies in Development, Johns Hopkins University Press, Baltimore, MD, pp. 503–530.

Evenson, R.E. (1989) Spillover benefits of agricultural research: Evidence from US experience, *Amer. J. Agricult. Econ.* **71**, 447–452.

Evenson, R.E. (1994) Analyzing the transfer of agricultural technology, in J.R. Anderson (ed.), *Agricultural Technology: Policy Issues for the International Community*, CABI, Wallingford, pp. 165–179.

Evenson, R. and Kislev, Y. (1973) Research and productivity in wheat and maize, *J. Political Econ.* **81**(6), 1309–1329.

Evenson, R., Waggoner, P.E. and Ruttan, V. (1979) Economic benefits from research: The case of agriculture, *Science* **205**, 1101–1107.

Falck-Zepeda, J.B. and Traxler, G. (1998) Rent creation and distribution from transgenic cotton in the U.S., Paper prepared for the Symposium on Intellectual Property Rights and Agricultural Research Impacts, sponsored by NC-208 and CIMMYT Economics Program, CIMMYT Headquarters, El Batan, Mexico, March 5–7.

Frantzen, D. (1998) R&D, international technical diffusion and total factor productivity, *Kyklos* **51**(4), 489–508.

Frantzen, D. (2000) Innovation, international technological diffusion and the changing influence of R&D on productivity, *Cambridge J. Econ.* **24**(2), 193–210.

Frey, K.J. (1996) National plant breeding study – I: Human and financial resource devoted to plant breeding research and development in the United States in 1994, Special Report 98, Iowa Agriculture and Home Economics Experiment Station, Iowa State University, IA.

Fuglie, K., Ballenger, N., Day, K. et al. (1996) Agricultural research and development: Public and private investments under alternative markets and institutions, AER-735, Economic Research Service, United States Department of Agriculture.

Göschl, T. and Swanson, T. (2000) Forecasting the implications of GURTs, *J. Internat. Develop.*, forthcoming.

Göschl, T. and Swanson, T. (2001) Genetic use restriction technologies and the diffusion of yield gains to developing countries, in T. Swanson (ed.), *Biotechnology, Agriculture and Developing Countries*, Edward Elgar, London, forthcoming.

Griliches, Z. (1957) Hybrid corn: An exploration in the economics of technological change, *Econometrica* **48**, 501–522.

Grossman, G. and Helpman, E. (1990) Comparative advantage and long-run growth, *Amer. Econ. Rev.* **80**, 761–773.

Grossman, G. and Helpman, E. (1997) *Innovation and Growth in the Global Economy*, The MIT Press, Cambridge, MA.

Hsaio, C. (1986) *The Analysis of Panel Data*, Econometric Society Monographs No. 11, Cambridge University Press, Cambridge.

Huffman, W.E. and Evenson, R.E. (1993) *Science for Agriculture: A Long-Term Perspective*, Iowa State University Press, Ames, IA.

Jaffe, W. and Van Wijk, J. (1995) *The Impact of Plant Breeders' Rights in Developing Countries: Debate and Experience in Argentina, Chile, Colombia, Mexico and Uruguay*, Inter-American Institute for Co-operation in Agriculture and University of Amsterdam, Amsterdam.

Jarrett, F.G. (1982) Location specificity, critical mass and the allocation of resources to agricultural research, *Agricult. Admin.* **11**, 49–65.

Kalton, R.R. and Richardson, P.A. (1983) Private sector plant breeding programmes: A major thrust in U.S. agriculture, *Diversity* (Nov.–Dec.), 16–18.

Krugman, P. (1979) A model of innovation, technology transfer, and the world distribution of income, *J. Political Econ.* **87**, 253–266.

Lai, E.L.-C. (1999) International intellectual property rights protection and the rate of product innovation, *J. Develop. Econ.* **55**, 133–153.

Lichtenberg, F.R. (1994) Testing the convergence hypothesis, *Rev. Econ. Statistics* **76**(3), 576–579.

Maredia, M. and Eicher, C.K. (1995) The economics of wheat research in developing countries: The one hundred million dollar puzzle, *World Develop.* **23**(3), 401–412.

McCunn, A. and Huffman, W.E. (2000) Convergence in the U.S. productivity growth for agriculture: Implications of interstate research spillovers for funding agricultural research, *Amer. J. Agricult. Econ.* **82**, 370–388.

Myren, D. (1969) The Rockefeller Foundation Program in corn and wheat in Mexico, in C. Wharton (ed.), *Subsistence Agriculture and Economic Development*, Aldine, Chicago.

Ordover, J.A. (1991) A patent system for both diffusion and exclusion, *J. Econ. Perspectives* **5**(1), 43–60.

Pardey, P.G., Roseboom, J. and Anderson, J.R. (eds.) (1991) *Agricultural Research Policy: International Quantitative Perspectives*, Cambridge University Press, Cambridge.

Perrin, R.K., Kunnings, K.A. and Ihnen, L.A. (1983) Some effects of the U.S. Plant Variety Protection Act of 1970, Economics Research Report No. 46, Department of Economics and Business, North Carolina State University.

Pingali, P. and Heisey, P. (1999) Cereal crop productivity in developing countries: Past trends and future prospects, CIMMYT Economics Program Working Paper No. 99-03.

Pioneer Hi-Bred International Inc. (1996) *Inventing and Building Seed Crop Genetic Improvements*, Research and Product Development.

Pray, C.E., Knudson, M.K. and Masse, L. (1993) Impact of changing intellectual property rights on US plant breeding R&D, mimeo, Department of Agricultural Economics, Rutgers University, New Brunswick, NJ.

Romer, P. (1990) Endogenous technological change, *J. Political Econ.* **98**(5), part II, S71–S102.

Schimmelpfennig, D.S. and Thirtle, C. (1999) Research spillovers between the European Community and the United States, *Contemp. Econ. Policy* **19**(4), 457–468.

Schmidt, J.W. (1984) Genetic contributions to yield gain in wheat, in W.R. Fehr (ed.), *Genetic Contributions to Yield Gains in Five Major Crop Plants*, Crop Science Society of America, Madison, pp. 89–101.

Schmitz, A. and Seckler, D. (1970) Mechanized agriculture and social welfare: The case of the tomato harvester, *Amer. J. Agricult. Econ.* **52**, 567–577.

Streissler, E. (1979) Growth models as diffusion processes: II, *Kyklos* **32**(3), 571–586.

Swanson, T. (1996) The reliance of northern economics on southern biodiversity: Biodiversity as information, *Ecol. Econ.*, 1–6.

Thirtle, C.G. (1985) Technological change and productivity slowdown in field crops: United States, 1939–1978, *Southern J. Agricult. Econ.* **17**, 33–42.

Traxler, G., Falck-Zapeda, J., Ortiz-Monasterio, J.I. and Sayre, K. (1995) Production risk and the evolution of varietal technology, *Amer. J. Agricult. Econ.* **77**, 1–7.

UNESCO (1993) *Statistical Yearbook*, Oxford University Press, New York.

Zilberman, D. and Heiman, A. (1997) The value of economic research, *Amer. J. Agricult. Econ.* **79**(5), 1539–1544.

PART E

POLICY CONCLUSIONS

11. Plants and Biotechnologies

PHILIP J. DALE

1. Introduction

Plant breeding has made a significant contribution to crop productivity. Wheat yields are now two to three times those of 80 years ago. Half of this improvement is genetic, using what we now term 'conventional' breeding methods; the other half is from the use of fertilisers and pesticides. The dwarfing character in cereals, for instance, has improved yield by the redistribution of plant resources from straw to grain. During the history of plant breeding, there have been steady advances and conventional plant breeding now uses many highly developed techniques to improve crops (Hayward et al., 1993).

2. Genetic Modification in the Context of Conventional Plant Breeding

Over the past 20 years, we have learnt how to isolate genes from different kinds of organisms and insert them into crop plants. We can now potentially isolate genes from any class of organism and insert them into our most important crop plants. This advance provides a wider choice of genes for use in crop improvement.

Since the beginning of scientific plant breeding in the early 1900s, plant breeders have continually searched for novel sources of genes for pest and disease resistance, for improved quality and for a range of other characters. The desire for a wider choice of genes for conventional plant breeding has led to the development of a range of sophisticated methods. Modern methods of embryo and ovary culture, for example, now make it possible to produce hybrids between plant species and genera that would not hybridise in nature. This has made it possible to move genes, in conventional plant breeding, across natural sexual barriers.

Crops are commonly improved by transferring genes from wild relatives, some of which carry undesirable toxins, or weedy characteristics. In these 'wide hybrids', there is usually the desire to transfer one or two genes that control disease resistance, for instance, but because of the way sexual hybridisation works, it results in the transfer of several thousands of unwanted genes. There is the potential for these linked genes to confer undesirable weediness or toxic characteristics and, because of this, it has been necessary to develop careful testing and selection procedures, to eliminate undesirable plant lines.

Timothy Swanson (ed.), The Economics of Managing Biotechnologies, 251–254.

Mutation plant breeding is another method widely used in conventional plant breeding. This exposes seeds or other plant parts to radiation or chemical mutagens which cause random and unpredictable genetic changes in the crop. The plant breeder has no control over the nature or the degree of the genetic change caused. The careful testing of mutagenised plant lines and the elimination of the vast majority of plants that do not have improvements has enabled mutation breeding to be surprisingly successful. Many of the crops that we now use each day, particularly the cereals, have induced mutation somewhere in their pedigree.

Genetic mapping and molecular marker assisted breeding techniques have reduced the number of unwanted genes transferred inadvertently in conventional breeding. However, the attraction of modern methods of genetic modification over more conventional methods is that it provides the ability to insert one or two genes that confer the desirable character, without the substantial and uncontrolled variation that is common in most conventional breeding methods. There is some variation between different genetically modified plant lines produced with the same introduced gene(s), so it is necessary to make about 100 plant lines for each genetic modification, in order to select those that have the most desirable characteristics. This contrasts with the many hundreds or thousands of plant lines (especially with mutation breeding) that usually have to be discarded in a conventional plant breeding programme.

An indication of some of the potential benefits of genetic modification are outlined in Table 1. Extensive international research on gene isolation and characterisation is ongoing. Genome sequencing in rice, *Arabidopsis* and other plants is now advanced and in the coming years will provide a wide variety of novel genes for testing in transgenic crop plants. The priorities for modification will, of course, depend on the needs of the developing country, but pest and disease resistance, and improved yield and nutritional quality are likely to be high on the list of priorities.

3. Assessing the Safety of GM Crops

Because it is now possible to introduce genes into crop plants from any class of organism, there is international agreement that we should carry out a thorough safety assessment on GM crops in addition to those followed for conventionally bred crops. The primary focus of the assessment is to determine the potential impact of the modified crop on human and animal health, and the environment. Table 2 lists some of the questions that must be addressed in safety assessment before a GM crop can be released experimentally or commercially.

Table 1. Potential benefits of transgenic crops.

1.	Resistance to pests and diseases
2.	Enhancement of nutritional qualities (e.g. vitamin A and iron in rice)
3.	Increased tolerance to environmental stresses including temperature, water, and saline soils
4.	Plant architecture (e.g. dwarfing) and flowering (e.g. flowering time)
5.	Modifications in fruit ripening and tuber storage
6.	Reduction in seed losses through shedding at harvest time
7.	Modification of oil, starch and protein to provide sustainable supplies of raw materials for food, biodegradable plastics, detergents, lubricants, paper making and packaging
8.	Herbicide tolerance so that crop varieties can tolerate specific herbicides
9.	Increase in the ability of certain plants to remove toxic metals from soils (bioremediation)
10.	The elimination of allergens from certain crops (e.g. rice) or the enhancement of certain plant products (eg anticancer substances)
11.	The production of pharmaceutical substances (e.g. edible vaccines)

Table 2. Questions addressed in biosafety assessment.

1.	What is the function of the gene in the donor organism?
2.	How does the introduced gene(s) modify the crop plant?
3.	Is there evidence of toxicity or allergenicity?
4.	Are there effects on non-target organisms within the environment?
5.	Are there changes in the weediness of the GM crop in subsequent crops, or in its invasiveness in natural habitats?
6.	What is the likelihood and consequence of gene transfer by pollination to crops or feral species?

4. Challenges for the Future

Advances in science and technology usually provide both benefits and challenges (Dale, 1999, 2000). The development and use of GM crops is no exception. It is important that the kinds of crops developed have a clear benefit to developing countries. The cost of chemical input in developing countries is often prohibitive and because there is often a desire to employ people for weed control, herbicide-tolerant crops are rarely high on the list of priorities. Improvement in pest and disease resistance, tolerance to environmental stresses, improvements in yield and nutritional quality could, however, provide clear benefits to low input farming systems. They can also make it possible to avoid the use of cheap and often hazardous chemicals for pest and disease control, that can present dangers to people handling the chemicals and to the environment.

It is important in developing novel crops and new varieties, whatever plant breeding method is used, that there is respect for traditional practices. It may also be desirable to use local landraces to maintain a broad genetic base in the crop. Genetic modification does frequently have the potential to insert genes into locally adapted landraces.

Intellectual property constraints can be a substantial challenge. Companies and laboratories investing money in research to isolate and test genes need to be able to pay for their research. There is a clear role for non-commercial organisations, like DFID, Rockefeller and others, in facilitating the development and adoption of crops to meet local needs.

There is the potential to significantly improve the productivity and nutrition of crops. A commonly quoted example is the improvement of vitamin A and iron properties of rice (Ye et al., 2000). Other applications being explored are the use of plants to produce vaccines and pharmaceutical substances (Walmsley and Arntzen, 2000). These issues present not insignificant challenges that require free and open discussion about their application. It is important that we use all the available methods to improve agricultural production and nutrition.

Modern biotechnology is not a 'magic bullet' that can be applied to all food production problems. But it is an important tool in the plant breeder's toolbox. The genetic improvement of crops, however, is only part of the solution; we need to engage open minded and progressive experts in relevant areas of sociology, economics, ethics and safety assessment. All products of plant breeding need to be developed and applied with sensitivity to social, cultural, agricultural and environmental needs.

References

Dale, P.J. (1999) Public concerns over transgenic crops, *Genome Res.* **9**(12), 1159–1162.

Dale, P.J. (2000) The GM debate: Science or scaremongering?, *Biologist* **47**, 7–10.

Hayward, M.D., Bosemark, N.O. and Romagosa, I. (eds.) (1993) *Plant Breeding*, Chapman & Hall, London, 555 pp.

Walmsley, A.M. and Arntzen, C.J. (2000) Plants for delivery of edible vaccines, *Curr. Opin. Biotechnol.* **11**(2), 126–129.

Ye, X.D., AlBabili, S., Kloti, A., Zhang, J., Lucca, P., Beyer, P. and Potrykus, I. (2000) Engineering the provitamin A (beta-carotene) biosynthetic pathway into (carotenoid-free) rice endosperm, *Science* **287**, 303–305.

12. Regulatory Harmony – Who's Calling the Tune?

Regulation of Crop Biotechnology and the Third World

TOM CROMPTON and GEORGE T. TZOTZOS

1. Introduction

Biotechnology enthusiasts foresee a pivotal role for transgenic crops in addressing the food requirements of a growing world population. And, as the agribiotechnology industry burgeons and trade in genetic commodities gathers pace, Third World countries are also being promised a slice of the cake, either through development of a domestic biotechnology industry, or (in the case of countries rich in genetic resources) through benefit-sharing. These economic and humanitarian prognoses are frequently uttered with the same breath, by those who portend neo-Malthusian calamity on the one hand and a patent transgenic cure-all (a remedy for both hungry people and ailing economies) on the other. Meanwhile, critics of biotechnology press the Green Revolution into service as evidence of the social havoc wreaked by past agricultural technofixes which, they suggest, reduced complex social problems to issues of photosynthetic efficiency or fertiliser use.

Somewhere on the spectrum between these two viewpoints, individual Third World countries try to formulate a national approach to biotechnology.[1] This approach will follow an assessment of the potential impacts of biotechnological developments in agriculture on the national economy. It will consider the effects of importation of the products of biotechnology following their manufacture elsewhere, the possible development of a domestic biotechnology industry, and the impact of biotechnology on patterns of global trade (through, for example, the substitution of imported products by domestically produced biotechnology products).

In the following overview we assume some natural grouping of countries as 'Third World'. Of course, this is to obscure a spectrum of national differences. Few countries currently enjoy a buoyant biotechnology export industry. The remainder range from nations with an embryonic export industry (including some European states) to those with no foreseeable prospect of producing commercially viable biotechnology products – either for exportation, or for domestic use. Differences of national interest will be most marked between these latter, and countries (the United States, for example) which have a well developed biotechnology sector. Whilst we retain the use of 'Third World', it should of course be borne in mind that this represents a generalisation which fails to take account of a range of differences

Timothy Swanson (ed.), The Economics of Managing Biotechnologies, 255–274.

in national interests and priorities.[2] Our juxtaposition of 'industrialised countries' with 'Third World countries' throughout this article is intended to underline the impossibility of mutually exclusive categorisation.

Competitive entry of Third World countries into the biotechnology industry is fraught with difficulties which, in view of the continuing coalescence of the industry under the control of a limited number of multinational companies, seem set to intensify. Biotechnology is becoming both increasingly proprietary, and diminishingly amenable to small-enterprise exploitation. For this reason, attention has been drawn to the assimilation and exploitation of existing knowledge (rather than the generation of new knowledge). Indeed, this is considered by some as the obligate state of new Third World technologies.[3] But whilst relying upon the absorption of established technologies may circumvent the need for heavy investment in the initial stages of research and development, it cannot obviate the need for technological expertise and institutional assets. Particularly with regard to agricultural biotechnology, many of the technological and economic problems are encountered far downstream from the innovative steps. Acquisition, or imitation, of the molecular biology can by no means guarantee a commercially viable product: significant barriers to entry into the agribiotechnology sector are associated with the transfer of the technology from laboratory to farm.[4]

The recommendation that non-industrialised countries should concentrate upon the exploitation of existing knowledge, rather than the pursuit of primary research, also overlooks the restrictions imposed by intellectual property rights (IPR) agreements; an important issue which lies beyond the scope of this paper. Thus it has been claimed that Third World countries are caught in a cleft stick. On the one hand, such countries lack the technological capacity for inventing around existing patents (or indeed the resources necessary for litigating against foreign infringements of their own patents), and on the other, under international trade agreements, they are prohibited from copying new technologies.[5]

Junne, in reflecting on the probable impact of biotechnology on the international commodity trade, argues that impacts in the agricultural sector will be much larger than those in the pharmaceutical sector. He suggests that world trade in agriculture is worth more than ten times trade in pharmaceuticals, and that many Third World countries rely almost exclusively upon such exports as a source of foreign exchange.[6] Substitution of these exports with biotechnological products developed for cultivation in developed countries will evidently have important economic ramifications. Although the result of technological developments in food processing rather than the exploitation of a transgenic crop, the replacement of sugar imports from Latin America by High Fructose Corn Syrup (HFCS) (produced enzymatically from maize) is cited as a prime example of the impact of biotechnology on the economies of non-industrialised countries. Another substitution, of tropical lauric oils through transgenic production of rapeseed (a temperate crop) with new oil qualities, seems imminent, though its commercial implications are not yet clear.[7]

Table 1. Asymmetries between the interests and concerns of industrialised and Third World countries over the introduction of genetically modified crops. These have been grouped according to whether they are (a) *conditional* upon the embryonic nature of a developing biotechnology industry (that is, there will be a convergence of interests or concerns between industrialised and non-industrialised countries as the biotechnology industry in the latter expands), or (b) *systemic* to either developing economies, or environmental conditions. Grouping these latter two factors together is not, of course, to suggest that none of these can be addressed during the course of economic development. However, convergence of interest between industrialised and Third World countries on these points will be contingent upon factors which are not directly influenced by a growth in the biotechnology industry.

Industrialised countries	Third World countries
(a) Conditional	
Existing commercially available transgenic crops are congruent with agricultural priorities	Imported transgenic crops may not directly reflect the current needs of farmers, requiring corresponding shifts in agricultural practice
Prolific domestic development: Regulators are likely to have a large number of new products to screen; pressure for simplification of the approval procedure may be intense	Little domestic development: Regulators may have relatively few products to screen. These are likely to have been approved by industrialised countries in earlier incarnations
Extensive, and rapidly expanding, experience of previous releases upon which to draw	May be minimal experience of release under local environmental conditions upon which to draw
Need for domestic industry to use developing countries as over-wintering grounds for crops	Reciprocal need less easily envisaged
Expertise available for the development of national biosafety frameworks without the intervention of foreign experts	Greater reliance upon foreign expertise (principally that of experts from industrialised countries) in the course of the development of biosafety frameworks
In the event of international legislation providing for compensation for the potential damaging effects of novel organisms, litigation is more likely to be against countries with a strong export market	Countries which are predominantly importers of novel organisms are more likely to require compensation
(b) Systemic	
Possible replacement of imports by endogenously produced substitute biotechnology products, boosting domestic industry	Possible negative socio-economic impact of replacement of exports by the biotechnological production of these in other countries
Effect on centres of origin of genetically modified crops unlikely	Environmental release at centres of origin may require special considerations
Effect of replacement of current crop varieties on crop diversity may be comparatively small	Widespread use of new crop varieties may threaten diversity of local landraces
Demographic effects of changing agricultural patterns may be less significant in the context of highly industrialised agriculture	Demographic effects of biotechnology in agriculture likely to be more significant in context of subsistence farming
Appeasement of peoples' organisations and non-governmental organisations may be important	May be little public or non-governmental concern about environmental issues, or otherwise little democratic accountability to such pressures

Table 1 summarises some of the possible national differences of interest *vis-à-vis* the development of a biotechnology industry. These have been classified as 'conditional' or 'systemic'. Those factors classified as 'conditional' will be directly affected by the growth of a national biotechnology industry. Those classified as 'systemic' transcend the specifics of the biotechnology sector, suggesting that these may persist even were an otherwise non-industrialised country to develop a buoyant biotechnology industry. Evidently, 'systemic' could have been further split into those factors which are an unavoidable consequence of environmental conditions (for example, ecological centres of crop diversity), and those which, whilst perhaps more typical of a developing economy, are nonetheless likely to change (for better or worse) with increasing industrialisation (for example, the vulnerability of subsistence farming to socio-economic change). The classification of a factor as 'systemic' is therefore intended to indicate that this does not arise simply as a transitory facet of a biotechnology industry in its infancy, but rather as a result of certain environmental, economic and social differences between nations. It is not intended to suggest that these factors are inescapable conditions of particular countries.

These asymmetries are present irrespective of whether one considers the possible importation of a product produced elsewhere, or the nurture of an infant domestic biotechnology industry. Whilst industrialised countries may well be considering the interests of a company based in that same country, to date most applications for field trials in Third World countries have been made by a foreign company, or at best a foreign transnational with local tie-ups.[8] This consideration will further skew the disparities. It is clear that an inchoate biotechnology industry may not be best managed on the model presented by an industrialised country.

National variations in regulatory interests will reflect, most obviously, an assessment of the social and economic implications of the import of biotechnology products, possible export substitution, and the prospects for development of a national biotechnology industry.[9] In addition however, other considerations may include an assessment of the potential impact of transgenic organisms upon local biodiversity; the domestic influence of pressure groups and the climate of public opinion; the historical constraints imposed by prior legislation, particularly where it is decided to use this as a basis for introduction of new biosafety legislation; the workload regulators are likely to face and the cost of implementing rigorous legislation; and the relative strength of the domestic public-sector industry as opposed to private-sector industry.[10]

This diversity of factors serves to underline the specificity of national interest in the development of legislation regulating biotechnology, which is itself required under the Convention on Biological Diversity (CBD).[11] However, there are certain considerations (global, regional and bilateral) which may intervene in the course of optimising a national framework for regulation. Here we will consider:

- *The implications of World Trade Agreements* for national biosafety legislation, specifically provisions concerning non-tariff trade barriers.
- *The Cartagena Protocol*, which opened for ratification by Parties to the CBD at CoPV in Nairobi from 15–26th May, 2000 (significantly, this does not at present include the United States), and the interaction of the Protocol with World Trade Agreements.
- *Harmonisation of biosafety regulation I*: regional trade agreements, membership of which may demand adoption of comparable biosafety legislation (we consider, for example, new biosafety legislation in Eastern European Countries intending to join the European Single Market).
- *Harmonisation of biosafety regulation II*: pressures from multinational companies to introduce regulations prior to local investment.
- *Bilateral governmental pressure to introduce regulations*; foreign direct aid for biotechnology capacity-building in Third World countries may be linked to development of regulatory structures in 'aid for regulation' deals.

It is the potential conflict between national interest and the harmonisation of biosafety legislation that we investigate here. It seems evident that the rate of expansion of the agribiotechnology industry has been underestimated in certain Third World countries, suggesting that this is a conflict which is set to intensify, rather than abate.[12]

There is general acceptance amongst representatives of multinational agribiotechnology companies, government representatives of both industrialised and non-industrialised countries, and environmental NGOs that there should, at some level, be a convergence of national biosafety frameworks – and this is a consensus with which we concur. Several arguments in favour of such convergence have been put forward:[13] Evidently, the environmental impact of genetically modified organisms is likely to extend beyond national borders, suggesting a corresponding need for trans-national regulation; disparities in national legislation may lead to competition between countries, exacerbating the so-called 'race-to-the-bottom' where environmental protection is compromised in an attempt to attract foreign investment; if national regulatory interests are shared (and this is a point which we explore critically here), convergence may avoid unnecessary repetition of legislative development; convergence will facilitate the exchange of regulatory experience between countries, where common environmental or socio-economic conditions apply; finally, convergence will lubricate free-trade, a desideratum of many governments and the *raison d'être* of the World Trade Agreements.

However, convergence of regulatory approaches belies disagreement on whether or not this should be addressed through international legislation, and a divergence of interests with respect to those aspects of biosafety that should be prioritised in the course of any such international agreement. The spectrum of national responses to biotechnology is encapsulated in viewpoints on the initial need for, and subsequent development of, a 'Biosafety Protocol', and also in the provisions of national biosafety legislation. The United States, for example, whilst realising the importance of regulatory convergence amongst trading partners in non-industrialised countries,[14] remained concerned that any international regulation would adversely affect its biotechnology exports. US representatives therefore adamantly opposed the inclusion of development of an international 'Biosafety Protocol', under the auspices of the CBD. The World Bank also rejected the need for a 'Biosafety Protocol'.[15]

Indeed 'harmonisation' has become a shibboleth of regulatory policy development in countries with a strong biotechnology base. This arises from a desire to see the adoption of compatible legislation in potential export markets, and the routine acceptance of products following their approval in their countries of origin.

Whereas industrialised countries have evolved biotechnology programmes which are congruent with existing economic and production frameworks, this is unlikely to be the case in Third World countries, where the history of the parallel development of technological and economic systems may be short. As a result, in Third World countries there may be little integration of science, technology and production.[16] Public-sector science, as mentioned above, may not be relevant to local problems, or may not address these in a realistic way. Whilst technology is imported, this contributes to the lack of correspondence between domestic science and technology programmes, and perpetuates an industrial dependency upon foreign technology.

2. The Implications of World Trade Agreements

The final round of the General Agreement on Tariffs and Trade (GATT),[17] completed in Marrakesh in 1993, is administered by the World Trade Organization (WTO). GATT obliges discrimination between trade restrictions which are imposed for reasons of environmental prudence ('trade related environmental measures', or TREMs) and restrictions that, whilst possibly masquerading as TREMs, are in fact tantamount to protectionist measures introduced purely to restrict commodity import. The implications of GATT evidently compromise the sovereignty of individual nations to develop specific biosafety legislation where this might be construed as a spurious non-tariff trade barrier. In other words, there is a requirement that national measures adopted for reasons of environmental protection are 'legitimate'. This legitimacy is established upon the basis of 'scientific justification'. Where national provisions deviate from international standards and guidelines, a country must, if challenged to do so, produce scientific evidence justifying such deviation. GATT is therefore one constraint that may lead to a convergence of national legislation according to international guidelines, except where countries can claim the legitimacy of more stringent regulations. Here, two factors will be considered further; the *development of international guidelines* accepted as standards by the WTO, and the conditions under which deviation from these might be upheld as legitimate *TREMs*.

2.1. International Guidelines

There are two GATT agreements which explicitly refer to the environment; the Agreement on Technical Barriers to Trade, and the Agreement on the Application of Sanitary and Phytosanitary Measures (SPS). Both rely upon a range of international standards and guidelines. The SPS Agreement requires that: "Members shall base their sanitary or phytosanitary measures on international standards, guidelines or recommendations, where they exist ...". But "Members may introduce ... measures which result in a higher level of ... protection than would be achieved by ... [such] measures, if there is a scientific justification ...".[18] The agreement on SPS defines 'international standards, guidelines and recommendations' as those established by the Codex Alimentarius Commission, the International Office of Epizootics, and the International Plant Protection Convention. (Provision is also made for inclusion of other relevant international organisations.)

2.2. TREMs

The Agreement on Technical Barriers to Trade requires that "Members shall ensure that technical regulations are not prepared, adopted or applied with a view to or with the effect of creating unnecessary obstacles to international trade".[19] However, the Article does itemise particular 'legitimate objectives', according to which trade restrictions may be permitted. These objectives include "... protection of human health or safety, animal or plant life or health, or the environment"; exceptions which were also made in Article XX of the General Agreement on Tariffs and Trade 1947.[20]

3. The Cartagena Protocol and World Trade Agreements

Deliberations over the inclusion of provision for development of a 'Biosafety Protocol' in the CBD, and negotiations during the subsequent stages of its development illustrated the conflict of interest between principle exporters of biotechnology products, and those countries which anticipated more modest domestic capacity building in biotechnology and a dependency upon import of biotechnological commodities. The United States was conspicuous in its opposition to a binding 'Biosafety Protocol'.[21] Although the US has not ratified the CBD, and was not therefore party to the 'Biosafety Protocol' negotiations, it continued to participate in the negotiations, both directly,[22] and by advising countries receiving US aid on the deliberations of the Conference of the Parties.[23]

Many Third World countries consistently supported the need for a protocol to cover not just the transboundary transfer of genetically modified organisms, but also their safe handling and use. There was no consensus amongst such countries however, with a more luke-warm response to stringent regulations from representatives of East Asian and Latin American countries.[24] The position of the European Union, representing the interests of both countries aspiring to export biotechnology products, and those with little foreseeable hope of developing a commercially viable biotechnology export industry, was intermediary to that of the US on the one hand, and Third World countries on the other.[25]

The Protocol's relationship with the WTO (embodied in the precautionary elements) can be applied 'appropriately', without violating SPS obligations. To what extent the Protocol offers a mandate for countries to impose precautionary import bans on specific products is still open to interpretation. As one commentator has observed: "The degree to which precaution is employed here is unprecedented in an operative portion of a binding multilateral environmental agreement. It goes beyond Principle 15 of Rio, although the more onerous aspects of this formulation are tempered by a reasonableness standard of 'as appropriate'".[26]

This new variation of precaution is contained in Article 10(6) stating:

> Lack of scientific certainty due to insufficient relevant scientific information and knowledge regarding the extent of the potential adverse effects of a living modified organism on the conservation and sustainable use of biological diversity in the Party of import, taking also into account risks to human health, shall not prevent that party from taking a decision, as appropriate, with regard to the import of the living modified organism in question as referred to in paragraph 3 above, in order to avoid or minimize such potential adverse effects.

Until the scope afforded by the Protocol for Parties to impose precautionary import bans is tested, its value in redressing the limitations on the freedom of Members of the WTO to implement biosafety legislation, where this could possibly be construed as a non-tariff trade barrier, is unknown.

4. Harmonisation of Biosafety Regulation I: Regional Trade Agreements

All 132 members of the World Trade Organisation (WTO) are also members of some form of regional trade agreement.[27] Membership of 'free trade areas' (or aspiration to membership) may place certain constraints upon the scope of national biosafety legislation. Several Eastern European countries have assimilated EU biosafety legislation, in anticipation of admission to the Union. Thus legislation which was designed by Member States of the EU (some, though by no means all, of which were considering the interests of emerging domestic biotechnology companies) is now in the process of being assimilated by Eastern European countries with less immediate hopes of building commercially viable biotechnology industries. In Bulgaria, for example, with regard to risk assessment, "the principles that have been developed by the OECD and the EC are to be applied".[28] In Poland, and until 1990, voluntary biosafety guidelines were adopted according to an American model. However, Poland aspires to membership of the EU, and a Polish 'Gene Law', currently proposed, is a 'national copy' of the European Council Directives 90/219/EEC and 90/220/EEC.[29]

Legislation has also been harmonised between Mexico, the US and Canada, each members of the North American Free Trade Agreement (NAFTA). Although there is a high degree of co-operation in the course of the regulatory approval of a new product, and its subsequent marketing,[30] there are clearly large differences between the biotechnology capacities in the three countries. To June 1996, there had been over 2000 trials in the US, as opposed to 14 as of 1995 in Mexico.[31] It seems likely that this disparity will be indicative of the relative flow of commercial biotechnology products between the two countries. A net import of biotechnology products from the US could of course be to the Mexican advantage, but this asymmetry reveals a potential difference of national interests in the course of harmonising regulatory approaches within NAFTA.

Malaysia (which sought advice from Australia, Japan, the EU, the US and Thailand) has adopted guidelines which most closely follow the Australian model. Although not a member of ASEAN, it is admitted that the assimilation of aspects of the Australian legislation was politically prudent. Here, however, it has also been suggested that the geographical proximity of Australia leads to ecological similarities which should be reflected in biosafety provisions.[32]

Nonetheless, scope apparently exists for diverse national interpretation of EU legislation. The European Council Directive on the Deliberate Release into the Environment of Genetically Modifed Organisms (90/220/EEC) provides that "Member States shall ensure that all appropriate measures are taken to avoid adverse effects on human health and the environment which might arise from the deliberate release or placing on the market of GMOs".[33] Previous work has drawn attention to the diverse interpretations placed upon this clause by different Member States. It has been suggested that the procedure established by the Directive, which was intended to accommodate potential concerns within a Europe-wide decision, has not achieved this level of harmony: "Regulators cannot avoid judgements about strategic advantages or disadvantages of a product; presumed benefits may influence how regulators define harm. Thus an implicit technology assessment enters their safety judgement".[34] Even in the comparatively homogeneous political and economic context of the European Union, it seems that individual countries have felt the need to place differing emphasis upon consideration of the socio-economic impact of GMO release. That they have done so may lead to conflicts within the Union. (Certainly, diverse national views of biosafety issues have lead to such conflict, with Austria and Luxembourg continuing to uphold national bans on

Novartis, Monsanto and AgroEvo modified maize, despite EU marketing approval being granted.) It remains to be seen whether such expressions of national interest will have to kow-tow to the demands of the Single Market.[35] Nor is it yet apparent whether those Eastern European states which are in the process of adopting legislation in line with the European Directives will find scope for interpretation of this with full expression of national interests.

Comparison of biosafety legislation in the EU Member States, with that in Norway (which is not a Member State of the EU) helps to reveal constraints which may impinge upon Eastern European states in the course of their seeking EU membership. Norway has comparatively far-reaching regulatory approval procedures. Under the Norwegian Gene Technology Act, there are explicit provisions for both consideration of broader socio-economic factors,[36] and public participation (at the discretion of the competent authority). However, there are fears that national Norwegian legislation, designed to meet specific national needs, would be compromised by subscription to a regional economic trading block.[37]

The issue of public participation presents another example of considerations to which different importance is attached by various member states of the European Union, despite the common regulatory frameworks imposed by the Directives 90/219/EEC and 90/220/EEC. Whereas diverse approaches to the consideration of socio-economic factors arose from different national interpretations of EU legislation by member states, public information and participation were left explicitly to the discretion of individual nations.[38] This freedom led some countries (Germany, for example) to require extensive public participation in reviewing applications for release.[39] As might be expected, such provisions drew criticism from the biotechnology industry, who viewed these as unnecessary protractions of the regulatory procedure. Indeed, the original German Law on Genetic Engineering[40] was amended in order to reduce the degree of public participation and remove the need for a public hearing prior to release approval.[41] However, as is the case with socio-economic considerations, a smaller number of release applications in non-industrialised countries may allow scope for public participation in this, particularly with regard to releases which have important aspects of novelty, or in the case of commercial introductions.[42] Furthermore, as most applications will be presented (and thus also financed) by foreign biotechnology companies, pressure from these to pare-down the approval process should, properly, be more easily resisted.

5. Harmonisation of Biosafety Regulation II: Pressures from Multinational Companies

Many multinational companies consider it imperative that a country has biosafety guidelines in place prior to the experimental introduction of transgenic crops. This arises in part from an anxiety to minimise the risks of liability in the event of environmental problems arising from the release, and in part from a concern to avert potential criticism from public interest groups that the country is exploiting a lack of regulation in choosing to develop their products in these countries.[43] Pioneer, although having research stations in Brazil was "unable to conduct research with transgenics until the recent establishment of regulations".[44] However, Asgrow (a division of Seminis) has allegedly conducted trials on transgenic vegetables in Guatemala, attracting criticism of the type other agribiotechnology companies apparently fear.[45]

With regard to the recent introduction of Genetically Modified Organisms legislation in South Africa,[46] Ralph Kirby has commented: "The incentive to place this Bill in the statute book in South Africa comes not really from within South Africa but from outside, in the form of multinational companies who are creating and marketing GMOs and wish to carry out these activities in South Africa".[47] There was apparently no bilateral pressure upon South Africa to adopt legislation (and bilateral aid for biotechnology in South Africa has not been contingent upon the implementation of a regulatory structure – voluntary guidelines were in place in the early 1980s). Rather, the development of a Genetically Modified Organisms Bill was seen as prerequisite for investment by multinational companies.

However, multinational pressure need not be directed towards the implementation of legislation which closely follows the US model. The South African legislation, for example, draws heavily upon the European Directives.[48] There is a recognition that where existing legislation cannot be easily adapted to cover biotechnology, conformity to the US model is unfeasible. In such circumstances, multinational companies are anxious to see legislation *per se*, irrespective of the system from which it is drawn.[49]

6. Bilateral Governmental Pressure to Introduce Regulations

The development of national biosafety legislation may be prerequisite for bilateral aid for biotechnology projects. The US Agency for International Development (USAID) requires the introduction of regulatory measures by Third World countries as a prior condition for aid for biotechnology capacity building.[50] A USAID sponsored project in Egypt provides a good example: A binding code of conduct for biosafety in Egypt, approved in February 1995, was developed by the Egyptian Agricultural Genetic Research Institute (AGERI) specifically to facilitate bilateral research projects.[51] However, although this code of conduct was produced with the collaboration of representatives from the USAID funded Agricultural Biotechnology for Sustainable Productivity (ABSP) Project, it does not closely follow a US model,[52] and it would therefore be misleading to suggest that the tying of bilateral aid packages to the development of biosafety regulations necessarily amounts to the transfer of legislative approaches from donor to recipient countries.

Indeed, it has been suggested that where regulatory frameworks have been adopted by Third World countries under bilateral aid agreements, these are less rigorous than might be hoped, facilitating the field-testing of foreign biotechnology products.[53]

Whilst it is doubtless in the best interests of Third World countries that they adopt some form of biosafety legislation, that this need follow a model presented by an industrialised nation, which has different (even, perhaps, conflicting) interests, seems more equivocal. In the course of the preceding analysis, we have stressed the importance attached to the harmonisation of biosafety legislation for facilitating free trade. The OECD's current advocacy for regulatory harmonisation amongst Member countries[54] began in June 1995 with the first meeting of the Expert Group on the Harmonization of Regulatory Oversight in Biotechnology. This group "instituted the development of consensus documents, which are mutually acceptable among member countries, as an initial step in efforts to facilitate harmonization".[55] The OECD advocates minimisation of international legislation whilst promoting harmonisation of regulatory approaches at the level of technical safety evaluation.

7. Deregulation

Hitherto we have explored the processes by which regulation is being harmonised. However, there are increasing industry pressures for a streamlining of the regulatory process, and there has recently been a relaxation of legislation in the US, with the strengthening of a notification system: Rather than apply for a permit for the importation, interstate movement or field testing of particular transgenic crops, researchers need simply notify APHIS of their intention to move the GMO, or conduct a field test. Provisions have also been made for the extension of the list of plants exempt from full regulatory control through petition.

The drive towards deregulation is based on the claim that as experience of previous environmental releases increases, there can be a relaxation of the rigour with which new applications for release are considered. However, regulation should properly reflect, on the one hand, the perception of risk, and on the other, the possible social and economic benefits of the new product. In the case of Third World countries, both these factors assume a different weighting: Risks will depend in part upon local biodiversity and the possibility of the crossing of novel crops with local varieties, and social and economic implications will evidently depend upon a plethora of local factors. Extensive experience of environmental releases in industrialised countries is not transferable, *in toto*, to non-industrialised countries, where different environmental conditions prevail – in particular where local biodiversity differs. With neither extensive experience of previous releases in the region, nor the direct transferability of the experience of other countries, regulators in Third World countries may feel the need to proceed with something of the caution once displayed by their counterparts in the US. This is a caution they can at present afford to exercise: With the regulatory authority having a far smaller number of applications to screen, rapid processing of each individual application may not accord the same priority as in industrialised countries, allowing a more rigorous review of each. Furthermore, in countries which lack a domestic biotechnology industry, most applications for field release or commercial importation will be made by foreign companies. These might reasonably be expected to meet the additional administrative costs of a more rigorous review of release applications: Costs which are likely to remain low by comparison to those of processing applications in industrialised countries.[56]

Important differences of national interest in the course of harmonising biosafety regulation arise from disparities in both the socio-economic milieu and the strength of national biotechnology programmes. Whilst there is clear need for the international co-ordination of mechanisms for the regulation of genetically modified organisms, the interests of Third World countries may not be best served by the straightforward assimilation of regulatory models taken from industrialised nations.

Notes

[1] For the purposes of this paper we restrict our use of 'biotechnology' to recombinant DNA technology.

[2] The World Trade Organisation lists 48 'least developed countries' (LDCs) (WTO Annual Report, 1997), which might be considered to be at one extremity of this spectrum. Although it has been suggested that many Third World countries may not consider biotechnology policy to be of primary importance, of these LDCs, 27 were represented at the Third Meeting of the Ad-Hoc Open-Ended Working Group on Biosafety (BSWG-3) at Montreal in October 1997. (This compares to 31 LDCs represented at the high profile UN Framework Convention on Climate Change at Bonn, later in the same month.)

[3] In the view of Goldstein, D.J. (Implications of biotechnology in the Third World, *Journal of Scientific and Industrial Research* **50**, 1991, 432–440), "[m]ost of the production of goods and services in the Third World does not depend on the endogenous generation of the original scientific results and/or technological innovations. Technologies are usually bought, or eventually copied. The innovative ceiling is adaptation and the few technological breakthroughs arising in the Third World are science-poor and of limited application. International competitiveness is attained by the reduction of production costs through low salaries, reduced ... taxation, and special state subsidies".

[4] Tzotzos, G.T. and Leopold, M., Commercial biotechnology: Developing world prospects, in D. Brauer (ed.), *Biotechnology, Volume 12, Legal, Economic and Ethical Dimensions*, VCH, Weinheim, Germany, 1995, pp. 339–367.

[5] See Goldstein (1991) (op. cit. 3). These are of course problems that are as inherent to the public sector as to the private. Indeed, one Indian government official, whilst bemoaning the poor performance of public-sector alternatives to commercially exploited Bt constructs, commented that such

public sector research "is driven not by serious attempts to develop a commercially viable product, but rather by academics vying with one another for publications in a trendy field" (pers. commun.).

[6] Junne concludes that: "Applications of biotechnology will first of all affect trade in agricultural products. It will make many importing countries more self-sufficient and increase trade conflicts among overproducing countries ... [B]iotechnology will ... contribut[e] to a stronger concentration of agricultural production for the world market on fewer developing countries" (Junne, G., The impact of biotechnology on international commodity trade, in E.J. DaSilva, C. Ratledge and A. Sasson (eds.), *Biotechnology – Economic and Social Aspects*, Cambridge University Press, 1992, pp. 165–188). An OECD publication, Biotechnology: Economic and Wider Impacts (1989) also provides an analysis of the potential impacts of biotechnology on international trade. It draws similar conclusions.

[7] In fact, high sugar prices and low quotas for sugar import to the US have exacerbated the trade implications of HFCS production (Junne, 1992, ibid.). Omvedt (Biotechnology: Miraculous or murderous?, *Economic and Political Weekly*, 22nd May, 1993, pp. 1033–1035) comments that "genuine 'free-trade' in this case ... [would] benefit the producers of natural products as compared to the high-cost industrial substitutes". See also Galhardi, R.M.A.A., Trade implications of biotechnology in developing countries: A qualitative assessment, *Technology in Society* **18**, 1996, 17–40. The example of lauric oils is discussed by Nichterlein, K., Biotechnology advances in oil crops affecting coconut production, *BINAS News* **3**(3–4), 1997, 2–4.

[8] For example, CTNBio lists 17 successful release applications for 2000 in Brasil. Of these, eight are from Monsanto, five from Novartis, two from AgrEvo, and two from the Cooperativa de Produtos de Cana e Alcool do Estado de Sao Paulo – Copersucar. In South Africa, of 15 release applications for genetically modified cotton, 13 are from large multinationals, of 10 release applications for maize, eight are from large multinationals (BINAS).

[9] Thus Munson notes that: "[The] need to incorporate different perspectives and priorities in risk assessment in each region and cultural area is highlighted by the different socioeconomic positions of different countries, as well as the different potential socioeconomic impact the same GMO may have on different countries' economies" (Munson, A., Should a biosafety protocol be negotiated as part of the biodiversity convention?, *Global Environmental Change – Human and Policy Dimensions* **5**, 1995, 17).

[10] Private sector industries may prove more difficult to regulate without appropriate legislation: In China, for example, regulations are nugatory – although there is a highly active domestic industry, this is centrally controlled. Given such central control, the drive to industrialise takes priority over the importance of stringent legislation.

[11] UNEP (1992) *Convention on Biological Diversity*, Environmental Law and Institutions Programme Activity Centre. Article 8 of the CBD states: "Each Contracting party shall, as far as possible and as appropriate: ... (g) Establish or maintain means to regulate, manage or control the risks associated with the use and release of living modified organisms resulting from biotechnology which are likely to have adverse environmental impacts that could affect the conservation and sustainable use of biological diversity, taking also into account the risks to human health". (The full text of the CBD is available at http://www.biodiv.org.)

[12] As late as 1995, P.K. Ghosh, at The Department of Biotechnology in the Ministry of Science and Technology in New Delhi, India, wrote: "The available data on Bt gene incorporated transgenic plant experiments seem to demonstrate that insects [sic] control through the development of transgenic plants is a distant dream". He continues, "... the extensive use of such unnatural engineered plants is expected to be delayed by the society for years to come, except in situations where transgenics have been developed by altering or moving *related genes* into plants ... (S)ignificantly greater exploitation of transgenic plant varieties over the existing natural ones by human kind is not foreseen in the near future" (Ghosh, P.K., Impact of industrial policy and trade related intellectual property rights on

biotech industries in India, *Journal of Scientific and Industrial Research* **54**, 1995, 217–230; his emphasis).

[13] See Virgin, I. and Frederick, R.J., The impact of international harmonisation on adoption of biosafety regulations, *African Crop Science Journal* **3**, 1995, 387–394; and Lesser, W. and Maloney, A.P., *Biosafety: A Report on Regulatory Approaches for the Deliberate Release of Genetically-Engineered Organisms – Issues and Options for Developing Countries*, Cornell International Institute for Food, Agriculture and Development (CIIFAD), Cornell University, Ithaca, 1993.

[14] Although the US opposed the need for a 'Biosafety Protocol', there is a recognition that harmonisation of biosafety legislation is in the economic interest of the US. A United States Office of Technology Assessment report foresees that: "The differences in approach [to biotechnology regulation] from nation to nation, particularly through their effects on investment and innovation, will influence the ability of the United States to remain competitive in biotechnology on the international scene" (US Congress, Office of Technology Assessment, Biotechnology in a Global Economy, OTA-BA-494; US Government Printing Office, Washington, DC, October, 1991, p. 14).

[15] A World Bank and International Service for National Agricultural Research report, published in May 1992 (immediately prior to the UN Conference on Environment and Development, the June 1992 'Earth Summit') specifically addressed the option for development of a 'Biosafety Protocol', rejecting this on the grounds that: "To establish international legal means of control for biosafety would be a departure from the experience acquired during the development of biotechnology in OECD countries. All OECD countries regulate biotechnology on the basis of guidelines or national legislation" (Persley, G.J., Giddings, L.V. and Juma, C., *Biosafety: The Safe Application of Biotechnology in Agriculture and the Environment*, The World Bank/International Service for National Agricultural Research, 1992, p. 20).

[16] Clark, N. and Juma, C., *Biotechnology for Sustainable Development: Policy Options for Developing Countries*, African Centre for Technology Studies, Nairobi, 1991.

[17] Raworth, P. and Reif, L.C., *The Law of the WTO: Final Text of the GATT Uruguay Round Agreements*, Oceana, New York, 1995.

[18] Agreement on the Application of Sanitary and Phytosanitary Measures, Paragraph 9.

[19] Raworth, P. and Reif, L.C. (1995) ibid., Agreement on Technical Barriers to Trade: Technical Regulations and Standards, Article 2.

[20] Article XX excludes, *inter alia*, measures "necessary to protect human, animal or plant life or health" and "relating to the conservation of exhaustible natural resources if such measures are made effective in conjunction with restrictions on domestic production or consumption" (Raworth and Reif, 1995, ibid.). Precedents have been taken to establish that living animals, according to GATT, can be regarded as an exhaustible natural resource (Andersson et al., 1995, op. cit. 17).

[21] The United States (then, as now, pre-eminent in the biotechnology industry) (Burrill, G.S. and Roberts, W.J., Biotechnology and economic development: The winning formula, *Bio/Technology* **10**, 1992, 647–653) consistently rejected the need for the development of a binding international 'Biosafety Protocol' during negotiations preceding the UN Conference on Environment and Development, or the 'Earth Summit' at Rio de Janeiro in 1992. At the meetings of the Agenda 21 Preparatory Committee, the "US delegation insisted that every reference on safety should be removed from the working documents. At PreCom 3, for example, the US wanted the phrase 'monitoring and evaluating the effectiveness and safety' of biotechnology changed to 'evaluating the success' thereof" (Munson, Genetically manipulated organisms: international policy-making and implications, *International Affairs* **69**, 1993, 497–517). Following such opposition, consideration of the need for a protocol was deferred, with Article 19.3 of the CBD reading "The parties *shall consider the need for* and modalities of a protocol" (Cantley, M.F., The regulation of modern biotechnology: A historical perspective, in H.J. Rehm and G. Reed (eds.), *Biotechnology*, Volume 12, 1995, pp. 629-630) (our emphasis). In the wake of the 'Earth Summit', an Expert Panel, convened by UNEP, met to advise on such

need. The panel did not reach consensus; a "minority of the panel, including the US representative, expressed themselves unconvinced of the need for a protocol" (Cantley, 1995, ibid. p. 631). This polarisation of opinion persisted in the following meeting of the Intergovernmental Committee on the Convention on Biological Diversity (ICCBD1) later in that same year (11–15 October, 1993, Geneva), with the US delegation expressing clear opposition whilst many Third World countries supported the need for a protocol (*Earth Negotiations Bulletin* **9**(6)). Again, at the second meeting of the ICCBD (Nairobi, 20th June–1st July, 1994), the US reiterated its position that a protocol on biosafety was unwarranted, whilst many Third World countries (along with Sweden, Norway and Denmark) affirmed their support for such a protocol (*Earth Negotiations Bulletin* **9**(17)). (The *Earth Negotiations Bulletin* are archived on the internet at http:www.mbnet.mb.ca/linkages.)

[22] With regard to the Fourth Meeting of the Open-Ended Ad Hoc Working Group on Biosafety (Montreal, 5–13 February, 1998), S. Burgiel comments that: "Countries that are not Party to the CBD, such as the US, cannot be Party to the Protocol until they have ratified the CBD. This provision for opting out of the 'Biosafety Protocol' by countries which have nonetheless ratified the CBD highlights the necessity of finding sufficient incentives to prompt as many countries as possible to ratify the Protocol. The process thereby presupposes that countries will buy into a Protocol by negotiating their particular interests, thereby encouraging ratification. On this point, the US delegation, which was one of the largest in Montreal, justified its involvement in the negotiations on the grounds of possibly becoming a Party to both the CBD and the Protocol at some point in the future", *BINAS News* **4**(1), 1998.

[23] Comment by B. Timmerman (University of Arizona) made at The Second Monroe Wall Symposium on Natural Products Discovery, Biodiversity and Biotechnology, Caracus, Venezuela, January 7–10, 1998.

[24] Masood, E. (Africa spearheads bid for strict rules on biosafety, *Nature* **387**, 1997, 326) writes: "Significantly, however, developing countries remain divided on the content and scope of the proposed biosafety protocol. Countries in the Far East and Latin America that are keen to grasp biotechnology to enhance trade and agriculture are broadly opposed to tough regulations. Less developed countries are in favour of such measures". And Hoyle (1997, ibid.) claims that "there is a growing realization, reportedly even among moderate G-77 countries in Latin America and elsewhere, that if the US withdrew from the biosafety talks, the outcome might be Draconian international biosafety standards that would effectively strangle the goose about to lay the golden egg".

[25] The European Union supported the preparation of an international protocol on biosafety during negotiations over Agenda 21 prior to the 'Earth Summit' (Munson, 1993; p. 500 op. cit. 20). Subsequently, they supported the development of interim guidelines. By the second Conference of Parties (COP2) in Jakarta, during November 1995, the European Union was promoting a twin-track approach, advocating the development of a protocol, and adoption of the UNEP International Technical Guidelines for Safety in Biotechnology. These guidelines were first developed through an Anglo-Dutch collaboration which critics have argued represented, from its inception, an attempt to deflect negotiations for a legally binding protocol to introduction of voluntary guidelines (Munson, A., 1994, Better biosafe than sorry, *New Scientist*, 25th June, 1994, 47–48). As Munson (1995, op. cit. note 9) notes, however, there were clearly differences of opinion within the EU: "The Spanish for example, subscribe to the majority opinion of the UNEP Expert Panel IV [that was, support for a biosafety protocol]. Norway and Sweden would also like to see a legally binding instrument on biosafety ... The UK intends to push for international guidelines, feeling that the negotiability of a protocol is unrealistic at present".

[26] T. Cors, in 'The Cartagena Protocol on Biosafety: Maintaining the Status Quo', *BINAS News*, forthcoming.

[27] All WTO Members are also members of either a customs union, free-trade agreement, or looser association with trade-related objectives, and this process of integration is continuing apace. For example, the European Commission has made a concrete proposal to admit Cyprus, the Czech Republic, Estonia, Hungary, Poland and Slovenia in a first phase of expansion. The North American Free Trade Agreement (NAFTA) is undergoing expansion, with negotiation of NAFTA-based agreements between Mexico, Canada and Chile. A Trade Protocol has been negotiated by the twelve members of the South African Development Community. In Asia the date for completion of the ASEAN (Association of South-East Asian Nations) Free Trade Area has been brought forward to 2005. The South Asian Free Trade Area (SAFTA) amongst countries in the Indian sub-continent aims for completion by 2001 (WTO Annual Report, Volume I, 1997, pp. 25–26).

[28] Petrinsky and Atanassov (1998), in G.T.Tzotzos and K.G. Skyrabin (eds.), *Biotechnology in the Developing World and Countries in Economic Transition*, CABI, 2000.

[29] Council Directive of 23rd April, 1990 on the contained use of genetically modified microorganisms (90/219/EEC) and Council Directive of 3rd April, 1990 on the deliberate release into the environment of genetically modified organisms (90/220/EEC). Twardowski et al. (in G.T. Tzotzos and K.G. Skyrabin (eds.), *Biotechnology in the Developing World and Countries in Economic Transition*, CABI, 2000) write "... during the period 1980–1990 the country [Poland] was transformed from a communist country to one determined to join the European Community ... Prior to 1990, regulations were adopted on the American model, though these were not legally enforceable. After 1990, Poland's aspiration for membership of the EU required acceptance of European legislation (with regard to biotechnology, EC Directives 90/219 and 90/220) ... Over the last three years there has been an intention to implement a new Polish law, which would be a 'national copy' of the EU Directives, drafted with the assistance of international organisations (OECD, UNEP, UNIDO). In November 1997 a team of experts submitted a proposal for a Polish 'Gene Law' to the government; the proposal is based on Directives 90/219 and 90/220".

[30] M. Schechtman, the USDA-APHIS Domestic Programmes Leader, comments: "This cooperation has been reflected in a variety of ways: in interactions between regulatory officials of the three countries, in the movement of products across national borders, and in the field testing and/or approvals of products developed by scientists or companies based in another of the three countries" (Schechtman, M., Harmonisation at the regional level – Americas: Case of the North American Plant Protection Organisation, in K.J. Mulongoy (ed.), *Biosafety Needs and Priority Actions for West and Central Africa*, International Academy of the Environment, 1997, p. 107).

[31] Ibid, p. 112.

[32] In commenting on the relevance of this geographical proximity, Zakri A. Hamid (Chairman, Genetic Modification Advisory Committee, Malaysia) comments: "the harmonization of biosafety regulations in the ASEAN region is not only imperative due to political considerations, but more so from the biological perspective (i.e. due to the commonality found in the ASEAN countries e.g. common centres of diversity)" (Zakri A. Hamid, pers. commun.).

[33] 90/220/EEC, Article 4. An instance of the need for such regional provisions is that of applications for the release of transgenic potatoes in Germany, where severe winters ensure that the germplasm cannot over-winter. In the UK, however, where winters are milder, and where the possibility of survival of germplasm from year-to-year was greater, more rigorous testing was demanded (Julian Kinderlerer, pers. commun.).

[34] 'Environmental risk disharmonies of European biotechnology regulation', in *AgBiotech News and Information*, 1997, 9(8)179N-183N. See also: An appraisal of the working in practice of directive 90/220/EEC on the deliberate release of Genetically Modified Organisms, a working document for the Scientific and Technological Options Assessment (STOA) Panel, by R. von Schomberg, European Parliament, January 1998.

[35] Von Schomberg (1998, ibid.) writes: "Only a ruling by the European Court of Justice could clarify the ambivalence between the obligatory lift in any trade barrier on products and the country's sovereignty to protect it's [sic] environment. The US have raised similar concerns about trade barriers and may appeal to the World Trade Organisation for an ultimate ruling".

[36] Chapter 3–10 (Approval) of The Norwegian Gene Technology Act (Act No. 38, 2 April, 1993) states, "In deciding whether or not to grant the application, significant emphasis shall also be placed on whether the deliberate release represents a benefit to the community and a contribution to sustainable development".

[37] Schenkelaars reports the view of a representative of the Norwegian Competent Authority (CA), "...that the Nordic approach to an open and democratic government would not be able to stand up against the way of decision-making in the European Union in the light of its democratic deficit" (Schenkelaars, P., Outlooks on public information and participation in the context of the European Biotechnology Directives 90/219/EEC and 90/220/EEC, in A. van Dommelen (ed.), *Coping with Deliberate Release: The Limits of Risk Assessment*, International Centre for Human and Public Affairs, 1996, pp. 168–169). This same representative of the Norwegian CA asks "...how one can secure participation on the national level in new legislation or changes in regulations, which are a result of changes in Community legislation" (Schneider, G., Public information and consultation in Norway, in Piet van der Meer (ed.), *Public Information and Participation in the Context of European Directives 90/219/EEC and 90/220/EEC*, MEBO Environmental Consultancy, 1994). It is interesting in this context to compare the EU Directives, and legislation in Sweden (a member of the EU) and Norway (which is not a Member State). Although there is no explicit suggestion that Swedish provisions for consideration of ethical aspects of release would be stronger were Sweden not subject to EU legislation, van der Meer (Public information and consultation in Norway, in Piet van der Meer (ed.), *Public Information and Participation in the Context of European Directives 90/219/EEC and 90/220/EEC*, MEBO Environmental Consultancy, 1994, p. 51) cites the comment that "...the Swedish Law on Biotechnology governing releases of GMOs is a copy of the EC Directives. But there is one big difference; there is a provision also to take ethical aspects of a GMO- release into account. However, this is a much weaker formulation than that in Norwegian law".

[38] Article 7 of 90/220/EEC directs that "Where a Member State considers it appropriate, it may provide that groups or the public shall be consulted on any aspect of the proposed deliberate release".

[39] Schubert, G., The situation in the Federal Republic of Germany, in Piet van der Meer (ed.), *Public Information and participation in the Context of European Directives 90/219/EEC and 90/220/EEC* MEBO Environmental Consultancy, 1994.

[40] Gentechnikgesetz, 1 July 1990.

[41] Schubert, G. (1994, ibid.).

[42] This is something that is recognised in Norway, where: "Public consultation would in the eyes of the [Norwegian] CA be important first and foremost in cases of marketing products, industrial production, and in cases where difficult ethical considerations have to be taken" (van der Meer, 1994, op. cit. note 34; p. 120).

[43] For example, Lawrence Zeph (manager, regulatory affairs, Pioneer Hybrid International Inc.) comments, "We are in the business of either testing or selling products for use in the environment and it is important that all relevant health and safety issues are considered before we proceed with these activities. If not, our company may incur liability risks or be the subject of criticism by environmental and consumer organizations who are interested in agricultural biotechnology. In addition, it is a benefit to all companies when a consistent, predictable regulatory system is in place which levels the playing field and does not allow one company to gain an advantage over other companies" (pers. commun.).

[44] Lawrence Zeph (ibid., pers. commun.).

[45] "Greenpeace says the lack of legal regulation of these activities is a major problem. Although Asgrow did not transgress any particular law, the pressure group says that the company acted with 'a lack of ethics', particularly by not informing authorities of the import and export of gene plasm" (Abbott, A., Greens attack transgenic plant trials, *Nature* **382**, 1996, 746). In fact, the same report offers an alternative interpretation of the liability of companies in the event of accidents, commenting that: "If there is damage to the environment, the Guatemala government [in the absence of national legislation] will not be able to hold companies liable".

[46] Republic of South Africa Genetically Modified Organisms Bill, 1997 (B 3-97).

[47] Kirby, R., A critical assessment of the South African Genetically Modified Organisms Bill of 1997, *BINAS News* 3(1–2), 1997, 12–14.

[48] Kirby, R. (1997, ibid.).

[49] "If a country has existing laws which could cover agricultural biotechnology (like the US model), then it is usually more efficient to build on those laws and the expertise of those responsible for the existing regulations. In general, we believe countries which are currently establishing their regulatory system should incorporate the experience with health and safety product reviews in the US, Canada, Europe and Japan" (Lawrence Zeph, op. cit. note 39, pers. commun.).

[50] The Agricultural Biotechnology for Sustainable Productivity (ABSP) Project, a US Agency for International Development (USAID) sponsored initiative based at Michigan State University, is unequivocal on this point: "Before any genetic transformation technology or materials may be transferred, the recipient country must have in place a regulatory approval mechanism to insure the safe transfer, handling and permitting of transgenic materials. ABSP works closely with USDA/APHIS to provide collaborators with consultation on the scientific, technical and regulatory issues encompassed in the exportation, importation, and safe handling of genetically engineered organisms, at both an individual and institutional level, for laboratory, greenhouse and field testing" (Policy Issues: IPR and biosafety, http://www.isp.msu.edu/ABSP). One representative of the USAID Biotechnology Programs lists the focus of these programmes "by order of priority: (a) research (b) policy (biosafety, intellectual property rights, plant variety protection) (c) human resources development" (Josette Lewis, pers. commun.).

[51] Madkour (in G.T. Tzotzos and K.G. Skyrabin (eds.), *Biotechnology in the Developing World and Countries in Economic Transition*, CABI, 2000) and Magdy Madkour (Director, AGERI) (pers. commun.).

[52] In fact, the code also draws upon the European Directives and the Australian regulations (Magdy Madkour, Director, AGERI, pers. commun.). See also Commercializing Agricultural Biotechnology Products in Egypt: Analysis of Biosafety Procedures. Madkour, M.A., Nawawy, A.S., Traynor, P.L. (ISNAR, 2000), which states: "Egypt's biosafety guidelines for greenhouse containment and field tests were drafted by AGERI scientists, based on existing guidelines from the US and Europe".

[53] Clark and Juma, 1991, p. 13, ibid.

[54] These include Poland (since 1996), Czech Republic (1995), Hungary (1996) and Mexico (1994).

[55] The rationale for such an initiative is given as follows: "OECD Member countries are moving rapidly towards the commercialisation and marketing of agricultural and industrial products of modern biotechnology. They have therefore identified the need for harmonization of regulatory approaches to the assessment of these products, in order to avoid unnecessary trade barriers" (OECD, 1997, *Consensus Document on Information Used in the Assessment of Environmental Applications Involving Pseudomonas*, OECD Environmental Health and Safety Publications, Series on Harmonization of Regulatory Oversight in Biotechnology, No. 6 [OCDE/GD(97)22]; Preamble).

[56] In suggesting that the South African Committee for Genetic Experimentation may be used as a model for regulatory committees in other African countries, J.A. Thompson (Biosafety in South Africa: The role of SAGENE, in K.J. Mulongoy (ed.), *Biosafety Needs and Priority Actions for West and Central Africa*, International Academy of the Environment, 1997, p. 85) comments that "Coun-

tries considering the establishment of similar bodies should realise that members of the committee will be required to put in considerable amounts of time, especially when it comes to assessing the risk involved in trial releases. SAGENE recently decided to draw on scientists with expert knowledge outside the committee to assist with risk assessments and that these people would in future be paid. As a result it is now charging companies which apply to carry out field trials in South Africa".

Economy & Environment

1. F. Archibugi and P. Nijkamp (eds.): *Economy and Ecology: Towards Sustainable Development.* 1989 ISBN 0-7923-0477-2
2. J. Bojö, K.-G. Mäler and L. Unemo: *Environment and Development: An Economic Approach.* 1990 ISBN 0-7923-0802-6
3. J. B. Opschoor and D. W. Pearce (eds.): *Persistent Pollutants: Economics and Policy.* 1991 ISBN 0-7923-1168-X
4. D.J. Kraan and R. J. in 't Veld (eds.): *Environmental Protection: Public or Private Choice.* 1991 ISBN 0-7923-1333-X
5. J.J. Krabbe and W.J.M. Heijman (eds.): *National Income and Nature: Externalities, Growth and Steady State.* 1992 ISBN 0-7923-1529-4
6. J. Bojö, K.-G. Mäler and L. Unemo: *Environment and Development: An Economic Approach* (revised edition). 1992 ISBN 0-7923-1878-1
7. T. Sterner (ed.): *Economic Policies for Sustainable Development.* 1994 ISBN 0-7923-2680-6
8. L. Bergman and D.M. Pugh (eds.): *Environmental Toxicology, Economics and Institutions.* The Atrazine Case Study. 1994 ISBN 0-7923-2986-4
9. G. Klaassen and F.R. Førsund (eds.): *Economic Instruments for Air Pollution Control.* 1994 ISBN 0-7923-3151-6
10. K. Uno: *Environmental Options: Accounting for Sustainability.* 1995 ISBN 0-7923-3513-9
11. K. Uno and P. Bartelmus (eds.): *Environmental Accounting in Theory and Practice.* 1997 ISBN 0-7923-4559-2
12. J.C.J.M. van den Bergh, K.J. Button, P. Nijkamp and G.C. Pepping: *Meta-Analysis in Environmental Economics.* 1997 ISBN 0-7923-4592-4
13. S. Faucheux, M. O'Connor and J. v.d. Straaten: *Sustainable Development: Concepts, Rationalities and Strategies.* 1998 ISBN 0-7923-4884-2
14. P. Kågeson: *Growth versus the Environment: Is there a Trade-off?.* 1998 ISBN 0-7923-4926-1
15. J.C.J.M. van den Bergh and M.W. Hofkes (eds.): *Theory and Implementation of Economic Models for Sustainable Development.* 1998 ISBN 0-7923-4998-9
16. J.N. Lekakis (ed.): *Freer Trade, Sustainability, and the Primary Production Sector in the Southern EU: Unraveling the Evidence from Greece.* 1998 ISBN 0-7923-5151-7
17. M. Boman, R. Brännlund and B. Kriström (eds.): *Topics in Environmental Economics.* 1999 ISBN 0-7923-5897-x
18. S.M. de Bruyn: *Economic Growth and the Environment.* An Empirical Analysis. 2000 ISBN 0-7923-6153-9
19. C. Kraus: *Import Tariffs as Environmental Policy Instruments.* 2000 ISBN 0-7923-6318-3
20. K. Uno (ed.): *Economy - Energy-Environment Simulation. Beyond The Kyoto Protocol.* 2002 ISBN 1-4020-0450-8

Economy & Environment

21. J. Beghin, D. Roland-Horst and D. Van der Mensbrugghe (eds.): *Trade and the Environment in General Equilibrium: Evidence from Developing Economies. 2002*
ISBN 1-4020-0479-6

KLUWER ACADEMIC PUBLISHERS – DORDRECHT / BOSTON / LONDON

MIX
Papier aus verantwortungsvollen Quellen
Paper from responsible sources
FSC® C105338

If you have any concerns about our products,
you can contact us on
ProductSafety@springernature.com

In case Publisher is established outside the EU,
the EU authorized representative is:
Springer Nature Customer Service Center GmbH
Europaplatz 3, 69115 Heidelberg, Germany

Printed by Libri Plureos GmbH
in Hamburg, Germany